Palaeoclimatology and Palaeoceanography from Laminated Sediments

Geological Society Special Publications
Series Editor A. J. FLEET

GEOLOGICAL SOCIETY SPECIAL PUBLICATION NO. 116

Palaeoclimatology and Palaeoceanography from Laminated Sediments

EDITED BY

A. E. S. KEMP
Department of Oceanography
University of Southampton
Southampton Oceanography Centre
UK

1996

Published by

The Geological Society

London

IUGS
UNESCO

IGCP 374

THE GEOLOGICAL SOCIETY

The Society was founded in 1807 as The Geological Society of London and is the oldest geological society in the world. It received its Royal Charter in 1825 for the purpose of 'investigating the mineral structure of the Earth'. The Society is Britain's national society for geology with a membership of around 8000. It has countrywide coverage and approximately 1000 members reside overseas. The Society is responsible for all aspects of the geological sciences including professional matters. The Society has its own publishing house, which produces the Society's international journals, books and maps, and which acts as the European distributor for publications of the American Association of Petroleum Geologists, SEPM and the Geological Society of America.

Fellowship is open to those holding a recognized honours degree in geology or cognate subject and who have at least two years' relevant postgraduate experience, or who have not less than six years' relevant experience in geology or a cognate subject. A Fellow who has not less than five years' relevant postgraduate experience in the practice of geology may apply for validation and, subject to approval, may be able to use the designatory letters C Geol (Chartered Geologist).

Further information about the Society is available from the Membership Manager, The Geological Society, Burlington House, Piccadilly, London W1V 0JU, UK. The Society is a Registered Charity, No. 210161.

Published by The Geological Society from:
The Geological Society Publishing House
Unit 7, Brassmill Enterprise Centre
Brassmill Lane
Bath BA1 3JN
UK
(*Orders*: Tel. 01225 445046
        Fax 01225 442836)

First published 1996

British Library Cataloguing in Publication Data
A catalogue record for this book is available from the British Library.

ISBN 1-897799-67-5
ISSN 0305-8719

Typeset by Aarontype Limited,
Easton, Bristol, UK.

Printed by The Alden Press,
Osney Mead, Oxford, UK.

Distributors

USA
AAPG Bookstore
PO Box 979
Tulsa
OK 74101-0979
USA
(*Orders*: Tel. (918) 584-2555
        Fax (918) 560-2652)

Australia
Australian Mineral Foundation
63 Conyngham Street
Glenside
South Australia 5065
Australia
(*Orders*: Tel. (08) 379-0444
        Fax (08) 379-4634)

India
Affiliated East-West Press PVT Ltd
G-1/16 Ansari Road
New Delhi 110 002
India
(*Orders:* Tel. (11) 327-9113
        Fax (11) 326-0538)

Japan
Kanda Book Trading Co.
Tanikawa Building
3-2 Kanda Surugadai
Chiyoda-Ku
Tokyo 101
Japan
(*Orders*: Tel. (03) 3255-3497
        Fax (03) 3255-3495)

Contents

Laminated sediments as palaeo-indicators

ALAN E. S. KEMP

Department of Oceanography, University of Southampton, Southampton Oceanography Centre, European Way, Southampton SO14 3ZH, UK

As society at large becomes increasingly concerned with the issues surrounding global climate change, so the pressure on the scientific community to produce models and predictions of climate variability increases. In many respects, however, that branch of science concerned with climate change is in its infancy. While recent meteorological and oceanographic studies have shed light on the processes and mechanisms of atmospheric and oceanic circulation, this has produced only a 'snapshot' perspective of global change, limited by the range of instrumental or historical records. On the other hand, palaeoclimatic and palaeoceanographic studies have been mainly on coarser (millennial) timescales that have a more academic and less immediate appeal. The palaeo-records which have the required temporal (interannual/ decadal) resolution are limited to tree rings, ice cores, coral records and laminated marine or lacustrine sediments. This volume is concerned with the wide-ranging application of lacustrine and marine laminated sediments as palaeo-indicators.

Environments of lamina formation and preservation

The two fundamental requirements for development of a laminated sediment sequence are: (1) variation in input/chemical conditions/biological activity that will result in compositional changes in the sediment; and (2) environmental conditions that will preserve the laminated sediment fabric from bioturbation. Within lakes, strong seasonal signals are dominant while preservation is effected by bottom water/ sediment anoxia resulting from stratification, high salinities or high sedimentation rates. In the marine environment, the dominant control on lamina preservation is reduced oxygen in anoxic silled basins (e.g. California Borderland basins) or marginal seas (e.g. Black Sea), or beneath regions of high primary productivity such as coastal upwelling zones (e.g. the Peru margin). A more recently recognized preservation mechanism applies to deep-sea laminated sediments, where rapid flux of strong diatom mats or giant diatoms overwhelms benthic activity in oxygenated bottom environments.

Approaches and methodology

The essential step prior to using laminated sediments as palaeo-indicators is to examine the sediment composition and micro-fabrics to develop a model for origins of the lamination. The examples given in this volume amply illustrate the range of lamina compositions, from clastic sediment through biological and chemical. The development of such sediments and the dominance of the annual cycle is discussed by **Anderson**, while an earlier review by Anderson & Dean (1988) gives an account of the composition of lacustrine lamina over geological timescales.

Given a thorough knowledge of the lamina components, a knowledge of the atmospheric and water column processes producing the lamina-forming flux is an important prerequisite to developing palaeoclimatic/palaeoceanographic models of lamina formation (**Sancetta**). With cores taken from existing marine basins or lakes the flux events recorded within the laminae may be directly compared with water column observations, meterological records and sediment traps. With ancient examples, appropriate modern analogues must be sought.

To ensure that individual flux events can be observed in samples of laminated sediments, the appropriate sampling and analysis techniques must be deployed. Recently, scanning electron microscope (SEM) methods have been increasingly used to resolve microfabrics and lamina components (**Pike & Kemp**). Such SEM-led approaches now facilitate study of seasonal-scale variability.

Defining varves: lamina origins

In this volume, the word 'varve' is used for the lamina or group of laminae that are interpreted to represent one year's deposition. Depending on the environment, the varve may be a couplet (two laminae, as is the case for the classical Swedish varves), triplet or have a larger number of laminae (see **Anderson; Pike & Kemp**).

From Kemp, A. E. S. (ed.), 1996, *Palaeoclimatology and Palaeoceanography from Laminated Sediments*, Geological Society Special Publication No. 116, pp. vii–xii.

Laminae defined by changes in terrigenous sediment grain size

The classical Swedish varves are clearly defined by a coarse terrigenous lamina (resulting from spring meltwater discharge) alternating with finer sediment (**Anderson**; **Petterson**). Lamination defined by similar changes in grain size of terrigenous sediment is common in marginal marine basins where seasonal climatic forcing produces a distinct coarser sub-lamina (e.g. Santa Barbara Basin; **Bull & Kemp**). Similar differences in grain size, however, may be also produced by a variety of mechanisms of particle sorting in sediment gravity flows or in sediment-bed interaction as well as by periodic changes in the grain size of settling sediment (see review by **O'Brien**). Thus, assessing the origins of lamination in ancient shales may not be straightforward and in some cases, differences in interpretation of lamina origins have arisen (e.g. for Silurian shales: Dimberline *et al.* 1990; Kemp 1991).

Biogenic lamination

Biogenic lamination may form either by episodic flux of plankton with hard parts or by in situ formation of organic structures within the sediment. The dominant record of biogenic flux within laminated sediments is the fall-out from surface-water algal blooms, either the opaline frustules of diatom algae, or the delicate calcite plates of coccolithophorid algae.

Diatom lamination. Diatom laminae are common in both marine and lacustrine sediments in which they record the seasonal productivity cycle, generally the spring bloom. **Pike & Kemp** show that more than one bloom episode may be recorded per year and further show, as do **Bull & Kemp** that multiple bloom episodes may be recorded within any one diatom ooze lamina. Such diatom laminae may also contain different sub-laminae containing a succession of species corresponding to the evolving bloom. Diatom laminae are generally parallel-sided and contain intact, unfragmented frustules and are thus interpreted to be mainly sedimented by rapid deposition as flocculated aggregates without the mediation of zooplankton.

Coccolith lamination. Coccolith laminae occur in settings ranging from Holocene sediments from the Black Sea (Hay 1988) through Oligo-cene shales and limestones (**Haczewski**) to Jurassic black shales (**O'Brien**). Individual coccolith laminae differ in form from diatom laminae in being more blebby and discontinuous, possibly due to sedimentation as faecal pellets (Pilskaln 1989). In sediments containing both diatoms and coccoliths (e.g. Black Sea), diatoms may form the deposit from the spring bloom while coccoliths are deposited from the Autumn bloom (Hay *et al.* 1990).

Lamination produced by algal or bacterial mats

Lamination produced by cyanobacterial mats in which organic filaments (which may be produced seasonally) alternate with clastic sediment is illustrated by **O'Brien**. Benthic bacterial mats (such as *Beggiotoa* or *Thioploca*) have been inferred to form laminae in the Santa Barbara Basin (see summary in **Schimmelmann & Lange**) although doubt is cast on this mechanism by **Bull & Kemp**.

Chemically induced lamination

Chemically induced lamination may form as water column precipitates which then settle or may develop within the surficial sediment due to early diagenesis. In some redox-sensitive cases, e.g. pyrite in marine environments and siderite in lacustrine environments minerals may form either within the water column or within the sediment.

Water-column precipitation. Many lakes precipitate calcium carbonate in the form of low-magnesium calcite during the summer when photosynthesis decreases dissolved CO_2 and temperatures increase (Kelts & Hsu 1978). In evaporitic basins, an annual cycle of evaporation may lead to alternating layers of halite/sulfate or calcite/anhydrite (Anderson; **Leslie** *et al.*).

Early diagenetic lamination. Early diagenetic changes related to variation of redox and the metabolisation of organic matter may influence lamina composition. For example, organic-rich laminae may form a locus of pyrite formation. Annual manganese carbonate laminae in sediments from the Baltic Sea and similar features in ancient black shales may result from seasonal changes in bottom water oxygenation (Huckriede & Meischner 1996).

Laminated sediments as geo-chronometers

Following the pioneering work of De Geer (see **Anderson**; **Petterson**) the development of varve chronologies has been a major research goal. This has led to the development of the Swedish Time Scale covering 13 527 varve years and to calibration of ^{14}C chronologies with varve years (Wohlfarth *et al.* 1995; references in **Petterson**). Where laminae are indistinct, however, or where complete couplets are not deposited every year, lamina-based timescales cannot be erected (e.g. Black Sea; Crusius & Anderson 1992).

Counting varves

Of course, the down side to having a thick laminated sediment sequence capable of use for chronology and generation of time series is that the varves must be counted! Until recently, such counting was exclusively manual. Automated varve counting by digital imaging and subsequent image analysis offers a solution to this. **Zolitschka** presents such analyses but emphasizes that with composite varves and the occurrence of thin varves complications arise that require additional resolution/examination. Ripepe *et al.* (1991) used automated image analysis from acetate peels for studies of cyclicity in Eocene oil shales. Schaaf & Thurow (1994) have developed a rapid method for laminated sediment core image acquisition and analysis but from the approach of using bulk or interpolated sedimentation rates rather than identifying separate years.

Non-annual laminae

In most environments of deposition of laminated sediments, the seasonal/annual signal is the strongest influence on sedimentation, as **Anderson** emphasizes. There are, however, other variations in climate/ocean dynamics which have a strong signal, the most prominent of which is the El Niño/Southern Oscillation (ENSO). Sequences that contain laminae interpreted to represent other than annual deposition are rare, but **Hagadorn** presents analysis of the Santa Monica Basin sediments, in which lamina couplets occur with 3–6 year periodicities characteristic of El Niño. Elsewhere, in Africa, based on records from Lake Turkana, Halfman & Johnson (1988) also suggest an El Niño periodicity for laminae.

Cyclicity recorded in annually laminated sediments

Varved sediments are a readily decipherable repository for records of interannual variability. In lake sediments, periodicites of 11 and 22 years, marking the solar (sunspot) cycles are common and longer period, decadal and century-scale periodicities are also recognized from spectral analysis (Glenn & Kelts 1991). ENSO signals (rare in lake sediments) are relatively common in marine sediments and 50–60 year cycles have also been identified (**Schimmelmann & Lange**; **Bull & Kemp**; **Hagadorn**). Increasing use of image analysis of laminated sediment sequences is producing more material for time series production, however, care is required in assessing the reliability and statistical validity of peaks produced in spectral analysis.

Laminated sediments as event correlators of palaeoseismicity and neotectonics

The chronological schemes derived from laminated sediments may also be applied to precisely date and correlate sedimentary or tectonic events. The reconstruction of geological events hinges on the ability to identify the relative timing of events recorded in the sedimentary record. It is here that the ability to correlate annual laminae over long distances can give precise information on the synchroneity of events. An illustration of this is derived from ancient Oligocene laminated limestones of the Polish Carpathians, allowing **Haczewski** to correlate palaeoseismicity over large distances. Holocene laminated sediments are being increasingly used for accurate dating in studies of neotectonics in tectonically active settings such as convergent plate margins (Brobowsky & Clague, 1990).

Laminated sediments as palaeo-oxygenation indicators

In settings where preservation of lamination may be confidently ascribed to reduced concentrations of dissolved oxygen (not deep-sea diatom ooze – see below), the degree of lamina disruption may be used as a palaeo-oxygenation index. A classification scheme based on lamination and the occurrence of trace-fossils was proposed by Savrda & Bottjer (1986) while a more refined scheme, based on the degree of disruption to laminae, integrated with benthic foraminiferal evidence has been employed in the Santa Barbara Basin by Behl & Kennett (1996).

Significance of deep-sea laminated diatom oozes

An exciting new development is the increasing recognition of the occurrence of laminated sediments composed of diatom mats or giant diatoms in open-ocean, deep-sea environments. Because of the laminated nature of these deposits, and given the existing preconceptions, origins have been ascribed previously to *ad hoc* and implausible occurrences of reduced oxygen conditions e.g. Muller *et al.* (1991). Work on laminated diatom mat deposits of the eastern equatorial Pacific (Kemp & Baldauf 1993) together with new insights into the oceanography of frontal systems (Yoder *et al.* 1994) have led to the development of integrated models for the origins of these enigmatic sediments (**Kemp et al.**; **Pearce et al.**). Analogous laminated diatom mat deposits have been described recently from the North Atlantic by Boden & Backman (1996) who ascribe their origins to a similar frontal zone origin to those of the equatorial Pacific.

The lamina-scale periodicities present within these deep-sea laminites are not as straightforward to interpret as those from lacustrine or marginal marine settings where an annually/seasonally-driven terrigenous sediment pulse provides a temporal control. **Kemp et al.** ascribe lamina-scale alternations in equatorial Pacific sediments to possible anti-El Niño perioditicies.

Laminated sediments as palaeoproductivity indicators

Biogenic laminae composed of diatoms commonly display thickness variation within individual sequences. Such thickness variation within Santa Barbara Basin sediments is related to variation in upwelling-driven productivity in the basin **Bull & Kemp** and new time series analysis of this reveals 4–7 year perioditicies. In laminated sediments of the Cariaco Basin **Hughen et al.** (2) also ascribe increased thickness of diatom-rich laminae to increased productivity.

Although variation in diatom lamina thickness (hence biogenic opal content) may be related to variations in primary production in marginal upwelling environments there is substantial evidence that variation in opal content in open-ocean, deep-sea settings may not be straightforwardly related to productivity (Kemp 1995; **Kemp et al.**).

Onset of laminations: monitoring anthropogenic effects

The recent onset of laminations in many marginal marine and lake environments is a direct record of anthropogenically induced eutrophication. **Petterson** illustrates the increasing incidence of laminae in Swedish Lakes within the last century which are a direct result of anthropogenic activity. In a marine environment, Jonsson *et al.* (1990) document the increasing incidence of laminae in surficial sediment samples from the Baltic since the end of the 1940s. However, care must be taken in oversimplifying such relationships. **Gorsline** summarizes the expansion of the laminated zone within the Santa Monica Basin over the last three hundred years: well before the incidence of significant anthropogenic influence. Such instances underscore the requirement to characterize natural trends in order to identify anthropogenically induced change and to separate local from regional and global-scale changes.

Laminated sediments as correlators of rapid global events

Some of the most intriguing questions in global change research concern the global extent and timing of change. How rapidly are variations in North Atlantic thermohaline circulation transmitted through the global ocean? Do the Dansgaard–Oeschger cycles have a global signature? Recent work from ODP Site 893 in the Santa Barbara Basin (Behl & Kennett 1996) has built on earlier work from the Gulf of California (Keigwin & Jones 1990) to show that changes in North Atlantic circulation (associated with Greenland Ice Core oxygen isotope variations) correlate with variations in preservation of laminae in basin sediments. Behl & Kennett (1996) relate these variations in lamina preservation to changes in the oxygenation of Pacific Intermediate Water controlled by variation in the volume of North Atlantic Deep Water production.

Another recent insight into transmission of high-latitude events has come from comparison of lamina thickness in sediments from the Cariaco Basin (**Hughen et al.**; Fig. 1) and the Greenland, GRIP ice core $\delta^{18}O$, which show a remarkable similarity in the timing and duration of events and in decadal-scale patterns of variability (Hughen *et al.* 1996).

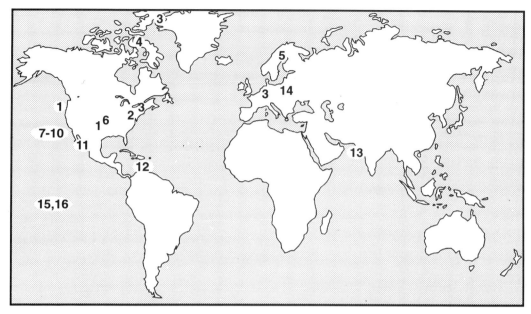

Fig. 1. Location of case studies of laminated sediments: (1) Anderson: evaporite basins, New Mexico & Texas; marine laminae, California Margin. (2) ÓBrien: Devonian shales, New York (& other miscellaneous Mesozoic-Proterozoic examples). (3) Zolitschka, Holzmaar, Germany; Fayetteville Green Lake (New York); Lake C2 (Canada). (4) Hughen *et al.*: tidewater lakes, Baffin Island (Canada). (5) Petterson: Swedish lakes, estuaries and seas. (6) Leslie *et al.*: evaporite basin, New Mexico. (7) Gorsline *et al.*: California Borderland Basins. (8) Hagadorn: Santa Monica Basin, California Borderland. (9) Schimmelmann & Lange: Santa Barbara Basin. (10) Bull & Kemp: Santa Barbara Basin. (11) Pike & Kemp: Gulf of California. (12) Hughen *et al.*: Cariaco Basin, offshore Venezuela. (13) von Rad & von Stackelberg: Northeastern Arabian Sea. (14) Haczewski: Polish Carpathians. (15, 16) Pearce *et al.*, Kemp *et al.*: eastern Equatorial Pacific Ocean.

Future research directions and initiatives

There is increasing focus on the acquisition of high-resolution records of environmental change and laminated sediment targets have featured prominently. One of the most notable developments in the production of highest-resolution data is the recent adoption of individual sites as targets of the Ocean Drilling Program (ODP). The ODP's *Joides Resolution* is the only platform capable of piston coring to greater than 50 m sediment depths. Several recent targets, including ODP Site 893 in the Santa Barbara Basin (200 m sediment to isotope stage 6); ODP Site 1002 in the Cariaco Basin and, most recently, the mini-Leg 169S to Saanich Inlet in British Columbia, have already started to produce fascinating new palaeoclimatic and palaeoceanographic data. The acquisition of increasingly longer, high quality cores from continental lakes (such as Monticchio) has also produced new high resolution records spanning the last glacial cycles and the new continental drilling initiatives promise more. Integration and correlation of multiparameter data-sets from these sites will provide key input to global change research.

This volume

Many of the papers included in this volume were presented at a conference on 'Palaeoclimatology and Palaeoceanography from Laminated Sediments' held at the Geological Society in September 1993. This meeting brought together workers on both marine and lacustrine laminated sediments. It emerged that the two communities were to a large extent quite separate and there had been only limited communication between terrestrial and marine researchers. As the meeting progressed, it became evident that the two communities had much in common yet, at the same time, had developed some quite separate sampling and analytical methods and approaches. The benefits from pooling the expertise from the two communities were readily apparent, both in terms of methodology and from a broader awareness of the combined approach of studying a global array of lacustrine and marine sites in the context of major initiatives such as the PAGES, PEP (Pole–Equator–Pole) transects. In order to initiate an ongoing method for interaction a proposal was submitted to initiate a IUGS/ UNESCO International Geological Correlation Programme. Subsequently IGCP 374 (Palaeoclimatology and Palaeoceanography from Laminated Sediments) has held a series of meetings designed as forums for interaction and to disseminate results of ongoing studies. The papers presented in this volume which cover approaches and methods as well as research results form a contribution to IGCP 374.

References

ANDERSON, R. Y. & DEAN, W. E. 1988. Lacustrine varve formation through time. *Palaeogeography, Palaeoclimatology, Palaeoecology*, **62**, 215–235.

BEHL, R. J. & KENNETT, J. P. 1996. Brief interstadial events in the Santa Barbara Basin, NE Pacific, during the past 60 kyr. *Nature*, **379**, 243–246.

BODEN, P. & BACKMAN, J. 1996. A laminated sediment sequence from the northern north Atlantic Ocean and its climatic record. *Geology*, **24**, 507–510.

BROBOWSKY, P. T. & CLAGUE, J. J. 1990. Holocene sediments from Saanich Inlet, British Columbia, and their neotectonic implications. *Current Research, Part E Geological Survey of Canada, Paper* **90-E1**, 251–256.

CRUSIUS, J. & ANDERSON, R. F. 1992. Inconsistencies in accumulation rates of Black Sea sediments inferred from records of laminae and ^{210}Pb. *Palaeoceanography*, **7**, 215–227.

DIMBERLINE, A. J., BELL, A. & WOODCOCK, N. H. 1990. A laminated hemipelagic facies from the Wenlock and Ludlow of the Welsh Basin. *Journal of the Geological Society, London*, **147**, 693–702.

GLENN, C. R. & KELTS, K. 1991. Sedimentary rhythms in lake deposits. *In*: EINSELE, G., RICKEN, W. & SEILACHER, A. (eds) *Cycles and Events in Stratigraphy*. Springer, Berlin, 188–221.

HAY, B. J. 1988. Sediment accumulation in the central eastern Black Sea over the last 5100 years. *Paleoceanography*, **3**, 491–508.

——, HONJO, S., KEMPE, S., ITTEKOT, V. A., DEGENS, E. T., KONUK, T. & IZDAR, E. 1990. Interannual variablity in particle flux in the southwestern Black Sea. *Deep-Sea Research*, **37**, 911–928.

HALFMAN, J. D. & JOHNSON, T. C. 1988. High resolution record of cyclic climatic change during the past 4 ka from Lake Turkana, Kenya. *Geology*, **16**, 496–500.

HUCKRIEDE, H. & MEISCHNER, D. 1996. Origin and environment of manganese-rich sediments within black-shale basins. *Geochimica et Cosmochimica Acta*, **60**, 1399–1413.

HUGHEN, K. A., OVERPECK, J. T., PETERSON, L. C. & TRUMBORE, S. 1996. Rapid climate changes in the tropical Atlantic duting the last deglaciation. *Nature*, **380**, 51–54.

JONSSON, P., CARMAN, P. & WULFF, F. 1990. Laminated sediments in the Baltic – a tool for evaluating nutrient mass balances. *Ambio*, **19**, 152–158.

KEIGWIN, L. D. & JONES, G. A. 1990. Deglacial climatic oscillations in the Gulf of California. *Paleoceanography*, **5**, 1009–1023.

KELTS, K. & HSU, K. J. 1978. Freshwater carbonate sedimentation. *In*: LERMAN, A. (ed.) *Lakes: Chemistry, Geology, Physics*. Springer, Berlin, 295–323.

KEMP, A. E. S. 1991. Silurian pelagic and hemipelagic sedimentation and palaeoceanography. *In*: BASSETT, M. G., LANE, P. D. & EDWARDS, D. (eds) *Special Paper in Palaeontology*, **44**, 261–300.

——1995. Laminated sediments from coastal and open ocean upwelling systems: what variability do they record? *In*: SUMMERHAYES, C. EMEIS, K.-C., ANGEL, M. V., SMITH, R. L. & ZEITZSCHEL, B. (eds) Upwelling in the Ocean: Modern Processes and Ancient Records. *Dahlem Workshop Report* ES18. Wiley, Chichester, 239–257.

—— & BALDAUF, J. G. 1993. Vast Neogene laminated diatom mat deposits from the eastern equatorial Pacific Ocean. *Nature*, **362**, 141–144.

MULLER, D. W., HODELL, D. A. & CIESIELSKI, P. F. 1991. Late Miocene to earliest Pliocene (9.8–4.5 Ma) paleoceanography of the subantarctic southeast Atlantic: stable isotopic, sedimentologic, and microfossil evidence. *In*: CIESIELSKI, P. F., KRISTOFFERSEN, Y. *et al. Proceedings ODP, Sci. Results*, **114**, College Station, TX (Ocean Drilling Program), 459–474.

PILSKALN, C. H. 1991. Biogenic aggregate sedimentation in the Black Sea Basin. *In*: IZDAR, E. & MURRAY, J. W. (eds) *Black Sea Oceanography*. Kluwer, Dordrecht, 293–306.

RIPEPE, M., ROBERTS, L. T. & FISCHER, A. G. 1991. ENSO and sunspot cycles in varved Eocene oil shales from image analysis. *Journal of Sedimentary Petrology*, **61**, 1155–1163.

SAVRDA, C. E. & BOTTJER, D. J. 1986. Trace-fossil model for reconstruction of paleo-oxygenation in bottom waters. *Geology*, **14**, 3–6.

SCHAAF, M. & THUROW, J. 1994. A fast and easy method to derive highest-resolution time-series datasets from drillcores and rock samples. *Sedimentary Geology*, **94**, 1–10.

WOHLFARTH, B., BJÖRK, S. & POSSNERT, G. 1995. The Swedish Time Scale – a potential calibration tool for the radiocarbon time-scale during the late Weichselian. *Radiocarbon*, **37**, 347–359.

YODER, J. A., ACKLESON, S., BARBER, R. T., FLAMANT, P. & BALCH, W. A. 1994. A line in the Sea. *Nature*, **371**, 689–692.

Seasonal sedimentation: a framework for reconstructing climatic and environmental change

Department of Earth and Planetary Sciences, University of New Mexico, Albuquerque, NM 87131, USA

*'Of the late glacial sediments the most important is the glaci-marine clay, the **varvig lera** (Hvarfvig lera) ... Already in my first field-work as a geologist, in 1878, I was struck by the regularity of these laminae, much reminding of the annual rings of trees ... The laminae were found to be so regular and so continuous that they could scarcely be due to any less regular period than the annual one ... I also succeeded in finding the first correlation between clay layers at three points, though not very far from one another ... Finally, in 1904, I happened to get a very good correlation between two clay sections 1 km apart from each other, and now I determined to make an earnest attempt to realize my old plan for a clay-chronology.*

I secured the assistance of a number of students from the Universities of Stockholm and Upsalla ... and after some training they all went out on a summer morning in 1905, each of them to his special part of a line about 200 km long ... going as nearly as possible in the direction of the ice recession. The main work was performed in four days.

I now finally have the conclusive proofs for the assumption that the individual 'varves' had a very wide distribution. This, together with their regular structure definitively showed, that they could not be due to any local or accidental cause of smaller importance or less pronounced periodicity than the climatic period of the year.'

Gerard de Geer (1912)

De Geer's intuition about the effects of seasonal climate forcing on sedimentation proved to be correct. His determination to leap-frog varve correlations from one site to another allowed him to recognize the power of the solstice cycle. In the late 1950s and early 1960s, my students and I performed a similar experiment by correlating much thinner, non-glacial laminae across single geological basins (Anderson & Kirkland 1966; Kirkland & Anderson 1970; Anderson *et al.* 1972). We relived the heady excitement of de Geer's discovery of varve correlation by extending his discovery into other environments. Long-distance, millimetre-scale correlations in single basins eventually were extended to around 1000 km in the Black Sea (Hay *et al.* 1991). It was de Geer's insight, the work of Sauramo (1923), of Richter-Bernberg (1960), and some of our own long-distance correlations that demonstrated that the annual cycle of climate forcing was strong enough to regulate the accumulation of sediment over entire basins, over broad geographic areas, and within a wide range of geological settings and environments.

Appreciation of the effects of the solstice cycle on sedimentation was reinforced by results from sediment trapping investigations carried out during the 1970s and 1980s. The design of sediment traps was improved to include time-series sampling (Anderson 1977), which revealed associations of specific sediment types with the seasons. It was found that sediment traps, even when deployed far from land and at great depths in the ocean, revealed a clear pattern of seasonal accumulation of sediment of varying composition. This information, when added to earlier detailed investigations of extant, seasonal organisms and their association with sedimentary laminations (Heim 1909; Nipkov 1927; Welten 1944; Kirkland & Anderson 1969), reduced resistance within the geological community to the idea that certain types of non-glacial laminations were also varves.

Seasonally generated accumulations of sediment were the rule, not the exception. Were it not for bioturbation, virtually the entire sediment pile, even in environments with very high accumulation rates, would bear evidence of seasonality in their laminated structure, as is the case for much of the Precambrian (Anderson *et al.* 1985b). Today, as the pendulum swings from skepticism about varves toward belief, a word of caution is in order. Laminations, even in quiet environments, are not necessarily varves and it is not safe to conclude that laminations are varves until one or more important criteria have been met. These criteria include known

From Kemp, A. E. S. (ed.), 1996, *Palaeoclimatology and Palaeoceanography from Laminated Sediments*, Geological Society Special Publication No. 116, pp. 1–15.

seasonal associations of major or minor components, an established chronology, and evidence for lateral continuity. However, when direct or indirect evidence for the annual cycle is not available, it is sometimes possible to conclude, through the cautious use of analogues, that laminations were generated seasonally and have faithfully recorded the annual cycle of climate forcing.

De Geer, in defining varves, recognized their persistence and noted that regularity of expression over time was characteristic of the process. Some non-glacial varve sequences display great regularity over time (Anderson 1982). In my opinion, however, it is not necessary that couplets or groups of seasonal laminations be precisely annual or continuously expressed to be considered varves. The key statement of de Geer is that the layering... 'could not be due to any local or accidental cause of smaller importance... than the climatic period of the year.' Thus, it is the relationship to the annual climate cycle and not regularity or persistence over time that defines the varve concept.

The majority of seasonally generated laminations in lacustrine environments, and especially in marine environments, are not beautifully preserved as regular varves. More commonly, a sequence of laminations is variable in expression, and continuity is interrupted by bioturbation, scour, turbidites, etc. For example, parts of a varve time series may be greatly expanded by the addition of sediment, to a point where the annual cycle of accumulation is not decipherable (Anderson 1992a, b; 1993). At the other extreme, rates of accumulation may be so slow that annual accumulations are unrecognizable and interannual and decadal changes in composition are mistaken for varves. There are probably many cases where there is but a single dominant seasonal component, leaving little physical evidence for seasonal accumulation. Other examples of less-than-perfect varving, in addition to disturbance by bioturbation, include cases where accumulations are interrupted or truncated by currents and downslope processes.

Limitations imposed on the use of varves by other processes can be seen, not as a deterrent to investigation, but as an important source of information about responses to environmental and climatic change. However, to obtain this information, and to use less-than-perfect laminations to reconstruct environmental and climatic variability, it helps to place the processes that generate seasonal laminations within a framework that facilitates environmental and climatic interpretation.

Seasonal sedimentation

Climatic effects on sedimentation

Joseph Barrell (1908) noted:

'It is natural that the influence of climatic change in producing shiftings of the sedimentary facies should be the last kind of action to reach a true appreciation.'

Now, many years later, and especially after the success of orbital forcing as a mechanism for explaining long-term changes in climate, Barrell's observation seems obvious. But it is not at all obvious that orbital forcing can account for the patterns of stratification found in deposits of clays and marls because the atmosphere, and especially the oceans, are very weakly coupled to the small, net changes in insolation that accompany orbital forcing. However, a sedimentation response within the Milankovitch band is more believable if most sediment is mobilized seasonally because orbital climate forcing works through seasonal changes in insolation. In the varved Permian Castile evaporite, for example, it is probably year by year changes in seasonal rates of accumulation of calcium sulphate that are responsible for the expression of cycles of precession and eccentricity (Anderson 1982, 1984).

An important conclusion drawn from large-scale changes in climate and stratification in the Milankovitch band is that the climate system must be in delicate balance. So delicate, in fact, that other subtle mechanisms of climate forcing probably are responsible for changes in sedimentation and stratification expressed on the time scale of decades to millennia. This conclusion is reinforced by the long, annual climate record from the Castile, which shows that variance within the Milankovitch band is not as strong as for millennial changes (Anderson 1982). The search is now on to find high-resolution climate records in lacustrine and marine environments and to understand climate variability at decadal to millennial timescales. Interpreting such records is aided by recognizing controls on seasonal sedimentation.

Framework for environmental and climatic reconstruction

Because seasonal changes in climate have a central role in regulating sedimentation, one can build a framework for sedimentology that is based on responses to the annual cycle of forcing.

My goal in this article is to present, in very general terms, a paradigm for sedimentology in which the centrepiece is not the environment of deposition. Rather, the focus is temporal, and the theme is the strong coupling that exists between seasonal climate forcing and the sedimentation response, as preserved in different sedimentary environments.

This brief sketch of a framework for seasonal sedimentation relies principally on examples

Table 1. Articles related to seasonal sedimentation published by R. Y. Anderson and co-workers

Category	Topics	Citations
Reviews	Evaporites, Orbital Forcing	Anderson (1984)
	Varve genesis. Meromictic Lakes	Anderson et al. (1985)
	Lacustrine Varves	Anderson & Dean (1988)
	Solar Variability in Varves	Anderson (1991)
Varve 'Theory'	Varves, Stratification	Anderson (1964; 1986)
	Marine Varves, ENSO	Anderson et al. (1990)
	Seasonal Sedimentation	This paper.
Methods, traps, seasonal processes	Time Series Methods	Anderson & Koopmans (1963; 1969; 1975)
	*	Anderson (1968)
	Sediment Trap Investigations	Anderson (1977)
	*	Dean & Anderson (1974)
	*	Anderson et al. (1984; 1985)
	*	Nuhfer & Anderson (1985)
	*	Nuhfer, et al. (1993)
Correlation and basin analysis	Varve Correlations	Anderson & Kirkland (1966; 1973)
	Turbidite Correlation	Dean & Anderson (1967)
	Evaporite Laminae	Dean et al. (1975)
	Folded Laminations	Kirkland & Anderson (1977)
	Brecciated Laminations	Anderson et al. (1978)
	Evaporites, Isotopic Shifts	Magaritz et al. (1983)
	Hydrologic Associations	Anderson & Dean (1995)
Palaeoclimatic time series Lacustrine	Rita Blanca, TX	Anderson (1969)
	*	Anderson & Kirkland (1969a, b)
	*	Kirkland & Anderson (1969)
	Elk Lake, MN	Dean et al. (1984)
	*	Bradbury et al. (1993)
	*	Anderson (1993)
	*	Anderson et al. (1993)
	*	Anderson, et al. (1993)
	Lake Estancia, NM	Allen & Anderson (1993)
	Florissant, CO	McLeroy & Anderson (1966)
Evaporite	Todilto Formation	Anderson & Kirkland (1960)
	Castile Formation	Anderson et al. (1972)
	*	Dean & Anderson (1973; 1978; 1982)
	*	Anderson (1982)
	*	Anderson & Kirkland (1987)
Marine, ENSO	Continental Slope, OMZ	Anderson et al. (1987)
	*	Anderson et al. (1989)
	*	Anderson et al. (1990)
	*	Dean et al. (1994)
	Isotopic Shifts	Linsley et al. (1990)
	ENSO Variability	Anderson, (1992b)
		Anderson et al. (1992)
Solar-climate associations	General	Anderson (1961; 1983; 1991)
	ENSO	Anderson (1990)
	Geomagnetic	Anderson (1992; 1993)

developed in a series of papers by myself and co-workers over nearly four decades. For the convenience of the reader I have assembled these investigations in Table 1. Articles in each category are cited chronologically and, along with other citations in the discussion to follow, provide the reader with examples of some of the methods used to reconstruct climate histories from laminated, partly laminated, and non-laminated sediments.

Earlier attempts to formulate a 'theory of sedimentation' based on varves drew on the observation that a varve was a 'microcosm' of stratification on a larger scale (Anderson 1986). That is, changes in sediment composition and geochemical and biological associations found during a single, annual cycle of sedimentation are repeated at all lower frequencies (longer periods), provided there has been no substantial change in the physical-biological system. Although this concept is an elaboration of the truism that the whole is the sum of the parts, it can be very useful when interpreting changes in composition and environmental associations under circumstances when laminations or varves are poorly preserved.

When examining the coupling between sedimentation and climate across the entire band of climate frequencies, I have found it useful to adopt a terminology developed for episodic processes, but less commonly applied in sedimentology.

Domains

Sedimentary processes are naturally separated into two domains according to their episodic behaviour and their resistance to forcing during the annual climate cycle. These domains were recognized as the most appropriate basis for rock classification (Walther 1897; Grabau 1904). Grabau based his genetic classification on agents or processes that produced rocks and he recognized that rocks were generated as the result of either mechanical (exogenetic) agents, or as a result of processes related to solution (endogenetic) agents. These two domains, here called mechanical and bio/chemical, (Fig. 1), can be used to categorize different types of seasonal laminations. Also, the concept is useful for recognizing the climatic implications of various

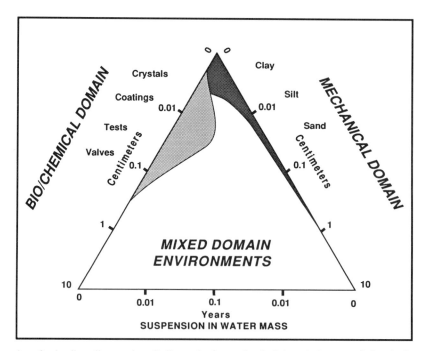

Fig. 1. Domains of episodic sedimentation. Sediment in the mechanical domain is suspended and advected after a forcing (climate) event exceeds a shear threshold. Sediment in the bio/chemical domian is generated in the water mass after a forcing event exceeds a weak chemical or vital threshold and accumulates as a 'sediment rain'. Fine particles generated in both domains remain suspended for long intervals, resulting in sedimentation in a mixed domain.

components within varves, especially when sediment is deposited in mixed domains, the usual case.

The mechanical domain corresponds to clastic sedimentary particles. Forces that move clastic particles at rest must first overcome resistance to movement by friction, adhesion, and cohesion. Stability in this domain and the resistance of fragmental materials to movement related to climate forcing is due to a strong shear threshold that exists between particles, resulting from the combined effects of gravity and frictional resistance. Within the mechanical domain, sediments tend to move laterally and episodically to lower elevations. Such particles are advectively transported by air and water currents, but also by gravitational sliding and flowage, as in the case of transport by ice, debris flows, or other processes related to gravity and mass wasting.

The bio/chemical domain includes a class of chemical-biological sediments, whose chief characteristic is production of particles directly or indirectly from a water body by biological fixation or chemical precipitation. In contrast to the mechanical domain, with its single, strong frictional threshold, thresholds for particle formation within the bio/chemical domain include various biological and chemical processes which extract particles from a watermass. Collectively, these processes have little 'resistance' to the annual cycle of climate forcing and are generally in harmony with the annual cycle through direct or indirect effects on physical properties such as temperature, pH, salinity, etc.). Within the bio/chemical domain, sedimentary particles, once created, also tend to move both vertically and laterally to lower elevations, often as a 'sediment rain.'

The two 'megadomains' are separated from each other by differences in the type and magnitude of resistance at thresholds. Such thresholds, or transition states, separate more stable states within each domain. Transitions between stable states in the mechanical domain where particles are at rest, tend to be abrupt, with particles set in motion or in suspension for short intervals of time (minutes to days) relative to time spent in stable states.

Stable states in the bio/chemical domain include materials in solution, and the same materials as organisms, organic fragments, valves, crystallites, etc. (Fig. 1). Threshold processes in the bio/chemical domain tend to be drawn out over days to weeks, corresponding to episodes of biologic growth or chemical precipitation. Once such particles are generated and accumulate on the bottom they fall into the mechanical domain. Examples of seasonal laminations formed entirely in the bio/chemical domain include evaporite and pure carbonate varves in which a seasonally precipitated component often is set off by a thin layer of organic matter, also derived from the water mass. Glacio-lacustrine layers and laminations accumulating in proglacial lakes represent sedimentation processes that fall almost entirely within the mechanical domain.

Mixed-domain environments

Mechanically or bio/chemically formed materials and laminae are end-members and most lacustrine and marine varves reflect deposition in environments that allow a mixing of processes in the two domains, *mixed-domain environments* (Fig. 1). Even in mixed-domain environments, separate components of varves or laminations accumulate as a result of mechanical or bio/chemical processes of sediment generation. In order to use varves effectively, whether for chronostratigraphy or for climate reconstruction, it is helpful to isolate the two markedly different types of sedimentation response associated with the two domains.

A good example of factors to be considered in reconstructing information from mixed domain environments is illustrated by varves that accumulated in an early Pleistocene lake basin on the high plains of Texas (Anderson & Kirkland 1969a, b). This small lake basin developed in the carbonate caprock of the Ogallala Formation, and carbonate that reached the floor of the lake was available as sediment both mechanically and as a bio/chemical precipitate. The other major component in the system was clay. We used the varves as a chronometer to obtain independent measurements of the flux of carbonate and clay to the lake bottom. Then we employed bivariate spectral analysis (Anderson & Koopmans 1969) to show that the flux of carbonate and clay were strongly anticorrelated for periods longer than around 60 years, but positively correlated for cycles at higher frequencies (< around 60 years, Fig. 2).

The shift in association at around 60 years can be explained as a change in relative dominance of bio/chemical and mechanical components and processes. Clay and detrital carbonate from the caprock margins of the basin were both available for episodic, advective, mechanical transport in surface runoff. Shear thresholds probably were exceeded each year, and large interannual runoff events flushed both clay and detrital carbonate into the small lake basin, explaining the positive association for abrupt,

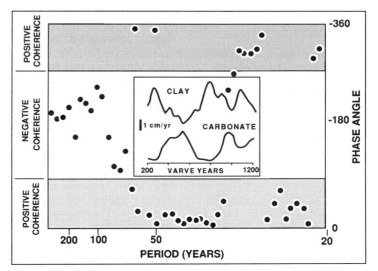

Fig. 2. Century-scale changes in rate of accumulation of clay and calcium carbonate in Rita Blanca varves and plot of phase angle. Note positive coherence for clay and carbonate in the range of interannual variability, and negative coherence for periods longer than around 60 years. The shift to negative coherence reflects the dominance of carbonate precipitated from bio/chemical domain during century-scale changes in climate.

interannual to decadal events (< around 60 years). However, finely divided carbonate was also being chemically and biogenically precipitated in the lake each year as an indirect result of the seasonal warming of lake water and biological activity during the warm season. Seasonal forcing of carbonate precipitation occurred each year during the annual cycle, but only longer and stronger changes in climate, cycles longer than around 60 years, precipitated enough carbonate to overcome the effect of episodic transport of carbonate.

The varve chronometer allowed us to isolate the effects of processes in each domain, and to show that anti-correlation of clay and carbonate, generated during the annual forcing cycle, was repeated again as an anticorrelation on the time scale of centuries. Because the relative contribution from each domain could be identified, the true association between carbonate and clay, over centuries, could be expressed in terms of seasonality. The seasonality expressed in century-scale changes, in turn, allowed us to infer sources of precipitation based on the regional climate setting of the lake basin (Anderson & Kirkland 1969a, b).

Many other examples of marls, marly clays, and calcareous silts, in basins of all sizes and types, could be cited to illustrate seasonal accumulations of sediment in mixed-domain environments. A particularly interesting example is the laminated calcareous siltstone of the Bell Canyon Formation that lies beneath the

varved Permian Castile evaporite (Anderson *et al.* 1972). The silt-organic couplets, which contain about 20–30% calcium carbonate, are transitional upward into evaporite varves. Some of the remarkable correlation over a distance of 24 km can probably be attributed to uniform precipitation of carbonate (seasonal 'pelagic' sediment rain). However, because quartz silt, probably aeolian, is the major component, the remarkable correlation (Fig. 3) shows that mechanically generated sediment also was distributed synchronously over a wide area in response to seasonal changes during the annual cycle.

Propagation of seasonality to longer-than-annual changes

Seasonality in sedimentation is re-expressed in the geological record as changes over decades, centuries, and millennia, often as stratification (Anderson 1964, 1986). The previous example from Rita Blanca lake sediments illustrates how seasonal processes are propagated to lower frequencies in a mixed-domain environment. In the pure bio/chemical domain of the Permian Castile Formation, the seasonal 'microcosm' consisted of carbonate, sulphate, and halite, organized as laminae. This triplet, on a millimetre to centimetre scale, was repeated to produce major units of stratification over thousands of millennia. The entire 1500 m thick geological formation, with limestone at the base, followed

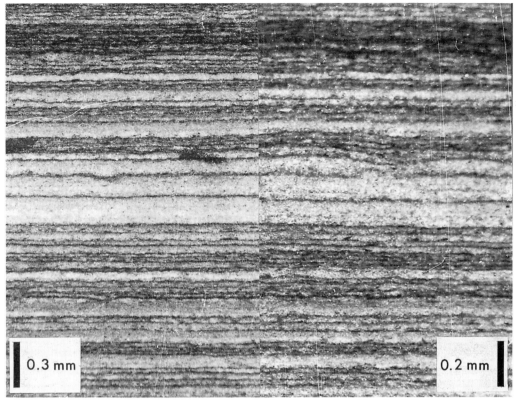

Fig. 3. Correlation of calcareous siltstone-organic laminations in cores from Permian Bell Canyon Formation over a distance of 24 km. Scale of core photos has been adjusted to align the laminations. Dominant component is fine silt, probably eolian. Uniformity of interannual response over 24 km distance testifies to efficiency of advection in the mechanical domain.

by sulphate and halite above, resembles a varve and reflects the importance of the annual cycle in regulating the accumulation of the entire deposit (Anderson 1986).

The Rita Blanca and the Castile examples, on very different temporal and spatial scales, show how the annual cycle of sedimentation resonates at longer time scales. Resonance of the annual cycle can be a powerful tool for interpreting the climatic and environmental significance of long term sedimentational and stratigraphic changes, even when varves are not preserved. This tool becomes even more important when it is recognized that some of the most responsive and informative climatic and environmental records are not varved. For example, in cases where varves are regular, continuously annual, and uninterrupted, preservation of such a record may mean that the basin was protected from extremes of variability. On the other hand, environments that are highly responsive to annual and interannual climate forcing tend to preserve laminations that are multiple,

interrupted, or otherwise poorly suited for use as a chronometer. Hence, there is a trade-off between varve continuity and responsiveness that calls for the reconstruction of high-resolution climatic and environmental records from partly laminated, and non-laminated sediments.

Resonance of the annual cycle offers a means for interpreting environmental and climatic change in responsive settings where varves are less than perfect, even where varves are totally lost to bioturbation or other processes. Application of the principle of seasonal propagation (resonance) will be illustrated with two examples, one lacustrine and one marine, in which varves have been obliterated or interrupted.

Estancia Basin, New Mexico

Local setting and sediment composition

Pluvial Lake Estancia had a large (1100 km²) surface area relative to its depth (40 m) at its maximum highstand during the last glacial maximum (LGM) (Allen & Anderson 1993). Near the centre of the lake, marly clay deposited during the LGM (Fig. 4) accumulated at an average rate of around 0.35 mm/year. Occasionally, thin, light–dark couplets less than 0.5 mm thick, in material similar to Fig. 4 are recognizable in fresh exposures, suggesting that a few varves may have been preserved. However, virtually all evidence for seasonal lamination has been destroyed by bioturbation of the top few millimetres of surface sediment, possibly in combination with gentle wave motion. What remains is a pattern of light and dark layering (Fig. 4).

Centimetre-scale couplets consist of alternations of light-coloured layers enriched in calcium carbonate (20–50%) and dark-coloured layers with high clay content (30–60%), slightly enriched in organic matter. Boundaries between light-coloured layers and dark layers are diffuse and transitional, owing to the mixing process (Fig. 4). Depth of mixing must have been less than the average thickness of a light or dark layer and probably was on the order of 0.5 cm. A third major component consists of layers, lenses, and laminations of pure detrital gypsum (Fig. 4). Gypsum laminae in the lake sediments are largely eolian, and the original gypsum particles grew as twin-crystals in shallow mud flats.

Even though the Estancia sequence is not varved it is assumed that sediment accumulated seasonally and that maxima in clay and carbonate accumulation occurred in different seasons. The regional setting plus other evidence for seasonal snowmelt and surface recharge of the lake (Allen & Anderson 1993) suggest that the clay component accumulated continuously during the year, but with more clay deposited near the end of the cool season and extending into late spring and summer. Carbonate occurs as finely divided particles or 'marl,' and as abundant valves of ostracodes. Carbonate and a small amount of dark-coloured disseminated organic matter reflect direct removal of soluble material from the water mass, mostly during the warm season. Of the major components, clay and detrital gypsum represent transport in the mechanical domain, and carbonate reflects particle generation in the bio/chemical domain.

The seasonal associations of clay and carbonate, defined above, although not directly known owing to a lack of varves, can also be deduced from a closely related analogue in the same region, the Rita Blanca lake deposits.

The probable relationship between laminae in the varved Rita Blanca analogue and the light and dark banding in Estancia sediment is shown

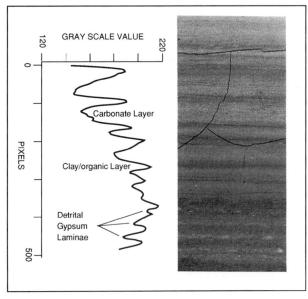

Fig. 4. Banded marl accumulated near the centre of Estancia lake basin at highstand during last glacial maximum and corresponding profile of gray scale values (3% smoothing). Light-coloured layers are enriched in calcium carbonate, dark layers contain more clay. Note gradational character of banding, believed to be the result of mixing and gentle bioturbation. Three, thin, light-coloured laminae near the base are composed of grains of eolian gypsum, partly disorganized from weak bioturbation.

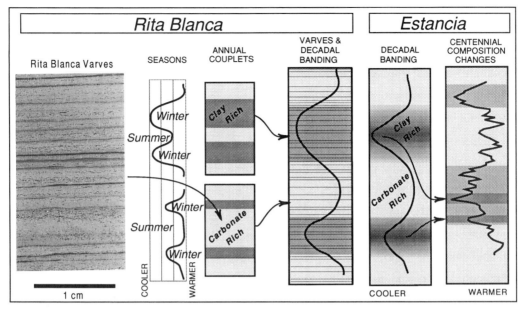

Fig. 5. Comparison of compositional changes in Rita Blanca varves with changes in non-varved analog from Lake Estancia. Winter and summer laminae at Rita Blanca occur in clay-rich and carbonate-rich intervals, with more clay reflecting winter and winter-like conditions, and more carbonate reflecting summer-like conditions over decades, and forming light- and dark-coloured banding (left panels). Varves are lost in the Estancia analogue due to sediment mixing, but interdecadal banding is preserved. Interdecadal changes accumulate over time to produce century-scale changes in composition and stratification (right panels).

in Fig. 5. Both deposits have clay and carbonate as major components. Both major components accumulated under a similar climate regime. In the Rita Blanca deposit one can readily show that banding on a larger-than-annual scale reflects incremental changes in seasonally generated laminae. The main difference between the two accumulations is the effect of weak bioturbation at Estancia and one can reasonably assume that seasonally incremental changes were responsible for banding at a larger scale in both deposits.

Reconstruction of decadal climate variability

The cm-scale banded structure of sediments in Estancia Basin is accurately depicted by relative changes in grey-scale values collected in profiles across the light-dark banding (Fig. 6). A 5-pixel-wide grey-scale plot profile was constructed from a CD-ROM photo image at a resolution of 6.4 pixels/cm, using Image 4.8 (NIH public domain software). Each dark band on the photographic image is marked on the profile plot and corresponds to a maximum in grey scale density (Fig. 6). Compositional changes in

the grey-scale profiles were linked to grey scale values by subsampling and analyzing for carbonate across several of the light and dark layers, as illustrated in Fig. 4 (inset in Fig. 6). Grey scale measurement was found to be an accurate estimator for relative proportion of carbonate and clay ($r = -0.97$) in those parts of the sequence where gypsum concentration is lower than around 5%.

Because varves were not available as a chronometer, AMS radiocarbon dating was used for temporal control. An age model for the sediment sequence, based on uncalibrated radiocarbon dates (Allen & Anderson 1993), assigns around 860 years to the illustrated interval (Fig. 6). The probable depth of mixing (<5 mm), when compared with rate of accumulation from the age model around 0.35 mm/year, indicates that sediment mixing filtered out cycles of less than 10–15 years. About 38 well defined dark laminae and grey-scale maxima occur in the *c*. 860-year interval, giving an average period of about 23 radiocarbon years for dark-light couplets. In addition, carbonate fluctuates systematically through about 4 to 5 longer cycles in the *c*. 860 year interval, for an average cycle period of around 200 years (Fig. 6).

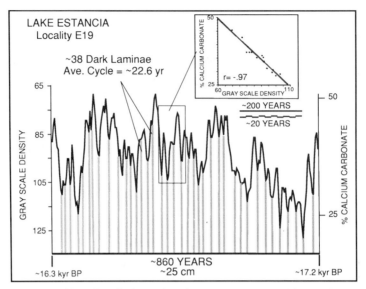

Fig. 6. Gray-scale density profile and equivalent values for percent percent calcium carbonate (inset) for a segment of Estancia sediment sequence that accumulated during the highstand of last glacial maximum. Note that dominant clay-carbonate cycles have periods of around 20 years and 200 years. Percent carbonate was estimated by corresponding gray scale values (see inset).

The continuous character of the grey-scale profile provides a continuity and resolution similar to that obtained from counting and measuring varves, and the Estancia example illustrates how patterns of interdecadal variability can be reconstructed from non-varved sediments. Even with inaccuracies inherent in the age model, temporal resolution is sufficient to characterize interdecadal variability. The continuity provided by continuous measurement offers a means to obtain near-annual resolution in sedimentary environments that were highly responsive to climate changes.

By applying the principle of resonance, one can assume that dark and light, clay and carbonate cycles in Lake Estancia sediment, as defined by continuous greyscale values with near-annual resolution, reflect systematic changes in the character of the seasons. Resonance with the annual cycle also implies that changes in climate, as reflected in compositional changes in sediments, can be projected backward to help interpret sedimentational responses during the annual cycle. For example, decade and century-scale alternations between clay and carbonate predict that Estancia varves, were they preserved, would probably be clay–carbonate couplets not unlike the interdecadal banding. Finally, the concept of resonance adds confidence that physical models that incorporate changes in lake level and composition, as

measured independently by physical, geochemical, and biologic proxies (Allen & Anderson 1993), can be linked to seasonal factors in regional climate.

Lamination–bioturbation cycles: offshore California

Laminations preserved in marine sediments at the depth of the oxygen-minimum-zone (OMZ, 700 m–1000 m) along the upper continental slope off northern and central California are sharply defined as dark and light-coloured laminae on a millimetre-scale (Fig. 7). Analysis of mineral and biogenic fractions from light–dark couplets, combined with data about seasonal flux of diatoms obtained from sediment traps (Anderson *et al.* 1987), has confirmed that couplets are seasonal and that series of light–dark couplets are annual. Dark laminae reflect suspension and transport of terrigenous materials to deeper waters by winter storms (mechanical domain). Light-coloured laminae are enriched in carbonate and summer-blooming species of diatoms associated with strong seasonal winds, Ekman transport, and upwelling (Anderson *et al.* 1987), representing the bio/chemical domain.

Interruptions in the processes of seasonal and annual deposition by the movement and

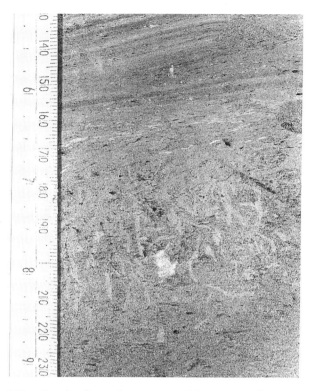

Fig. 7. Laminated and bioturbated sediment from oxygen minimum zone of upper continental slope off central California. Vertical borings of trace organisms change to horizontal borings and then to zones of undisturbed laminations (varves) during a shift to lower concentrations of dissolved oxygen. Large borings from a larger and later varve–bioturbation cycle are superimposed on laminations.

burrowing habits of benthic organisms provide indirect evidence for changes in concentration of dissolved oxygen (DO) at the sediment–water interface. Such interruptions range from a slight disturbance of a few recently deposited milli-metre-laminae (Fig. 7) to deep burrowing by organisms to depths of several decimeters. Data on concentrations of DO in the OMZ from oceanographic surveys permit a rough calibra-tion of DO and an index of bioturbation (Anderson *et al.* 1989). The temporal pattern and degree of disturbance of the primary varve structure shows that, by inference, the OMZ was highly dynamic, with DO fluctuating in response to changes in the ocean-atmosphere system (Anderson *et al.* 1989).

In the California slope example, marine varves, even though preservation was discon-tinuous over time, provide a reference point against which to measure the temporal response of the OMZ and permit an interpretation of that response in terms of ocean–atmosphere dynamics. For example, higher concentrations of both biogenic silica and organic matter in varved sediment (Dean *et al.* 1994) suggest that lower concentrations of DO and com-plete preservation of varves probably resulted from higher productivity, increased upwel-ling, stronger coastal winds, and stronger flow of the California Current under La Niña-like conditions.

Conditions that prevailed during bioturbated intervals were probably related to a weakening of the California Current, leading to a higher concentration of DO and a return of bioturbat-ing organisms. Weakening of the California Current is associated with El Niño conditions. Hence, the alternation of varves and bioturba-tion, in half-metre-scale lamination-bioturbation cycles (Fig. 8), suggests that El Niño-like and La Niña-like climate and circulation regimes per-sisted over intervals ranging from decades to several millennia.

In the offshore California marine example, even though varves are discontinuous and occur in relatively small intervals of the cores, their

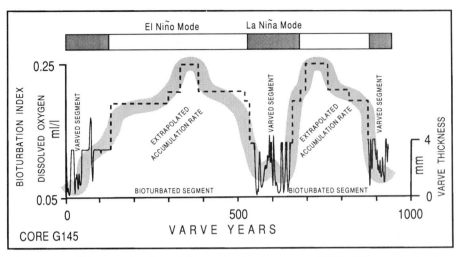

Fig. 8. Plot of varve thickness and estimated changes in dissolved oxygen (DO) within two half-metre-scale lamination-bioturbation cycles. Sediments accumulated in the oxygen minimum zone off central California. Systematic changes in DO and bioturbation culminated in zones of varves and small-scale bioturbation, reflecting persistent changes in El Niño- and La Niña-like conditions over time scales of hundreds to several thousand years (the extrapolated sedimentation rate in bioturbated zones is only approximate and may be higher owing to downslope sediment transport or lower owing to minor discontinuities).

preservation, along with the aid of seasonal data from a sediment trap, identifies a long-term pattern in ocean–atmosphere circulation related to El Niño, and also provides a rough estimate of the timing of those events. As was the case for the Estancia lacustrine example, the annual cycle of climatic forcing, and its relation to sediment type and domain, provides a means for reconstructing and interpreting past changes in climate and the environment in settings where varves are imperfectly developed.

Power of the solstice cycle

It is difficult to imagine two environmental settings of greater contrast than the closed lake of Estancia Basin in the continental interior, and the open-marine system along the upper continental slope. Yet, in spite of these differences, the light–dark structure of the seasonal laminations, their association with the seasons, and their expression of domains of sediment production are remarkably similar. Also similar is the fact that both examples lose their laminated character as a result of bioturbation and climate-driven changes in circulation. These genetic similarities, expressed in radically different environments, are indirectly the result of having a strong solstice cycle.

The Rita Blanca and Estancia lacustrine examples, and the California marine example, as well as other examples documented by many other investigators, follow logically from Mitchell's (1976) observation that the effects of seasonality outrank other climatic variability by two orders of magnitude. Climate proxies, once identified at any frequency can, based on the evidence for resonance, be used to help interpret changes anywhere in the time series, provided the environment has not changed.

As our understanding of the processes of seasonal sedimentation improves, the need for detailed investigations of the effects of seasonality may be partly replaced by usable analogues and generalizations. Without varves, it may eventually even be possible to reconstruct regional patterns of seasonality from large-scale changes in stratigraphy. For now, it seems safe to use the annual climate cycle and varves, even when imperfectly preserved, as both a chronometer and as an interpreter of environmental and climatic variability.

This paper incorporates work done under National Science Foundation Grant EAR90–19134. I am deeply indebted to co-workers B. D. Allen, J. P. Bradbury, W. E. Dean, J. V. Gardner, E. Hemphill-Haley, D. W. Kirkland, B. K. Linsley, and E. B. Nuhfer for their original ideas and contributions to the study of seasonal sediments.

References

ALLEN, B. D. & ANDERSON, R. Y. 1993. Evidence from western North America for rapid shifts in climate during the last glacial maximum. *Science,* **260,** 1920–1923.

ANDERSON, R. Y. 1961. Solar-terrestrial climatic patterns in varved sediments. *Annals New York Academy of Science,* **95,** 424–439.

—— 1964. Varve calibration of stratification. *In:* MERRIAM, D. G. (ed.) *Symposium on Cyclic Sedimentation.* Kansas Geological Survey Bulletin, **169,** Vol. 1, 1–20.

—— 1968. Sedimentary laminations in time-series study. *In:* MERRIAM, D. G. (ed.) *Computer Applications in the Earth Sciences: Colloquium on Time-series Analysis.* Kansas Geological Survey Computer Contribution, **18,** 68–72.

—— 1969. Systematic list of fossils in the Rita Blanca lake deposits: *In:* ANDERSON, R. Y. & KIRKLAND, D. W. (eds) *Paleoecology of an Early Pleistocene Lake on the High Plains of Texas.* Geological Society of America, Memoir, **113,** 77–81.

—— 1977. Short-term sedimentation response in lakes in western United States as measured by automated sampling. *Limnology and Oceanography,* **22,** 423–433.

—— 1982. A long geoclimatic record from the Permian. *Journal of Geophysical Research,* **87,** C9, 7285–7294.

—— 1983 (1961). Solar-terrestrial climatic patterns in varved sediments. *In:* SCHOVE, D. J. (ed.) *Sunspot Cycles,* Benchmark Papers in Geology, **68,** Dowden, Hutchinson, and Ross, Stroudsburg, Pa.

—— 1984. Orbital forcing of evaporite sedimentation. *In:* BERGER, A. L. *et al.* (eds) *Milankovitch and Climate,* Part 1, Reidel, Dordrecht, 147–162.

—— 1986. The varve microcosm: Propagator of cyclic bedding. *Paleoceanography,* **1,** 373–382.

—— 1990. Solar-cycle modulation of ENSO: A possible source of global climatic change. *In:* BETANCOURT, J. L. & MACKAY, A. M. (eds) *Proceedings, Sixth Annual Pacific Climate (PACLIM) Workshop.* California Department of Water Resources, Interagency Ecological Studies Program, Technical Report, **23,** 77–81.

—— 1991. Solar variability captured in climatic and high-resolution paleoclimatic records: A geologic perspective. *In:* SONETT, C. P., GIAMPAPA, M. S. & MATHEWS, M. S. (eds) *The Sun in Time.* Tucson, University of Arizona Press, Space Science Series, 543–561.

—— 1992a. Possible connection between surface winds, solar activity, and the earth's magnetic field. *Nature,* **358,** 51–53.

—— 1992b. Long-term changes in the frequency of occurrence of El Niño events. *In:* DIAZ, H. F. & MARKGRAF, V. (eds) *El Niño: Historical and Paleoclimatic Aspects of the Southern Oscillation.* London, Cambridge University Press. 193–200.

—— 1993. The varve chronometer in Elk Lake, Minnesota: Record of climatic variability and evidence for a solar/geomagnetic ^{14}C-climate connection. *In:* DEAN, W. E. & BRADBURY, J. P. (eds) *Elk Lake, Minnesota: Evidence for Rapid Climate Change in the North-Central United States.* Geological Society of America Special Paper, **276,** 45–67.

—— & DEAN, W. E. 1988. Lacustrine varve formation through time. *Palaeogeography, Palaeoclimatology, Palaeoecology,* **62,** 215–235.

—— & —— 1995. Filling the Delaware Basin: hydrologic and climatic controls on the Permian Castile Formation varved evaporite. *In:* SCHOLLE, P. A., PERYT, T. M. & ULMER-SCHOLLE, D. S. (eds) *The Permian of Northern Pangea. Vol. 2: Sedimentary Basins and Economic Resources,* Springer, New York, 61–78.

—— & KIRKLAND, D. W. 1960. Origin, varves, and cycles of Jurassic Todilto Formation, New Mexico. *AAPG Bulletin,* **44,** 37–52.

—— & —— 1966. Intrabasin varve correlation: *Geological Society of America Bulletin,* **77,** 241–256.

—— & —— 1969a. Paleoecology of the Rita Blanca lake deposits – a synthesis. *In:* ANDERSON, R. Y. & KIRKLAND, D. W. (eds) *Paleoecology of an Early Pleistocene Lake on the High Plains of Texas.* Geological Society of America Memoir, **113,** 141–157.

—— & —— 1969b. Geologic setting of the Rita Blanca lake deposits. *In:* ANDERSON, R. Y. & KIRKLAND, D. W. (eds) *Paleoecology of an Early Pleistocene Lake on the High Plains of Texas.* Geological Society of America Memoir, **113,** 3–13.

—— & —— 1973 (1966). Intrabasin varve correlation. *In:* KIRKLAND, D. W. & EVANS, R. (eds) *Marine Evaporites: Origin, Diagenesis, and Geochemistry,* Benchmark Papers in Geology, Dowden, Hutchinson, and Ross, Stroudsburg, Pa, 220–239.

—— & —— 1987. Banded Castile evaporites, Delaware Basin, New Mexico. *Geological Society of America Centennial Field Guide,* **2,** 455–458.

—— & KOOPMANS. L. H. 1963. Harmonic analysis of varve time series. *Journal of Geophysical Research,* **68,** 877–893.

—— & —— 1969. Statistical analysis of the Rita Blanca varve time-series. *In:* ANDERSON, R. Y. & KIRKLAND, D. W. (eds) *Paleoecology of an Early Pleistocene lake on the High Plains of Texas,* Geological Society of America Memoir, **113,** 59–75.

—— & —— 1975 (1963). Harmonic analysis of varve time series. *In:* Benchmark Papers in Geology, Dowden, Hutchinson, and Ross, Inc., Stroudsburg, Pa.

——, DEAN, W. E., KIRKLAND, D. W. & SNIDER, H. I. 1972. Permian varved Castile evaporite sequence, west Texas and New Mexico. *Geological Society of America Bulletin,* **83,** 59–86.

——, KIETZKE, K. K., & RHODES, D. J. 1978. Development of dissolution breccias, northern Delaware Basin, New Mexico and Texas. *New Mexico Bureau Mines and Mineral Resources Circular,* **159,** 47–52.

——, NUHFER, E. B. & DEAN, W. E. 1984. Sinking of volcanic ash in uncompacted lake sediment in Williams Lake, Washington. *Science,* **225,** 505–508.

——, ——, BRADBURY, J. P. & LOVE. D. 1985a Meromictic lakes and varved lake sediments in North America. *US Geological Survey Bulletin*, 1607.

——, —— & ——1985b. Sedimentation in a blast-zone lake at Mount St. Helens, Washington: Implications for varve formation. *Geology*, **13**, 348–352.

——, HEMPHILL-HALEY, E. & GARDNER, J. V. 1987. Persistent late Pleistocene-Holocene seasonal upwelling and varves off the coast of California. *Quaternary Research*, **28**, 307–313.

——, GARDNER, J. V. & HEMPHILL-HALEY, E. 1989. Variability of the late Pleistocene-early Holocene oxygen minimum zone off northern California. *In*: PETERSON, D. H. (ed.) *Aspects of Climate Variability in the Pacific and Western America*. American Geophysical Union, Geophysical Monograph, **55**, 75–84.

——, LINSLEY, B. K. & GARDNER, J. V. 1990. Expression of seasonal and ENSO forcing in climatic variability at lower than ENSO frequencies: Evidence from marine varves off California. *In*: MEYERS, P. A. & BENSON, L. V. (eds) *Paleoclimates: The Record from Lakes, Ocean, and Land. Palaeogeography, Palaeoclimatology, Palaeoecology*, **78**, 287–300.

——, SOUTAR, A. & JOHNSON, T. C. 1992. Long-term changes in El Niño/Southern Oscillation: evidence from marine and lacustrine sediments. *In*: DIAZ, H. F. & MARKGRAF, V. (eds) *El Niño: Historical and Paleoclimatic Aspects of the Southern Oscillation*. Cambridge University Press, 419–433.

——, BRADBURY, J. P. & DEAN, W. E. 1993a. Elk Lake in Perspective. *In*: BRADBURY, J. P., & DEAN, W. E. (eds) *Elk Lake, Minnesota: Evidence for Rapid Climate Change in the North-Central United States*. Geological Society of America Special Paper, **276**, 1–6.

——, ——, —— & STUIVER, M. 1993b. Chronology of Elk Lake sediments: Coring, sampling, and time-series construction. *In*: BRADBURY, J. P. & DEAN, W. E. (eds) *Elk Lake, Minnesota: Evidence for Rapid Climate Change in the North-Central United States*. Geological Society of America Special Paper, **276**, 37–43.

BARRELL, J. 1908. Relations between climate and terrestrial deposits. *Journal of Geology*, **16**, Nos. 2, 3, 4.

BRADBURY, J. P., DEAN, W. E. & ANDERSON, R. Y. 1993. Holocene climatic and limnologic history of the north central United States as recorded in varved sediments of Elk Lake, Minnesota – a synthesis. *In*: BRADBURY, J. P. & DEAN, W. E. (eds) *Elk Lake, Minnesota: Evidence for Rapid Climate Change in the North-Central United States*. Geological Society of America Special Paper, **276**, 309–328.

DEAN, W. E., JR & ANDERSON, R. Y. 1967. Correlation of turbidite strata in Pennsylvanian Haymond Formation, Marathon region, Texas. *Journal of Geology*, **75**, 59–75.

—— & ——1973 Trace and minor element variations in the Permian Castile Formation, Delaware Basin, Texas and New Mexico, revealed by varve calibration. *Fourth Symposium on Salt*, Northern Ohio Geological Society, Inc., Cleveland, Ohio, 275–285.

—— & ——1974. Application of some correlation coefficient techniques to time-series analysis. *Mathematical Geology*, **6**, 363–372.

——, DAVIES, G. R. & ANDERSON, R. Y. 1975. Sedimentological significance of nodular and laminated anhydrite. *Geology*, **3**, 367–372.

DEAN, W. E. & ANDERSON, R. Y. 1978. Salinity cycles – evidence for subaqueous deposition in the Castile Formation and lower part of the Salado Formation, Delaware Basin, Texas and New Mexico. *New Mexico Bureau Mines and Mineral Resources Circular*, **159**, 15–20.

—— & ——1982. Continuous subaqueous deposition of the Permian Castile evaporites, Delaware Basin, Texas and New Mexico. *In*: HANFORD, C. R., LOUCKS, R. G. & DAVIES, G. R. (eds) *Depositional and Diagenetic Spretcra of Evaporites*, Society of Economic Paleontologists and Mineralogists, Core Workshop, Calgary, Canada, 324–353.

——, BRADBURY, J. P., ANDERSON, R. Y. & BARNOSKY, C. J. 1984. Variability of Holocene climatic change: Evidence from varved lake sediments. *Science*, **226**, 1191–1194.

——, GARDNER, J. V. & ANDERSON, R. Y. 1994. Geochemical evidence for enhanced preservation of organic matter in the oxygen minimum zone of the continental margin of northern California during the late Pleistocene. *Paleoceanography*, **9**, 47–61.

DE GEER, G. 1912. A geochronology of the last 12,000 years. *Compte Rendus du XI Congrs Géologique International (Stockholm 1910)*, 241–253.

GRABAU, A. W. 1904. On the classification of sedimentary rocks. *American Geologist*, **33**, 228–247.

HAY, B. J., ARTHUR, M. A., DEAN, W. E., NEFF, E. D. & HONJO, S. 1991. Sediment deposition in the late Holocene abyssal Black Sea with climatic and chronological implications. *Deep Sea Research*, **38**, S2, S1211–1235.

HEIM, A. 1909. Einige Gedanken über Schichtung. *Verteljahrssachr. naturf. Gesell. Zurich*, 330–342.

KIRKLAND, D. W. & ANDERSON, R. Y. 1969. Composition and origin of the Rita Blanca varves. *In*: ANDERSON, R. Y. & KIRKLAND, D. W. (eds) *Paleoecology of an Early Pleistocene Lake on the High Plains of Texas*. Geological Society of America Memoir, **113**, 15–46.

—— & ——1970. Microfolding in the Castile and Todilto evaporites, Texas and New Mexico. *Geological Society of America Bulletin*, **81**, 3259–3282.

LINSLEY, B. K., ANDERSON, R. Y. & GARDNER, J. V. 1990. Rapid $\delta^{18}O$ and $\delta^{13}C$ isotopic shifts preserved in late Pleistocene marine varves on the California margin. *In*: BETANCOURT, J. L. & MACKAY, A. M. (eds) *Proceedings, Sixth Annual Pacific Climate (PACLIM) Workshop*, California Department of Water Resources, Interagency Ecological Studies Program, Technical Report **23**, 83–86.

MAGARITZ, N., ANDERSON, R. Y., HOLSER, W. T. & SALTZMAN, E. S. 1983. Isotope shifts in the late Permian of the Delaware Basin, Texas, precisely timed by varved sediments, *Earth and Planetary Science Letters*, **66**, 111–1124.

MCLEROY, C. A. & ANDERSON, R. Y. 1966. Laminations of the Oligocene Florissant Lake deposits. *Geological Society of America Bulletin*, **77**, 605–618.

MITCHELL, J. M., JR 1976. An overview of climatic variability and its causal mechanisms. *Quaternary Research*, **6**, 481–493.

NIPKOV, H. F. 1927. Uber das Verhalten der Skelette planktonischer Kieselalgen im geschichteten Tiefenschlamm des Zurich und Baldeggersees. *Rev. Hydrobiologie*, 4th Ann., 71–120.

NUHFER, E. B. & ANDERSON, R. Y. 1985. Changes in sediment composition during seasonal resuspension in small shallow dimictic inland lakes. *Sedimentary Geology*, **41**, 131–158.

——, ——, BRADBURY, J. P. & DEAN, W. E. 1993. Modern sedimentation in Elk Lake, Clearwater County, Minnesota. *In*: BRADBURY, J. P. & DEAN, W. E. (eds) *Elk Lake, Minnesota: Evidence for Rapid Climate Change in the North-Central United States*, Geological Society of America Special Paper, **276**, 75–96.

RICHTER-BERNBERG, G. 1960. Zeitmessung geologischer Vorgange nach Warven-Korrelation im Zechstein. *Geologische Rundschau*, Heft 1 Bd. **49**, 132–148.

SAURAMO, M 1923. Studies on the Quaternary varved sediments of southern Finland: Helsinko. *Bulletin de la Commission Gologique de Finlande*, No. 60.

WALTHER, J. 1897. Versuch einer Classification der Gesteine auf Grund der vergleichenden Lithogenie. Congres Géologique International, *Compte Rendu de la 7th Session*, St. Petersburg, **3**, 9–25.

WELTEN, M. 1944. Pollenanalytische, stratigraphische, und geochronologische Untersuchungen aus dem Faulenseemoos bei Spiez. *Geobot. Inst. Rubel Veroff.* (Zürich), **21**, 1–201.

Laminated diatomaceous sediments:
controls on formation and strategies for analysis

CONSTANCE SANCETTA

Room 725, National Science Foundation, 4201 Wilson Blvd, Arlington, VA 22230, USA

Abstract: This paper summarizes observations and ideas developed over a decade studying a variety of laminated sediments. Laminations of alternating lithic (dark) and biogenic (light) particles occur near land masses, where the lithic laminae may be interpreted as a record of precipitation, and the biogenic laminae as a record of primary production resulting from wind stress. The seasons corresponding to each type can be predicted from regional climatology. Biogenic laminae frequently consist of sublaminae representing a sequence of production events, so that several seasons of information may be included. Laminations of two types of biogenic particles are unusual, but have been reported from deep-sea settings. The physical control on these laminations is very indirect, and their occurrence is more probably a result of complex ecologic interactions. Although many laminar couplets are true varves, problems can occur when the normal cycle of alternating conditions is disturbed. Examples include the occurrence of a heavy rainstorm producing a dark lamination within the light lamina of a productive season, juxtaposition of two dark laminae from separate rainy seasons if the intervening productive events did not occur, and continuous rainfall throughout a productive season such that the light biogenic material is masked by the dark lithic material. Such cases, while complicating the use of laminae for time control, may provide useful information on interannual variability if they can be accurately identified. As an alternative to generating very long time series, it is suggested that selected clusters of samples combined with the use of analyses of variance may provide such information. Another strategy is the use of factor analysis to define seasonal proxies from the top part of a section; the proxies may then be projected downcore to derive information on seasonal trends of the past.

In the last ten years I have studied several types of laminated sediments, as well as following the literature on others. In addition to studying sediment cores, I have participated in two multi-year projects involving the use of biweekly or monthly sediment traps to examine the processes controlling the deposition of laminated diatomaceous sediments. The results of these studies have been published in detail elsewhere. However, over the years I have also evolved some general concepts which do not appear in the publications. I suspect that in some cases they are restating the obvious, but I offer them here in the hope that others may find them useful. They fall generally into three categories, dealing with the kind of seasonal information that may be derived from different lithologies, the problem of distinguishing true varves from other laminae, and strategies for extracting the unique information that laminated sediments can provide.

Lithology and seasonal forcing

Laminar pairs almost always consist of alternating lithologies which may be of three types: both biogenic, both non-biogenic (lithic) or one of each. Examples with which I am familiar contain at least one biogenic lamina, and I will confine my remarks to these.

Lithic/biogenic couplets

Cases of paired lithic and biogenic laminae occur on or close to land masses in lakes, estuaries, marginal seas and the continental shelf to upper slope. The lithic component almost always results from runoff, although conceivably aeolian transport could play a role as suggested for the Gulf of California (Baumgartner *et al.* 1991). The runoff may be derived directly from rainfall, or from melting of ice and snow. In the former case, thickness of the lamina is an indicator of precipitation; in the latter it may reflect temperature (hence, volume of ice melted) but this interpretation is less straightforward, since it may depend upon other elements such as variable concentration of detritus within the ice. Interpretation of these sequences requires a knowledge (or reasonable estimate) of the local climatology which may, very roughly, be inferred from latitude. Saanich Inlet, at 48° N on a western continental margin, lies within the temperate zone characterized by winter rainfall, and the sediment-trapping program showed that the dark laminae do result from deposition of lithic material between November and March (Sancetta 1989*a*, *b*). The central basins of the Gulf of California (25–27° N) lie within the subtropical zone where precipitation occurs primarily during

From Kemp, A. E. S. (ed.), 1996, *Palaeoclimatology and Palaeoceanography from Laminated Sediments*, Geological Society Special Publication No. 116, pp. 17–21.

the summer and fall. Sediment-trap samples for the months of July through October were dominated by lithic particles (Thunell *et al.* 1992), although it remains an open question whether they were derived directly from runoff, from winds blowing over the local deserts, or from resuspension of shelf sediments by internal waves. Composition of the diatom assemblage, which contains a variety of shallow-water benthic taxa, suggests that the latter may be the case. In any case, the resulting dark laminae can be interpreted as reflecting the occurrence of summer/fall storms.

The light biogenic lamina is usually dominated by phytoplankton microfossils (calcareous in the Black Sea, diatomaceous in most other locales) and derives from the season(s) of maximal production. It is therefore an indication of increased nutrient or (rarely) light availability, resulting from wind mixing followed by stratification. Variations in thickness or composition of this lamina type most likely depend upon intensity and frequency of mixing versus stratification, and may be interpreted as a proxy for wind stress (Bull & Kemp 1996). A thick layer implies several mixing events, in which case there may be taxonomic variations within the lamina as different taxa successively dominate the assemblage. Saanich Inlet and the Gulf of California both contain examples of sublamina sequences. In Saanich Inlet our sediment-trapping program identified a recurrent sequence of spring bloom, moderate summer production, and a small early fall bloom, each characterized by distinctive diatom taxa (Sancetta 1989*a*). Examination of lamina sequences from different depths in several cores consistently revealed the same sequence within each light lamina (Sancetta 1989*b* and unpublished observations). In the Gulf of California a two-year sediment-trapping program suggests that the typical sequence is a winter bloom of diverse large diatoms, followed by a spring bloom of smaller colonial forms (Sancetta 1993, 1995) and the same sequence has been identified in sediments from Guaymas Basin (Pike & Kemp 1996). In contrast, a thin lamina of uniform composition would reflect a single bloom event during an otherwise unproductive period.

Biogenic/biogenic couplets

Although less common, cases of alternating biocalcareous and biosiliceous laminae have been reported from deepwater sediments far from land, such as the Miocene sediments of the eastern equatorial Pacific (Sancetta 1983; Kemp & Baldauf 1993) and some of the Mediterranean

sapropels (Schrader & Matherne 1981; Sancetta 1994). In these cases the relationship to climate is less easy to interpret; while physical forcing may be the ultimate control, there is also implicit a strong effect related to plankton dynamics rather than to a single climatic variable. Generally, diatoms have a competitive advantage when nutrient levels and supply rates are high and silica is not limiting. Such conditions occur when stratification is first established after a period of upwelling or wind mixing. The calcareous phytoplankton dominate when nutrient supply is low and silica limiting, conditions that often occur after prolonged stratification, often with recycled nutrients accounting for a large proportion of production. Both the Mediterrean sapropels and the Miocene Pacific sediments consist of mats of almost monospecific diatoms alternating with laminae of mixed calcareous microfossils and diatoms. The mats presumably represent single bloom events while the mixed-composition laminae represent 'normal' pelagic production. The Pacific mats may have been produced by aggregation along an oceanic frontal zone (Kemp *et al.* 1995, 1996), which implies a unique combination of atmospheric and surface-ocean conditions. The Mediterranean mats were probably triggered by unusual density stratification resulting from fluvial influx (Sancetta 1994). In both cases 'normal' pelagic production resumed only after the mats had settled out of the surface waters, either from nutrient exhaustion, cellular senescence, or a wind event that destroyed the stratification. Although not directly demonstrated, it is inferred that during the time the mats were growing they suppressed production of other phytoplankton groups by any of a variety of mechanisms: shading, excretion of noxious compounds, or simply much higher growth rates.

Lithology thus can be used as a reflection of cyclic forcing, probably seasonal, which may be driven by temperature, runoff (precipitation or melting), or winds. Thickness of each lamina implies the intensity or duration of the respective forcing mechanism; sublaminae may record even shorter-scale events. If the climatic setting is known, it may be possible to draw inferences concerning palaeoseasonality of climatic variables. Purely biogenic lithologies are less easy to interpret, because the effect of physical forcing is complicated by the physiologic requirements of the different plankton organisms.

Laminae or varves?

The occurrence of visually distinct laminae necessarily implies two conditions or processes which alternate regularly for very long periods (centuries or millennia). Anderson (1986) has provided a cogent discussion of the implications of cyclic laminae, stressing the fact that the seasonal cycle is usually of such strong amplitude that it is likely to dominate the frequency domain. That is, in many cases the laminar couplets will be true varves, reflecting the contrast between two seasonal conditions. In Saanich Inlet the dark lamina represents winter and the light one comprises the sequence spring-summer-fall (Sancetta 1989a, b). In the Gulf of California the dark lamina corresponds to summer-fall and the light lamina to winter-spring (Thunell et al. 1992).

However, exceptions may be found to every rule, and even in a primarily varve-dominated system not all laminae are varve-halves. If conditions usually associated with one season occur during another, a 'false varve' will be created which may be indistinguishable from the 'true' ones. For example, in the Gulf of California the late winter and spring trap samples for 1992 contained a high proportion of lithic material that would probably produce a dark lamina, although this period is usually the season of high diatom production and light lamina formation (Pride et al. 1993). 1992 was an ENSO year, and ENSO events typically produce greater-than-normal storms and rainfall in the eastern Pacific (Bray & Robles 1991). Periods of high rainfall during the productive season, therefore, can produce dark laminae which might be misidentified as varve-halves. Conversely, one can imagine that an unusual sunny period during the rainy season might result in a small bloom and a light lamina which might be taken as a varve-half representing a very short year. In these cases it may be possible to distinguish varves from single-event laminae (false varves) if the taxonomic sequence of the normal seasonal cycle is known. For instance, if in Saanich Inlet one found a light lamina with spring bloom species overlain by a dark lamina and then a light lamina with sequential summer and fall taxa, it would be possible to determine that the dark lamina represented a heavy rainfall event in late spring, rather than the usual winter rains.

In the examples above the effect of an unusual event is an expansion of the varve chronology by one year. The opposite is also possible: for instance if a year were unusually rainy there might be a high proportion of lithic material throughout, resulting in a single thick dark lamina which would comprise two of the 'normal' rainy seasons plus the intervening anomalous wet season. The result would be a condensation of the varve chronology by one year. Again, it may be possible to use taxonomic composition of the microfossils to determine that a bloom occurred, even though

it is visually masked by the dark lithic material. Another example is the case in which the productive event that produces the light lamina fails to occur, resulting in immediate juxtaposition of two dark laminae. An extreme case of this kind of condensation occurs in the Black Sea, where 'varve' counting yields an age a factor of 2–3 younger than radioisotope analyses (Crusius & Anderson 1992).

False varves are even more likely in biogenic/biogenic couplets because the processes controlling the alternating lithologies may not be dependent upon a systematic annual cycle at all. In the case of the Miocene tropical Pacific laminae, it is impossible to determine the time scale on which the putative mechanism occurred; conceivably it could be anywhere from several times per year to once every few years or decades.

These complications are annoying if one is interested in the laminae primarily for their utility in chronologic control. However, they are probably the most interesting part of the section for the study of palaeoclimate, because they are the record of interannual variability. The frequency of false varves, if they can be identified as such, may provide a record of variability such as the ENSO cycle or the elusive sunspot cycle. It may be worth the effort to investigate a few suspicious cases in any given section – unusually thin or thick laminae, for instance – to determine whether false varves are common. This would, of course, require establishing the 'normal' pattern to either side of the suspicious section, and is most likely to be successful if another age control can be applied to test alternative varve-countings. Since sublamina composition may be the key to the problem, it would also require examination of an intact section by light- or scanning electron microscope.

Strategies for studying laminated sediments

Laminated sediments represent an embarrassment of riches. A variety of types of data can be generated (sedimentology, geochemistry, microfossils) and potentially thousands of samples if each lamina is studied individually. There is an understandable temptation to generate huge data sets merely because it is possible to do so, and it is very easy to yield to this temptation. But more knowledge can be achieved for considerably less time and expense if we first identify the key questions we wish to answer about environmental variability and design our sampling scheme accordingly. Some good scientific problems do require continuous annual samples, but many others do not. Even those which require annual intervals may not require them for hundreds of consecutive years. For instance, suppose we have a proxy for ENSO events and we want to know if the frequency of ENSOs has changed over time. Identification of the events themselves requires semi-annual sampling, since each event may last only for one year and occurs every 4 – 7 years. But if we wish to compare, say, the modern world with the Medieval Warm Epoch it is not necessary to study the full section in between. A 50- or at most 100-year section from each interval would be sufficient to establish whether there is a statistically significant difference. One would need to study the intervening centuries only if one found a difference and then wished to study the time evolution of ENSO frequency.

The idea of selecting time slices for comparison is not new, of course, and this is the most common strategy people have chosen for reducing the workload of studying laminated sequences. However, the assumption that one must do large numbers of consecutive samples continues to dominate our thinking. I would like to suggest that there are more efficient strategies to address the question of seasonal to interannual variability and how it may have changed over time. The key lies in using clusters of samples arranged in a chronologic hierarchy, and in the appropriate use of analyses of variance (ANOVA).

Suppose that we have a varved record spanning the interval from the present back through Isotope Stage 5 (125 000 ka BP), with accumulation of 2 mm/year so that each lamina (half-varve) is 1 mm. Clearly we are not going to analyze 250 000 samples, unless we plan to bequeath this project to our grandchildren! But with a much smaller number of samples we can address questions such as: which season(s) have been more variable (or more constant) at different parts of the glacial/interglacial cycle? Was interannual variability greater during, say, deglaciation (Termination I) or Stage 3 than during the mid-Holocene or the full glacial? How does interannual (or seasonal) variability during Stage 5e compare with that of the Holocene? What is the amplitude of interannual variability as compared with that of decadal, century, or millennial variability at different parts of the cycle?

As an example, let's say we want to compare the annual scale of variability for the Holocene and Stage 5e, and we want to study both light and dark laminae to determine the variability of winters (dark) and spring/summers (light) in the two time intervals. We decide that 15 years should be enough to capture the interannual range. We then sample 15 varves (30 laminae, 15 each of light and dark) for the Holocene and Stage 5e, yielding a

total of 60 analyses. ANOVA can then tell us whether the variance of the Holocene samples as a group is significantly different from that of Stage 5e as a group. Alternatively, we could compare the dark laminae against the light ones within each time interval to answer the question: How variable is winter compared to spring/summer during each time? If one is interested in century-scale variability, a different set of sample clusters would be used, say ten-year clusters every hundred years for some section. It would probably be a wise strategy early on to test how much incremental benefit is gained by adding one year to a cluster. For example, do we really need 15 years to capture the interannual variability, or would ten be enough?

The second strategy I suggest is best used on microfossil assemblages, and is probably practicable only for systems that are forming laminae at present. The idea is to establish a reference data set based upon the most recent laminae and then interpret older samples in terms of the reference set. It is analogous to the strategy proposed by Imbrie & Kipp (1971), substituting the uppermost laminae for core tops. One must first establish that the laminae are (usually) varves, using radioisotope control, and determine which seasons and conditions produce the light and dark laminae respectively. In cases like Saanich Inlet and the Gulf of California, the light laminae may be composed of sublaminae representing a sequence of several seasonal events. If so, one has the choice of subdividing each light lamina (more work, but potential for greater detail) or not (easier, but integrates all the production for each year). One then generates a reference data set in which each lamina (or sublamina) is a separate sample, working downward from the sediment surface until the number of samples is sufficiently large to capture the variability. If instrumental records from the region are available, one can link each sample to an environmental variable, using the average value for the season represented by each lamina.

One then performs a factor analysis upon the species composition data, just as in the Imbrie/Kipp technique for core tops. In most circumstances, the first few factors will correspond reasonably well to certain seasons or conditions. For instance, for diatoms in Saanich Inlet one might obtain factors corresponding to winter (rainfall, minimum production), early spring bloom (cold water, maximum production, no grazing), late spring/summer (warm stratified surface water, moderately low production, high grazing) and fall bloom (moderate wind mixing and production, no grazing). Factors can be related to climatic variables by regression analysis. One then quantifies taxonomic composition in older samples downcore, and expresses these data in terms of the factors by matrix multiplication (Imbrie & Kipp 1971). The result is a downcore plot of each factor and, if desired, of the climatic variables estimated from the regression equations.

It is generally agreed that laminated sediments have great potential to provide insights into short-term climate variability. The trick will be to develop creative ways to exploit this potential without becoming bogged down in masses of data. The approaches suggested here are essentially modifications of existing methods used by palaeoceanographers. Another promising direction is the adaptation of techniques used in tree-ring analyses, such as spectral analyses of varve thickness.

I thank Robert Thunell (University of South Carolina) and Steve Calvert (University of British Columbia) for their generous sharing of sediment trap material and data from their unpublished work.

References

ANDERSON, R. Y. 1986. The varve microcosm: propagator of cyclic bedding. *Paleoceanography*, **1**, 373–382.

BAUMGARTNER, T. R., FERREIRA-BARTRINA, V. & HENTZ-MORENO, P. 1991. Varve formation in the central Gulf of California: a reconsideration of the origin of the dark laminae from the 20th century varve record. *In*: DAUPHIN, J. P. & SIMONEIT, B. R. T. (eds) *The Gulf and Peninsular Province of the Californias*. American Association of Petroleum Geologists, Memoir, **47**, 617–635.

BRAY, N. A. & ROBLES, J. M. 1991. Physical oceanography of the Gulf of California. *In*: DAUPHIN, J. P. & SIMONEIT, B. R. T. (eds) *The Gulf and Peninsular Province of the Californias*. American Association of Petroleum Geologists, Memoir, **47**, 511–553.

BULL, D. & KEMP, A. E. S. 1996. Composition and origins of laminae in late Quaternary and Holocene sediments from the Santa Barbara basin. *This volume*.

CRUSIUS, J. & ANDERSON, R. F. 1992. Inconsistencies in accumulation rates of Black Sea sediments inferred from records of laminae and [210]Pb. *Paleoceanography*, **7**, 215–227.

IMBRIE, J.& KIPP, N. G. 1971. A new micropaleontological method for quantitative paleoclimatology: application to a late Pleistocene Caribbean core. *In*: TUREKIAN, K. K. (ed.) *The Late Cenozoic Glacial Ages*. Yale University Press, New Haven, 71–181.

KEMP, A. E. S. & BALDAUF, J. G. 1993. Vast Neogene laminated diatom mat deposits from the eastern equatorial Pacific Ocean. *Nature*, **362**, 141–143.

——, —— & PEARCE, R. B. 1995. Origins and paleoceanographic significance of Neogene laminated sediments from the eastern equatorial Pacific. *In*: PISIAS, N. G., MAYER, L. A., JANECEK, T. R., PALMER-JULSON, A., & VAN ANDEL, T. H. (eds) *Proceedings of the Ocean Drilling Program, (Scientific Results)*, **138**, College Station, TX (Ocean Drilling Program), 641–645.

PIKE, J. & KEMP, A. E. S. 1996. Preparation and analysis techniques for studies of laminated sediments. *This volume*.

PRIDE, C., THUNELL, R. & TAPPA, E. 1993. Laminated sediments of the Gulf of California: how many laminae make a varve? *EOS, Transactions, American Geophysical Union*, **74**, 372.

SANCETTA, C. 1983. Biostratigraphic and paleoceanographic events in the eastern equatorial Pacific: Results of DSDP Leg 69. *Initial Reports of the Deep Sea Drilling Project*, **69**, 311–320.

——1989a. Spatial and temporal trends of diatom flux in British Columbian fjords. *Journal of Plankton Research*, **11**, 503–520.

——1989b. Processes controlling the accumulation of diatoms in sediments. *Paleoceanography*, **4**, 235–251.

——1993. Seasonal succession of diatoms in the Gulf of California derived from two years of moored sediment traps. *EOS, Transactions, American Geophysical Union*, **74**, 371.

——1994. Mediterranean sapropels: seasonal stratification yields high production and carbon flux. *Paleoceanography*, **9**, 195–196.

——1995. Diatoms in the Gulf of California: Seasonal flux patterns and the sediment record for the last 15,000 years. *Paleoceanography*, **10**, 67–84.

SCHRADER, H. J., & MATHERNE, A. 1981. Sapropel formation in the eastern Mediterranean Sea: Evidence from preserved opal assemblages, *Micropaleontology*, **27**, 191–203.

THUNELL, R., PRIDE, C., TAPPA, E., MULLER-KARGER, F., SANCETTA, C., & MURRAY, D. 1992. Seasonal sediment fluxes and varve formation in the Gulf of California. *4th International Conference on Paleoceanography*, Kiel, Germany, 280–281.

Shale lamination and sedimentary processes

NEAL R. O'BRIEN

*Geology Department, State University of New York, College at Potsdam, Potsdam
New York 13676, USA*

Abstract: Lamination is a common feature of shale and other argillaceous rocks. The type
of shale lamination provides an important clue in identifying sedimentary processes when
combined with other geological evidence. Presented here is an overview of some common
lamination signatures in shale and the sedimentary processes responsible for their
formation. Macro-and microfabric and textures of Proterozoic, Palaeozoic, and Mesozoic
shales from various sedimentary environments are revealed by X-radiographs, thin-section
photomicrographs, and scanning electron microscope micrographs. Characteristic fabrics
associated with the following sedimentary processes are presented: (1) bottom flowing
currents produce features of fine to thick laminae (approx 0.5–1.0 mm thick) in association
with small-scale features of crossbedding, ripple marks, cut and fill and flame structures;
(2) deposition from low-density bottom currents and turbidity currents leaves a fabric
signature of graded bedding; (3) suspension settling is shown by alternating very fine
(<0.5 mm thick) non-graded silt and clay laminae; (4) microbial mats are recognized by
wavy lamination.

Shale is a 'fine-grained detrital sedimentary rock
which is characterized by its finely laminated
structure' whereas mudstone has the composition and texture of shale, but lacks its fine
lamination (Bates & Jackson 1987). This paper
illustrates some of the more common lamination
features and fabric signatures found in shales
and discusses how they may be used as clues in
interpreting sedimentary processes. Certain sedimentary processes seem to be very influential in
the development of lamination types. The most
important involve the individual or combined
effects of suspension settling, traction, and/or
sediment gravity flow (Stanley 1983). In general,
fine lamination could relate to one or a combination of (1) turbidite deposition; (2) contourites;
(3) nepheloid flow and drift deposits; (4) suspension settling; (5) variations in sediment delivery
or in the sedimentary environment; (6) plankton productivity variations; and (7) diagenesis
(Robertson 1984).

Lamination in fine-grained rocks has been
studied in an attempt to classify argillaceous
rocks or to use lamination as a clue to origin.
For example, petrographic classification of black
shales based upon lamination types has been
developed by O'Brien & Slatt (1990). Others
have used features seen in X-radiographs as a
basis for argillaceous rock classification (Nuhfer
et al. 1979; Cluff 1980; Nuhfer 1981; O'Brien &
Slatt 1990). Descriptive terms for laminae or
stratification types used by Campbell (1967) and
Cole & Picard (1975) are parallel and non-
parallel laminae which are further subdivided
into groups such as even, discontinuous, wavy,
or curved. Their work illustrates very well the
diversity of lamination types that actually exist
in fine-grained rocks.

Some have attempted to determine the reason
for lamination diversity by relating it to processes. Hallam's (1967) study of shale with
bituminous laminae indicated formation of
shale in shallow water under quiet conditions.
O'Brien (1989) found that the fine lamination in
Devonian shales of New York State was
produced by a repetitive alternate deposition of
hemipelagic and detached turbid layer sediments
in a density stratified basin. Three types of
lamination signatures are present in the Toarcian (Lower Jurassic) black shales from Yorkshire, Great Britain (O'Brien 1990). The fine
lamination is a clue of suspension settling; thick
lamination resulted from bottom flowing current
deposition; and wavy lamination is produced by
benthic microbial mats. Reineck & Singh (1980)
present a very comprehensive survey of rhyth-
mites (i.e. rocks or sediment with alternate beds
of laminae usually less than 3 or 4 mm thick and
of different composition, texture and colour).

This study shows the utility of the fabric
and other textural features in interpreting the
dominant sedimentary processes forming shale.
Presented here are examples revealed by X-radiography, thin-section petrography, and scanning
electron microscopy of common macro- and
microfabric features found in shales from various
sedimentary environments and geographic and
stratigraphic positions.

The focus of this paper is on identifying rock
features and relating them to the important shale
forming sedimentary processes such as those
associated with (1) bottom flowing currents, or

From Kemp, A. E. S. (ed.), 1996, *Palaeoclimatology and Palaeoceanography from Laminated Sediments*,
Geological Society Special Publication No. 116, pp. 23–36.

(2) low-density turbidity currents, or (3) suspension settling, or (4) microbial mat development.

Methods

The samples mentioned in this paper are listed in
Table 1. Pictured are selected examples from over
200 grey and black shales studied from various
stratigraphic and geographic locations. Standard
X-radiography, thin-section petrography, and
SEM techniques were used (refer to O'Brien &
Slatt 1990 for preparation details). All samples
show pictures of sections cut (X-radiographs and
thin-sections) or broken (SEM) perpendicular to
bedding.

Bottom flowing current processes

Kuehl *et al.* (1988) studied the fine-grained
sediments of the Amazon subaqueous delta and

observed that silt lamination forms as a result of
changes in sediment supplied from suspension
and from particle-sorting processes operating
near or on the seabed. Shown here are fabric
examples found in shales associated with the
latter process.

Cross and parallel lamination

Large or small scale cross-bedding found in
coarse clastic sediment provides good evidence
of unidirectional bottom flowing currents. In
shale outcrops cross-lamination is not conspicuous. Rine & Ginsburg (1985), however, have
observed in recent argillaceous sediments off
the Suriname coast the presence of micro cross-
lamination and parallel to subparallel lamination formed by wave generated currents. This
lamination type is interpreted to result from
boundary-layer shearing within bottom muds

Table 1.

Fabric Signature	Fig.	Geographic Location	Stratigraphic Location
cross-lamination	1a	Ontario County, New York (outcrop)	Devonian-Genesee Fm. Penn Yan Member
cross-lamination	1b	Steuben County, New York (EGSP well#3, 1208.7) (core)	Devonian-West Falls Fm. Rhinestreet Member
cross and parallel lamination	1c	Livingston County, New York (outcrop)	Devonian-Sonyea Fm. Cashaqua Member
ripple marks	2	Allegany County, New York (EGSP well#1, 2727) (core)	Devonian-Genesee Fm. West River Member
cut and fill	3	Conneaut, Ohio (outcrop)	Devonian-Ohio Shale Fm. Huron Member
flame structure	4	Steuben County, New York (EGSP well#3, 1208.7) (core)	Devonian-West Falls Fm. Rhinestreet Member
convolute bedding	5a, b	Mason County, West Virginia (EGSP well#146, 3169.4) (core)	Devonian-West Falls Fm. Angola Member
graded bedding	6a	Ontario County, New York (outcrop)	Devonian-Genesee Fm. Geneseo Member
graded bedding	6b	Wyoming County, New York (outcrop)	Devonian-Genesee Fm. Geneseo Member
graded bedding	7a, b	Ontario County, New York (outcrop)	Devonian-Genesee Fm. Geneseo Member
suspension settling	8a	Seneca County, New York (outcrop)	Devonian-Genesee Fm. Geneseo Member
suspension settling	8b	Ontario County, New York (outcrop)	Devonian-Genesee Fm. Penn Yan Member
suspension settling	9a, b	Seneca County, New York (outcrop)	Devonian-Genesee Fm. Geneseo Member
microbial mat	11	Meagher County, Montana (outcrop)	Mid-Proterozoic Newland Formation
microbial mat	12a, b	Port Mulgrave, Yorkshire Great Britain (outcrop)	Jurassic, Toarcian Jet Rock Formation Falciferum Zone
microbial mat	13a, b	Kimmeridge Bay, Dorset Great Britain (outcrop)	Jurassic, Kimmeridgian Kimmeridge Clay Fm. White Stone Band

produced by currents. Similar features are also found in some shales when viewed in X-radiographs or thin sections (Fig. 1a, b, c). The Penn Yan shale from the Appalachian basin of New York State is dominantly dark grey to black and is interpreted to have formed under anoxic conditions in quiet water. The presence of fine-scale cross-lamination in this unit indicates deposition not from absolutely stagnant water but instead under the influence of a very low velocity bottom flowing current under anoxic conditions (Fig. 1a). The grey

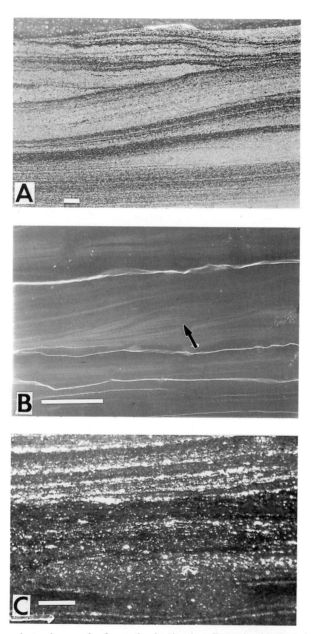

Fig. 1. (a) Thin section photomicrograph of cross-lamination in a Devonian black shale (Genesee Fm., New York). (b) X-radiograph of cross-lamination (see arrow) in a Devonian grey shale (West Falls Fm., New York). Scale = 1 cm. (c) Fine, parallel lamination (thin-section photomicrograph) in a Devonian grey shale (Sonyea Fm., New York). Scale = 1 mm

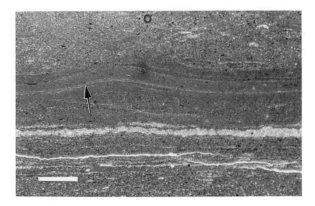

Fig. 2. Thin-section photomicrograph of ripple mark in a Devonian grey shale (Sonyea Fm., New York). Arrow points to fine laminae in the ripple. Scale = 1 mm.

Fig. 3. Cut and fill structure shown in a thin-section photomicrograph in a Devonian laminated black shale (Ohio Shale Fm., Ohio). (current lamination) in zone Y. Scale = 1 mm.

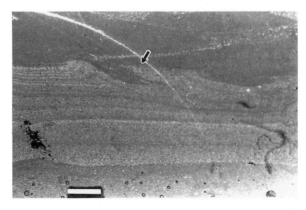

Fig. 4. Flame structure (arrow) in a Devonian shale shown in a thin-section photomicrograph (West Falls Fm., New York). Scale = 1 mm.

Rhinestreet shale from the Appalachian Basin in New York also shows cross-lamination indicative of a similar sedimentary process although the grey colour of the unit suggests a more oxygenated environment. (Fig. 1b). Current lamination is also indicated in the parallel laminae found in the West River shale (Fig. 1c). Minor fluctuation in current velocity is indicated by the change in laminae thickness vertically in the section.

Ripple marks

Another indication of bottom flowing currents is found in preserved ripple marks (Fig. 2). Notice the ripple is constructed from increments of sediment forming fine parallel laminae (see arrow Fig. 2). An increase in bottom flowing current velocity results in the following three shale fabric signatures.

Cut and fill and flame structures

Rine & Ginsburg's (1985) study of the modern mud shoreface in Suriname revealed scour and fill and discontinuity features that indicate high-energy events. These features are also found in the shales of this study.

The cut and fill feature is found in thin-section analysis of the black Ohio Shale of the Appalachian Basin in Ohio (Fig. 3). The feature is not apparent in outcrop. The microstratigraphic sequence represented by the 1 cm thick sequence shown in thin-section (Fig. 3) indicates a series of events starting with an interruption of existing background sedimentation by an erosion event (storm?) at zone X. Bottom scouring produced the undulating contact at X. That the current velocity of bottom flowing currents decreased during deposition is suggested by the fine parallel layers in zone Y. A similar erosion event is recorded in Fig. 4 which shows a rip-up zone or flame structure at arrow.

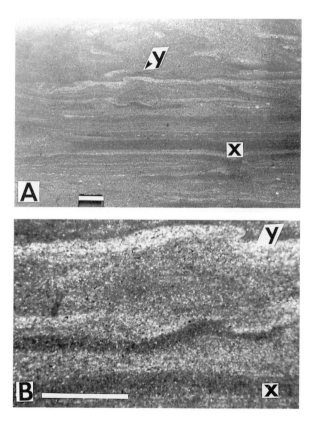

Fig. 5. Convolute lamination shown in a thin-section photomicrograph of a Devonian shale (Angola Fm., West Virginia). Scale = 1 mm.

Convolute lamination

Convolute bedding results from a variety of sedimentary processes which include differential compaction of sediment or from the shearing action of current flowing over bottom sediment (Reineck & Singh 1980). These authors state liquefaction is the important factor in the genesis of convolute bedding which can be found in turbidite sequences, intertidal flats, and fluvial environments. One sedimentary process common in these environments which relates to shales is that associated with bottom flowing currents. In Fig. 5 are examples of convolute bedding found in the dark grey Angola shale of the Appalachian Basin. Notice the presence of fine continuous uncontorted parallel laminae at Zone X (Fig. 5a, b) which becomes contorted in Zone Y. This fabric change suggests a change in sedimentary process probably produced by an increase of velocity of flow of bottom current. Note: since evidence of bioturbation is lacking in this shale, this structure is primary and indicative of original sedimentary processes.

Low-density bottom currents (dilute turbidity currents)

Graded bedding

Deposition from low density bottom flowing currents resulting from bottom flowing turbidity currents also may be responsible for lamination in shales. The distinctive fabric signature of graded bedding in shale laminae should be combined with other stratigraphic features normally associated with the turbidite mechanism to confirm this process, however. This author has noticed in a study of the Middle and Upper Devonian dark grey and black shales of the Appalachian Basin (New York State) that certain portions of the units display thick (>0.5–1.0 mm) laminae which are graded. Others investigating siltstones associated with these Devonian shales have proposed a turbidity current mechanism for the siltstone emplacement (Sutton 1963; Walker & Sutton 1967; McIver 1970). Fabric evidence also supports this sedimentary process for the origin of some of the Appalachian Basin shales.

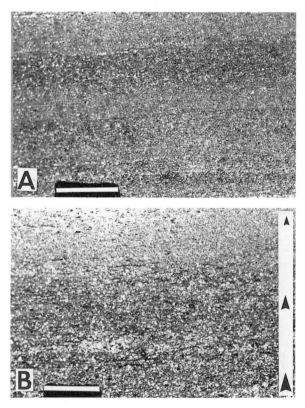

Fig. 6. (a) Graded bedding shown in a thin-section photomicrograph of a Devonian shale (Genesee Fm., New York). (b) Same as Fig. 6a. Arrows show particle size gradation in a lamination zone. Scale = 1 mm.

Figure 6(a, b) shows lamination in a portion of the shale member (Genesee Fm) from New York believed to have formed from a turbid flow. Most apparent is that laminae are thicker (>0.5+ mm) than those formed by other processes (Figs. 1–5 & Fig. 8). Graded bedding from fine silt to clay size occurs in adjacent lamina (Fig. 6B). Viewing by SEM reveals details which show particle size segregation vertically in a 1 mm thick lamina (Fig. 7). Fine silt and randomly oriented clay is found in lower parts, grades vertically to totally clay size material at the top of a typical lamina. The randomness of the clay and silt fabric has been interpreted to indicate rapid deposition of flocculated clay and silt which would occur in turbidite deposition (O'Brien *et al.* 1980).

The turbidity current mechanism has been proposed also by others to explain the origin of

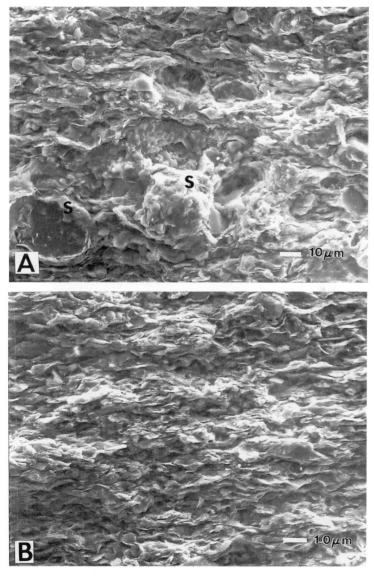

Fig. 7. SEM micrograph of graded bedding in a laminated shale (Devonian, Genesee Fm., New York). (**a**) Lower portion of a laminae showing coarse grains. S, Silt. (**b**) Upper finer, more clay rich portion of the same lamina. Total couplet thickness is 1 mm.

lamination in fine grained mudstones and sediment (Piper 1972; Stow & Bowen 1978, 1980). The latter (1978) state that depositional sorting by increased shear in the boundary zone (of a turbidity flow) separates clay flocs from silt grains and results in regular mud/silt lamination. The alternation of silt and clay laminae (and the associated graded bedding) in the shales of the Lower Jurassic Fjerritsler Formation is interpreted by Pedersen (1985) as evidence of rapid deposition from a storm generated cloud of suspended sediment. He said the units formed in a manner similar to fine-grained turbidites.

Suspension ettling

Fine, parallel, non-graded lamination

Some shales lack those fabric features commonly associated with the previously mentioned processes. These shales, for example, are characterized by alternating very thin (tenths of a millimetre thick) laminae of silt and clay which are not graded. They also show more couplets per unit of rock thickness than the other types. It is proposed that this very thin lamination type represents suspension settling.

Certain portions of the Devonian Genesee Formation in the Appalachian Basin, New York also illustrate features associated with suspension settling. The individual lamina are thin (commonly <0.5 mm) (Fig. 8a, b) and not graded (Fig. 9a, b). Laminae are parallel and continuous and contacts are abrupt. Some layers (Fig. 9b) contain only a few silt grains. No features similar to those produced by bottom currents are present.

There are three possible mechanisms to produce this type of fabric from suspension settling: (1) hemipelagic settling, (2) deposition from higher in the water column from a detached

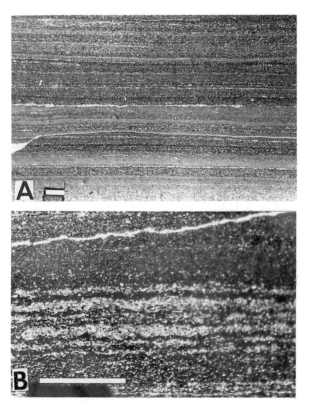

Fig. 8. Typical fine, parallel, ungraded laminae seen in thin-section Devonian shale (Genesee Fm., New York). (**a**) Notice thin, parallel laminae Scale = 1 mm. (**b**) Ungraded laminae, light layers are quartz-rich, darker layers contain organics and clays. Scale = 1 mm.

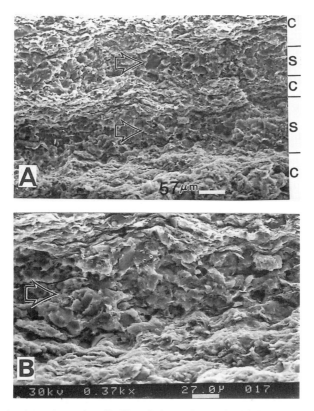

Fig. 9. SEM micrographs show alternating silt (S) and clay and organic rich layers (C) in a finely laminated Devonian shale (Genesee Fm., New York). (a) Thickness of a silt (quartz rich) lamina varies from 0.06–0.09 mm. See arrow. (b) Magnified view of an ungraded silt lamina (arrow) between two clay-organic rich zones.

turbid layer, and (3) deposition from dust clouds blown out into the basin and subsequent settling out.

Hemipelagic settling through the water column has been proposed as a mechanism of deposition for some of the organic-rich sediments in the Mediterranean (Anastasakis & Stanley 1984). It is suggested that the examples shown in Figs 8 & 9 show shale fabric signatures indicating deposition in the more distal portions of a basin where finer and better sorted sediments would be found. Episodic sedimentation also is indicated by the regular alternation of different grains in the adjacent lamina.

Another likely mechanism for the formation of this fabric may be that associated with detached turbid layers. Such layers have been described in sediment deposited off the coast of California in the Santa Barbara channel (Drake 1971; Drake *et al.* 1973). A detached turbid layer model is proposed by Pierce (1976) to explain how suspended sediment is transported over the outer continental margin. Stanley (1983) uses a

similar model to account for some laminated muds in the Mediterranean Sea. He states the laminated facies could indicate deposition in a stratified water column. Concentrations of particles could spread broadly on density interfaces in the water column as continuous or detached turbid layers and with time produce a progressive release of silt and clay over large areas of the sea floor (Stanley 1983). The Devonian shale described in Figs 8 & 9 has formed in a density stratified sea (Ettensohn & Elam 1985). Thus, the presence of lamination in this shale may be a clue to the presence of a pycnocline which facilitated sediment segregation into laminae. It is proposed that the silty laminae were deposited mainly from silt-rich detached turbid layers which interrupted the background sedimentation of the hemipelagic organic-rich clays (O'Brien 1989). Sedimentation at the distal margins of the turbid flow is suggested by the sorting, fine grain size and thickness (i.e. fine, well sorted silt would be more likely carried out to the distal portions of such a flow). Silt from the detached layer rained down onto the organic-rich clayey (of hemipelagic origin) bottom sediment that formed in the density stratified basin. The dark colour of the shale and sparse benthic fossil content suggest the presence of an oxygen deficient bottom condition which would occur within a strongly stratified basin water column. This type of laminated signature of the shale indicates frequent episodes of intrusion of a silty turbid layer followed by longer periods of hemipelagic organic-rich clay sedimentation.

Thirdly, an alternate mechanism, involves an aeolian source for the silt (and/or clay?). Suspension sedimentation would still be the dominant final process producing lamination; however, some or all of the sediment could be delivered out into the distal portions of the basin by dust storms. Microlaminations (approximately 1 mm thick) in a core sequence have been described as representing a yearly fluctuation in the supply of aeolian material to the Gulf of Aden (Olausson & Olsson 1969). Aeolian origin is also ascribed to terrigenous laminae in Gulf of California sediments – see Pike & Kemp (1996). Once the silt and clay settled into the water, the sedimentation process would be the same as previously described.

Microbial mat

The process of formation of dark grey to black shales in association with an organic mat (referred to by some as an 'algal-fungal' mat and in this paper as a microbial mat) has been discussed by numerous investigators (Kaufmann 1981; Brenner & Seilacher 1978; Reigel et al. 1986; Schieber

MICROBIAL MAT

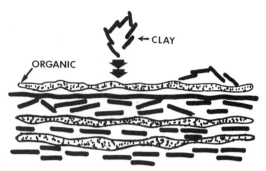

Fig. 10. Model illustrating microbial mat development and formation of wavy lamination.

1986; Loh et al. 1986; O'Brien 1990). The most distinguishing characteristic of shale lamination produced by microbial mats is that it is wavy and consists of alternating light and dark laminae (commonly <1 mm in thickness) (Fig. 10). The dark layers represent organic matter whereas light coloured laminae are usually composed of silt and/or clay. Schieber (1986) states the association of the sediment with carbonaceous matter can be explained by sediment particle trapping on the sticky mucilaginous mat surface. The alternating layers thus represent repeating conditions of mat growth and sediment trapping (Figs 10 & 11). This wavy or fine lamination is present in the Toarcian (Jurassic) black shales of Great Britain and Germany (Fig. 12a, b). The significance of wavy lamination in interpreting sedimentary processes is well illustrated in Kauffman's (1981) study of the Posidonienschifer of Germany. He

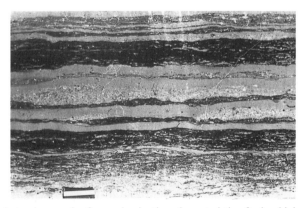

Fig. 11. Thin-section photomicrograph of wavy lamination characteristic of microbial mats. Sample is the Newland shale Fm. described by Schieber (1986) as of mat origin and given to the author for this study by Schieber.

Fig. 12. Wavy lamination of a microbial mat signature. (**a**) Thin-section photomicrograph of wavy lamination in a black shale (Jet Rock Fm., Great Britain) Scale = 1 mm. (**b**) SEM micrograph lamination features of shale shown in Fig. 12a. Notice alternating textures in lamina. Scale = 1 mm.

indicates that this shale (assumed by many to have formed in an anoxic stagnant basin) formed under conditions of a fluctuating oxic–anoxic boundary above and below the sediment–water interface. He supports a model 'that proposes an extensive-fungal mat situated a few centimetres above the interface, entrapping anaerobic waters below it even in the face of currents ....' The wavy lamination signature found in organic rich shales thus suggests alternate conditions of mat development and silt/clay deposition.

The wavy laminated portion of the White Stone Band of the Kimmeridge Clay (Dorset, England) is related in part to mat development (Fig. 13a, b). Organic-rich layers alternate with coccolith rich lamina (Fig. 13a). A similar fabric has been found in Kimmeridge facies of the Southern Jura (Tribovillard *et al.* 1992). They interpret the origin of the laminated fabric as that produced by conditions of organic or

carbonate sedimentation alternating by periods of nondeposition marked by microbial mat development. Gallois (1976) interpreted the coccolith layers in the Kimmeridge Clay of Dorset as indications of season blooms. The alternation of sediment types shown in Fig. 13b could be produced by episodes of mat development followed by an influx of coccolith fragments which signifies the time of a bloom. Regardless of the details of events, the significance is that this type of lamination is an indicator of the role played by microbial mats in the sedimentary process.

Conclusion

X-radiography, thin-section microscopy, and scanning electron microscopy reveal small scale features of shales useful as clues in interpreting sedimentary processes. Lamination types and

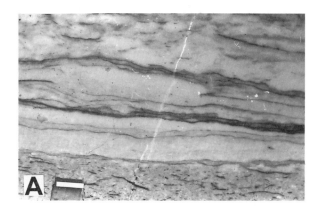

Fig. 13. Wavy lamination fabric: (**a**) Thin-section photomicrograph of wavy lamination in the White Stone Band (Kimmeridge Clay, Dorset, Great Britain). Dark layers are organic-rich; white layers are coccolith rich. Scale = 1 mm. (**b**) SEM micrograph showing alternating organic-rich and coccolith lamina in the White Stone Band

associated features may be used in identifying these processes. Fine-grained laminated argillaceous rocks formed by bottom flowing current processes possess cross-lamination, fine to thick (approximately 0.5–1.0 mm) parallel lamination, ripple marks, cut and fill structure, flame structure, or convolute lamination. Graded bedding in a lamina of shale suggests deposition by low density bottom or turbidity currents. The fine-grained, thin (<0.5 mm thick), non-graded and parallel lamination in other shales is a signature of a suspension settling process associated with any of three possible mechanisms – hemipelagic

settling, deposition from detached turbid layers or from dust clouds blown out over the depositional site. Wavy lamination results from microbial mat development. It is proposed that in any study of the origin of shales one should consider the important feature of lamination type because it offers clues to sedimentary processes.

The author is grateful for financial assistance from the Donors of the Petroleum Research Fund, administered by the American Chemical Society, the National Science Foundation (grant EAR-8611608), and the

Potsdam College Mini-grant program. Thanks is also given to R. W. Dalrymple, Jurgen Schieber, Michael House, Douglas Patchen, William Kirchgasser, Gordon Baird, The West Virginia Geological Survey, and New York Geological Survey who provided samples or technical support. The author also thanks Janet Bullis for important technical and editorial assistance.

References

ANASTASAKIS, G. C. & STANLEY, D. J. 1984. Sapropels and organic-rich variants in the Mediterranean: sequence development and classification. *In*: STOW, D. A. V. & PIPER, D. J. W. (eds) *Fine-Grained Sediments: Deep Water Processes and Facies*. Geological Society, London, Special Publication, **15**, 497–510.

BATES, R. J. & JACKSON, J. A. 1987. *Glossary of Geology*. American Geological Institiute, Alexandria, Va., 788.

BRENNER, K. & SEILACHER, A. 1978. New aspects about the origin of the Toarcian Posidonia Shales. *Neues Jahrbuch für Geologie und Palaeontologie Abhandlungen*, **157**, 11–18.

CAMPBELL, C. V. 1967. Lamina, laminaset, bed, and bedset. *Sedimentology*, **8**, 7–26.

CLUFF, R. M. 1980. Paleoenvironment of the New Albany Shale Group (Devonian-Mississippian) of Illinois. *Journal of Sedimentary Petrology*, **50**, 767–780.

COLE, R. D. & PICKARD, M. D. 1975. Primary and secondary sedimentary structures in oil shale and other fine-grained rocks, Green River Formation (Eocene), Utah and Colorado, *Utah Geology*, **2**, 49–67.

DRAKE, D. E. 1971. Suspended sediment and thermal stratification in Santa Barbara Channel, California. *Deep-Sea Research*, **18**, 763–769.

——, KOLPACK, R. L. & FISHER, P. J. 1972. Sediment transport on the Santa Barbara–Oxnard Shelf, Santa Barbara Channel, Calif. *In*: SWIFT, D. J. P., DUANE, D. B. & PILKEY, O. H. (eds) *Shelf Sediment Transport: Process and Pattern*. Dowden, Hutchinson and Ross, Stroudsburg, PA, 307–331.

ETTENSOHN, F. F. & Elam, T. D. 1985. Defining the nature and location of a Late Devonian–Early Mississippian pycnocline in eastern Kentucky. *Geological Society of America, Bulletin*, **96**, 1313–1321.

GALLOIS, R. W. 1976. Coccolith blooms in the Kimmeridge Clay and origin of North Sea oil. *Nature*, **259**, 473–475.

HALLAM, A. 1967. The depth significance of shales with bituminous laminae. *Marine Geology*, 5 481–94.

KAUFMAN, E. G. 1981. Ecological reappraisal of the German Posidonienschiefer (Toarcian) and the stagnant basin model. *In*: GRAY, J., BOUCOT, H. J. & BERRY, W. B. N. (eds), *Communities of the Past*. Dowden, Hutchinson and Ross, Stroudsburg, PA, 311–381.

KUEHL, S. A., NITTROUER, C. A. & DEMASTER, D. J. 1988. Microfabric study of fine-grained sediments: observations from the Amazon subaqueous delta. *Journal of Sedimentary Petrology*, **58**, 12–23.

LOH, H., MAUL, B., PRAUSS, M. & RIEGEL, W. 1986. Primary production, maceral formation and carbonate species in the Posidonia Shale of NW Germany. SCOPE-UNEP, Sonderband, **60**, 397–421.

McIVER, N. L. 1970. Appalachian turbidites. *In*: FISHER, G. W., PETTIJOHN, F. J., REED, J. C. JR. & WEAVER, K. N. (eds) *Studies of Appalachian Geology: Central and Southern*. Wiley, New York, 69–81.

NUHFER, E. B. 1981. Mudrock fabrics and their significance: discussion. *Journal of Sedimentary Petrology*, **51**, 1027–1029.

——, VINOPAL, R. J. & KLANDERMAN, D. S. 1979. X-radiograph atlas of lithotypes and other structures in the Devonian shale sequence of West Virginia and Virginia. METC/CR-79/27, NTIS, Springfield, Virginia.

O'BRIEN, N. R. 1989. Origin of lamination in Middle and Upper Devonian black shales, New York State. *Northeastern Geology*, **11**, 159–165.

——1990. Significance of lamination in Toarcian (Lower Jurassic) shales from Yorkshire, Great Britain. *Sedimentary Geology*, **67**, 25–34.

——, NAKAZAWA, K. & TOKUHASHI, D. 1980. Use of clay fabric to distinguish turbiditic and hemipel-agic siltstones and silt. *Sedimentology*, **27**, 47–61.

—— & SLATT, R. M. 1990. *Argillaceous Rock Atlas*. Springer, New York.

OLAUSSON, E. & OLSSON, I. V. 1969. Varve stratigraphy in a core from the Gulf of Aden. *Palaeogeography, Palaeoclimatology, Palaeoecology*, **6**, 87–103.

PEDERSEN, G. K. 1985. Thin, fine-grained storm layers in a muddy shelf sequence: an example from the Lower Jurassic in the Stenlille 1 well, Denmark. *Journal of the Geological Society, London*, **142**, 357–374.

PIERCE, J. W. 1976. Suspended sediment transport at the shelf break and over the outer margin. *In*: STANLEY, D. J. & Swift, D. J. P. (eds) *Marine Sediment Transport and Environmental Management*. Wiley, New York, 437–458.

PIKE, J. & KEMP, A. E. S. 1996. Records of seasonal flux in holocene laminated sediments, Gulf of California. *This volume*.

PIPER, D. J. W. 1972. Turbidite origin of some laminated mudstones. *Geological Magazine*, **109**, 115–126.

REINECK, H. E. & SINGH, I. B. 1980. *Depositional Sedimentary Environments*. Springer, Berlin.

RIEGEL, W., LOH, H., MAUL, B. PRAUSS. M. 1986. Effects and causes in a black shale event – the Toarcian Posidonia shale of NW Germany. *In*: WELLISER, O. (ed.) *Global Bio-Events*. Lecture Notes in Earth Sciences, **8**, Springer, Berlin, 267–276.

RINE, J. M. & GINSBURG, R. N. 1985. Depositional facies of a mud shoreface in Suriname, South America – a mud analogue to sandy shallow-marine deposits. *Journal of Sedimentary Petrology*, **55**, 633–652.

ROBERTSON, A. H. F. 1984. Origin of varve-type lamination, graded claystones and limestone-shale couplets in the lower Cretaceous of the western North Atlantic. *In*: STOW, D. A. V. & PIPER, D. J. W. (eds.) *Fine-Grained Sediments: Deep Water Processes and Facies*. Geological Society, London, Special Publication, **15**, 437–452.

SCHIEBER, J. 1986. The possible role of benthic mats during the formation of carbonaceous shales in shallow Mid-Proterozoic basins. *Sedimentology*, **33**, 521–536.

STANLEY, D. J. 1983. Parallel laminated deep-sea muds and coupled gravity flow-hemipelagic settling in the Mediterranean. *Smithsonian Contributions to the Marine Sciences*, **19**.

STOW, D. A. V. & BOWEN, A. J. 1978. Origin of lamination in deep sea, fine-grained sediments. *Nature*, **274**, 324–328.

—— & ——1980. A physical model for the transport and sorting of fine-grained sediment by turbidity currents. *Sedimentology*, **27**, 31–46.

SUTTON, R. G. 1963. Correlation of Upper Devonian strata in south-central New York. *In*: SHEPPS, V. C. (ed.) Symposium on Middle and Upper Devonian stratigraphy of Pennsylvania and adjacent states. *Pennsylvania Geological Survey Bulletin*, G-39, 87–101.

TRIBOVILLARD, N. P., GORIN, G. E., BELIN, S., HOPFGARTNER, G. & PICHON, R. 1992. Organic-rich biolaminated facies from a Kimmeridgian lagoonal environment in the French Southern Jura Mountains – a way of estimating accumulation rate variations. *Palaeogeography, Palaeoclimatology, Palaeoecology*, **99**, 163–177.

WALKER, R. G. & SUTTON, R. G. 1967. Quantitative analysis of turbidites in the Upper Devonian Sonyea Group, New York. *Journal of Sedimentary Petrology*, **37**, 1012–1022.

Preparation and analysis techniques for studies of laminated sediments

JENNIFER PIKE & ALAN E. S. KEMP

Department of Oceanography, University of Southampton,
Southampton Oceanography Centre, European Way, Southampton SO14 3ZH, UK

Abstract: Laminated sediments commonly represent the highest resolution continuous records available of marine and lacustrine variability. To exploit this information effectively careful sampling and preparation procedures must be followed, thus the method of analysis chosen must have sufficient resolution to recover the data. The scanning electron microscope, because of its good spatial resolution, is an ideal tool for analyzing laminated sediment fabric. Backscattered electron imagery produces preserved flux data on intra-annual time-scales, which enables high-resolution fabric studies, palaeoenvironmental reconstructions and compilation of data for time-series analysis. This contribution summarizes the techniques used for laminated sediment fabric investigation and discusses the use of the scanning electron microscope, including the best preparation method for thin sections and the procedure for backscattered electron imagery analysis.

Laminated sediments occur intermittently throughout the geological record and have contained within them the highest resolution sediment record of ancient ocean/atmosphere variability. In order to facilitate palaeoenvironmental reconstructions on millennial to intra-annual scale, it is essential to exploit fully the record of past climate and hydrography recorded in these sediments. A wide range of methodologies, from visual description of exposures cropping-out on the Earth's surface for facies analysis (De Greer 1912; Domack 1984), to the application of geophysical logging for investigating millennial-scale cyclicity (Fischer & Roberts 1991), have been employed for this purpose. Techniques such as X-radiography and optical microscopy have been used to retrieve the high-resolution intra- to inter-annual record of variability recorded by laminae. Recently, scanning electron microscope (SEM) techniques, particularly backscattered electron image analysis (Table 1), have also been used for this purpose.

Techniques for the study of laminated sediments

The most common laboratory techniques used to investigate laminated sediments include X-radiography, core surface photography, digital imaging, optical and scanning electron microscopy (Figs 1 & 2). Core surface photography is a standard method of imaging cores and requires no description. X-radiography and digital imaging are discussed below.

X-radiography

This technique was introduced into sedimentology by Hamblin (1962), and its potential in the study of laminated sediments was immediately realized (Calvert & Veevers 1962). A comprehensive description of X-radiography is given by Bouma (1969). Briefly, X-radiography is based on the differential passage of X-radiation, through an heterogenous media, onto X-ray sensitive photographic film. Dense, lithogenic laminae will produce light negative images and more penetrable, biogenic laminae produce darker negative images. X-radiography thus provides useful information on the gross structure of the sediment (Soutar & Crill 1977).

Digital imagery

This is a relative new technique for producing core photographs as previously, traditional photographs/X-radiographs were digitized rather than digital images being taken directly from the core (Bond *et al.* 1992). In digital imagery, a digitizer is scanned down the core surface and the colour/grey-scale of the sediment are recorded producing an image similar to a standard photograph (e.g. Schaaf & Thurow 1994).

Time-series analysis of laminated sediment sequences may be carried out at any scale using either photographic, digital or electron microscope images. Grey-scale and colour density analyses are most commonly used. For example, high-resolution grey-scale variation can provide time-series for spectral analysis. This analysis

From Kemp, A. E. S. (ed.), 1996, *Palaeoclimatology and Palaeoceanography from Laminated Sediments,*
Geological Society Special Publication No. 116, pp. 37–48.

Table 1. *Studies of laminated sediments using backscattered electron imagery*

Paper	Study
Pike & Kemp 1996*a, b*	Unconsolidated Holocene laminated sediments, Gulf of California (sediment fabric analysis, palaeoceanography, palaeoclimatology)
Kemp 1990; Patience *et al.* 1990; Aplin *et al.* 1992; Brodie & Kemp 1994, 1995	Unconsolidated Quaternary laminated sediments, Peru continental margin (sediment fabric analysis, palaeoceanography, palaeoclimatology)
Lange & Schimmelmann 1994; Bull & Kemp 1996	Unconsolidated Quaternary laminated sediments, Santa Barbara Basin, California Borderland (sediment fabric analysis, palaeoceanography, palaeoclimatology)
Grimm 1992*a, b*	Unconsolidated Quaternary laminated sediments, Japan Sea (sediment fabric analysis, palaeoceanography)
Macquaker & Gawthorpe 1993; Belin & Kenig 1994; Macquaker 1994	Jurassic mudstones/silty mudstones, southern England (facies analysis, depositional environments)
Kemp 1991	Silurian pelagic/hemipelagic mudstones, Southern Uplands (UK) (palaeoceanography)
Krinsley *et al.* 1983	General – shales (petrology)
Primmer & Shaw, 1987	Shales (diagenesis)
Pye & Krinsley 1984	General – sedimentary rocks (petrology)

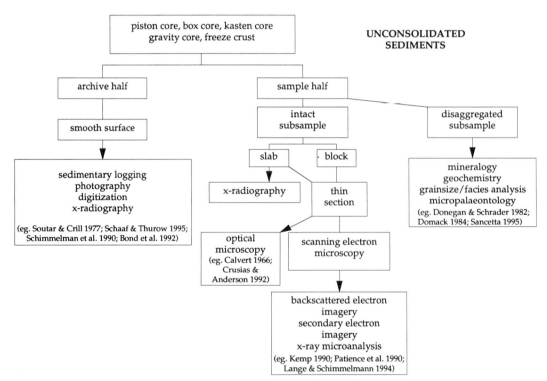

Fig. 1. Summary of the analytical techniques applied to unconsolidated and semi-consolidated laminated sediments.

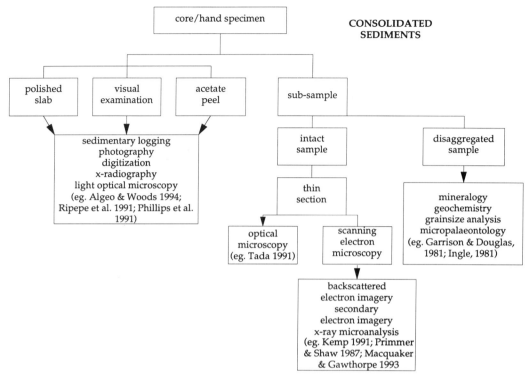

Fig. 2. Summary of the analytical techniques applied to consolidated laminated sediments (laminites).

may yield information about the cyclicity of such variables as sediment total organic carbon (TOC) content (Schaaf & Thurow 1994).

Optical microscopy

This has been used to analyse sediment fabric elements such as microfossils and lamina boundary relationships, ie. whether they are gradational or sharp (Byrne & Emery 1960; Domack 1984; Crusius & Anderson 1992). Preparation of thin sections from rock is straightforward (Macquaker & Gawthorpe 1993), however, for either unconsolidated and/or wet sediments the procedure is more complicated, as prior to thin section manufacture samples need to be dried and hardened (Bouma 1969; Murphy 1986).

Resolution of X-radiography, digital imagery and optical microscopy

The techniques discussed above all have failings when used for obtaining high-resolution records

of palaeoenvironmental variability. X-radiography is incapable of clearly recording very thin ($<200\,\mu$m) laminations, and cannot distinguish between microfossil assemblages preserved *within* biogenic laminae. Likewise, time-series analysis of grey-scale variability can give insights into yearly variation of, for example, TOC but lacks the sensitivity to detect intra-annual variations in biogenic assemblages. High-magnification study of thin sections is the only way to fully investigate these variations, preserved in many laminated sediments. Historically, optical microscopy has been used for this purpose, however, this lacks the resolution needed for high magnification studies. Optical micrographs have poor contrast between components (e.g. siliceous microfossils) and resin; the colour of detrital minerals, such as clays, often obscures constituents, as does interference due to the depth of view through the thin section.

Scanning electron microscopy

The scanning electron microscope (SEM) (including low-vacuum SEM) is an ideal tool for high-resolution analysis of laminated sediments. Both secondary electron imagery (SEI) (e.g. Reynolds & Gorsline 1992) and, more recently, backscattered electron imagery (BSEI) (Kemp 1990; Macquaker & Gawthorpe 1993; among others, Table 1) have been used for fabric studies, providing hydrographic reconstructions, high-resolution proxies for climate, and information about depositional environment. The remainder of this paper discusses thin section preparation for SEM analyses, and the SEM techniques used for investigating laminated sediments.

Preparation of thin-sections for SEM analysis

Very little preparation is required to produce thin sections from indurated rocks (Krinsley *et al.* 1983). Friable lithologies (e.g. diatomite), however, may require resin-impregnation to hold them together. Once this has been completed either impregnated blocks, or saw-cut samples of rock are glued to a glass microscope slide, the major portion of the block cut away, and the remaining sample ground and polished (avoiding the use of water-based lubricants which may cause rearrangement of the clay minerals and fabric disturbance).

More involved preparation is needed to produce thin sections from wet, unconsolidated sediments. Careful sampling is required to ensure as little fabric disturbance as possible; pore-fluids are removed and the sample is resin-impregnated before thin section production.

Sampling wet, unconsolidated or semi-consolidated sediment

Sampling from water-saturated sediment cores has been improved by the development of a sediment slab cutter (Schimmelmann *et al.* 1990) (Fig. 3). The cutter, which uses the osmotic knife principle, provides undisturbed sediment slabs which should then be X-radiographed for a permanent record of internal structure (Bull & Kemp 1996). Where strong sediment fabrics preclude the use of the slab cutter (e.g. many mesh-like diatom mats (Kemp *et al.* 1995); slabs of sediment should be cut using either a large scalpel or 'autopsy' blade. Great care must be taken not to lacerate oneself during this process. Slabs are divided for thin section preparation, micropalaeontological, geochemical and bulk sedimentological analyses. For thin sections,

sediment blocks are initially cut from the slab using a scalpel, then dried and finally resin-impregnated.

Critical point, freeze and vacuum drying, and resin-impregnation of wet sediment

There are a number of different techniques that have been used to dry wet sediments prior to resin impregnation.

Critical point drying. Above a critical point of temperature and pressure, the phase boundary between liquid and gas disappears (Bouma 1969). The critical point of water is very high so pore-fluids are usually replaced by an alcohol (e.g. ethanol), followed by liquid CO_2. After the sample has been subjected to the appropriate temperature and pressure, gaseous CO_2 is removed and the dry sample is ready for resin-impregnation (Bennett *et al.* 1981; Reynolds & Gorsline 1992).

Freeze drying. Freeze drying is carried out using liquid nitrogen (Bouma 1969). Samples must be in contact with the nitrogen at one face only so that it can move through the sediment block as a front. Quick freezing turns water into ice without ice-crystal growth. Frozen samples are placed into a freeze-drier immediately, and water is removed by sublimation (Crevello *et al.* 1981; Kuehl *et al.* 1988).

Vacuum drying. During vacuum drying, samples are evacuated, removing the pore-fluids. This is carried out using a vacuum impregnation unit (Kemp 1990; Patience *et al.* 1990).

Resin-impregnation of dry sediment. Critical point, freeze and vacuum dried sediments are usually resin-impregnated using a vacuum impregnation unit. Evacuated resin is added to evacuated sediment, the resin filling the pore spaces. After releasing the vacuum, the resin-impregnated block is then cured in an oven (Kemp 1990).

Discussion. For high-resolution SEM analysis of sediment fabric, the techniques of thin section preparation outlined above have a number of drawbacks. Vacuum drying of very wet and/or clay-rich samples causes severe desiccation of the sediment (Fig. 4). This micro-cracking prevents reliable measurements of lamina thicknesses and can obscure lamina boundaries. Critical point

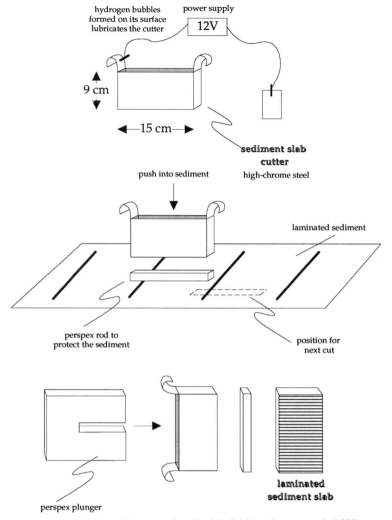

Fig. 3. The Schimmelmann sediment slab cutter, described in Schimmelmann *et al.* (1990).

and freeze drying of sediments causes little disturbance of the sediment fabric. Both methods, however, involve a stage where fragile sediment blocks need moving, and are not ideal for routine procedures as they require specialized equipment for both drying and resin-impregnation.

Fluid displacive drying and resin-embedding of wet sediments

This technique of drying and resin-embedding provides the best quality thin sections for microscopy. The method outlined in the Appendix is adapted after Jim (1985) and Polysciences Inc. (1986) and uses low-viscosity Spurr epoxy

resin (Spurr 1969). This resin is toxic, as are some of the component chemicals, and should be handled with care.

Fluid displacive resin-embedding of wet sediment is a passive procedure; samples are never physically dry and the fabric is supported by fluid throughout. Chemical dehydration prevents the cracking which results from vacuum drying, and the technique requires no specialized drying, or resin-impregnating equipment. Water-saturated samples are more successfully embedded using fluid replacement than those which are drier, and finer-grained samples are more successful than coarser sediment. Increasing the number of replacements of solvent and resin, however, facilitates successful embedding of silty and fine-sandy sediments. Thin sections produced by this

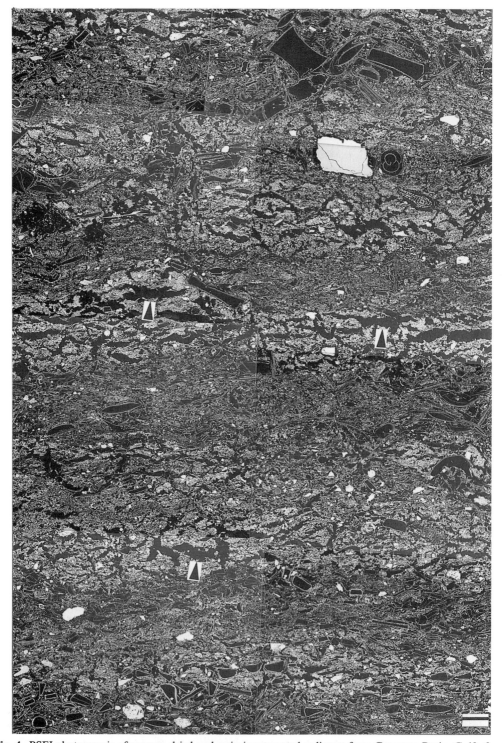

Fig. 4. BSEI photomosaic of vacuum dried and resin-impregnated sediment from Guaymas Basin, Gulf of California. Arrows indicate severe micro-cracking in the bright lithogenic laminae, disturbing the sediment fabric. Scale bar is 100 μm.

Fig. 5. BSEI photomosaic of fluid displacive resin embedded sediment from Guaymas Basin, Gulf of California. There is no cracking of the lithogenic laminae, which are very bright, in contrast with the diatom ooze laminae which are very dark (low backscatter coefficient resin filling frustules). Scale bar is 1 mm.

Table 2. *Summary of scanning electron microscopy techniques applied to laminated sediments*

Scanning electron microscopy	Backscattered electron imagery (thin sections)	Porosity and sediment fabric analysis (20× photomosaics) High-magnification imaging – component identification (e.g. microfossil types, pellets)
	Secondary electron imagery/low vacuum scanning electron imagery (strew sildes, sediment fracture surface)	High-magnification imaging – component identification (binocular microscopy analogue) (e.g. microfossil species, clay)
	X-ray microanalysis (thin sections)	Point analysis/line scans/X-ray maps – qualitative chemical composition analysis (e.g. elemental distribution)

technique show no sediment fabric disturbance and preserve the most detail for SEM-scale studies (Fig. 5).

Thin section preparation

Thin sections of resin-impregnated blocks are prepared in the same way as saw-cut rock samples. Oil- rather than water-based lubricants should be used for cutting. For SEM analysis, thin sections should be highly polished, using a range of grits down to $1\,\mu m$, particularly for BSEI (Krinsley *et al.* 1983). If thin sections are to be used for optical, as well as electron, microscopy they should be polished down to at least $35\,\mu m$. Thin sections are carbon-coated before being analysed in the SEM (Goldstein *et al.* 1981).

SEM analysis of laminated sediment

A comprehensive summary of scanning electron microscopy, and types of analyses possible, is provided by (Goldstein *et al.* 1981). The following is a description of BSEI, SEI and X-ray microanalysis; methods used for analysing laminated sediments (Table 2).

Backscattered electron imagery (BSEI)

Analytical technique. Backscattered electrons are the result of elastic collisions between energetic beam electrons and atoms within the specimen (Goldstein *et al.* 1981). The number of backscattered electrons generated (backscatter coefficient, η), is primarily related to the average atomic number of the target (when using polished thin sections), and is recorded on a photograph as image brightness. Mineral grains such as pyrite, quartz and carbonate have relatively high average atomic numbers, and therefore have higher backscatter coefficients and produce brighter images than organic matter and carbon-based

resin, which have a low average atomic numbers, low backscatter coefficients, and produce dark images, i.e. black. BSEI photomosaics of laminated sediments, provide compositional data, but may also be regarded as porosity maps which give sediment fabric information (Table 2). Biogenic laminae comprising diatoms, for example, have high porosity and are recorded as dark layers on the photomosaic (carbon-based resin fills the frustules). Lithogenic laminae, comprising silt and clays, have low porosity and are characterised by bright layers. This contrast between components and resin, and the shallower sampling depth within the thin section provides greatly improved resolution images over optical micrographs.

Image analysis. Analysis of thin sections, using BSEI, begins with the production of a 20× magnification photomosaic, used as a base-map for higher magnification work (Fig. 6). Photomosaics are time-consuming to produce, and single images are often distorted around the edges at low magnifications (SEMs are designed for high-magnification imaging). The advent of new pc-windows SEMs, however, allows the production of mosaics to be fully automated. The mosaic may be stored as a single image, removing the problem of unfocused photograph edges. If the thin section has been made from an X-radiographed slab, the photomosaic is located on the X-radiograph (Bull & Kemp 1996). Individual flux events are identified, the thin section logged, lamina thicknesses measured and data collected for fabric analysis. Where long sections of sediment have been impregnated (e.g. 50 cm), time-series and lamina transition analysis may be carried out using both thickness and compositional data. Higher magnification imagery (i.e. up to 1500×) can be used to characterize components, for example, microfossil identification (in association with optical microscopy).

Sediment fabrics, such as bioturbation, may be investigated using BSEI photomosaics (e.g. Brodie & Kemp 1995). Porosity contrasts highlight biogenic and lithogenic material

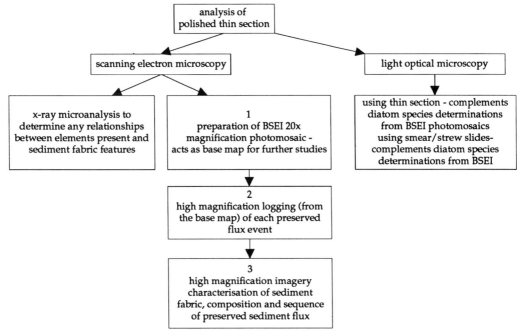

Fig. 6. Summary of the analysis of polished thin sections using the SEM and optical microscope.

redistributed by burrowing. BSEI can also be used for facies analysis of fine grained rocks (e.g. the Jurassic Kimmeridge Clay Formation (UK); Macquaker & Gawthorpe 1993).

Secondary electron imagery (SEI)

Analytical technique. Blocks of sediment are fractured along laminations, mounted on SEM stubs using epoxy glue, and gold-coated for SEI analysis (Table 2). Secondary electrons are produced by inelastic collisions between high-energy beam electrons and atoms within the specimen (Goldstein *et al.* 1981). The number of secondary electrons emitted, and the resulting the secondary electron coefficient, is affected by the topography of the specimen, therefore the images produced are topographic and may be considered analogous to binocular optical microscopy images.

Image analysis. High-magnification SEI images of fractured sediment surfaces are used to aid characterization of laminations, e.g. component and microfossil taxa identification (Table 2). If polished thin sections are thin enough, optical microscopy (utilising the depth of view within the section) can be used to reduce the amount of SEI imagery required (Fig. 6).

Low vacuum scanning electron microscope (LVSEM)

Analytical technique. The LVSEM uses back-scatter electrons which are energetic enough to move through the low vacuum (~6%) in the specimen chamber/column to the detector. Thus, either wet sediment blocks or strew mounts may be analysed directly and topographic images are produced without the need for carbon/gold coating.

Image analysis. High magnification LVSEM images have the same uses as traditional SEI images.

X-ray (electron microprobe) microanalysis

Analytical technique. X-radiation is produced by interaction of the electron beam with the specimen (Goldstein *et al.* 1981). Using energy dispersive spectrometry (EDS) X-ray analysis, point analysis, line scans and X-ray maps provide qualitative and quantitative chemical analyses of minerals in the thin sections (Table 2; Fig. 6).

Applications. Samples may be impregnated with osmium tetroxide vapour, before vacuum resin-impregnation, to stain reactive organic matter

(Patience *et al.* 1990; Aplin *et al.* 1992). The electron microprobe can be used to study diagenesis in shales (Primmer & Shaw 1987), and White *et al.* (1984) used EDS X-ray mapping to delineate grains of organic matter in oil shales and map out areas of, and intergrowths of different clay minerals.

Conclusions

The scanning electron microscope is one of the best tools to fully characterize palaeoenvironmental variability, as recorded in laminated sediments. Other techniques, such as X-radiography and optical microscopy, are not so useful because they lack the resolution of SEM backscattered electron imagery. Preparation of unconsolidated sediment using fluid displacive resin embedding prevents fabric disturbance, and produces high quality thin sections. It is crucial to keep the sediment as wet as possible for this preparation technique. Thickness of thin sections is also an important consideration if sections are to be used for optical, in conjunction with electron, microscopy.

JP wishes to acknowledge the receipt of NERC Research Studentship GT4/92/263/G. We are grateful to Bob Jones, John Ford, Konrad Hughen and Carina Lange for many useful discussions about resin embedding soft sediments. We would also like to thank Joe Macquaker and Andy Fleet for their thorough reviews of this manuscript.

Appendix

Fluid displacive, low viscosity resin embedding procedure for wet, unconsolidated sediment (adapted after Jim, 1985 and Polysciences Inc., 1986).

Acetone-resistant containers are placed on a balance, tared, and filled with water to the depth required to cover the sample. This will be the amount of resin/acetone needed per container. Proportions of resin chemicals required are calculated using the following ratio:

vinylcyclohexene dioxide (VCD)	10.0 g
diglycidol ether of polypropyleneglycol (DER 736)	6.0 g
nonenyl succinic anhydride (NSA)	26.6 g
dimethylaminoethanol (DMAE)	0.2 g

Samples of wet sediment are placed into close-fitting, well-perforated aluminium foil boats. These are placed into the containers, into a glass desiccator over silica gel and submersed in high-purity acetone. Acetone is replaced three times daily, until 15 replacements have been made. This chemically dries the sediment by replacing the aqueous pore fluids with the solvent. *Samples must not be allowed to desiccate at any stage.* After the final soaking in acetone, this is removed and the first batch of resin is added. It is a good idea to dilute the first two additions of resin with high-purity acetone in the proportions 90 : 10, and 95 : 5 resin : acetone, respectively. Resin is changed every 12 hours (to prevent viscosity increase) until 15 replacements have been carried out. After each addition of resin, gently lift the foil boats to ensure fresh resin is drawn up through the base of the sample, and prevent the boats sticking to the bottom of the container. Samples are left to soak for up to four weeks (in the desiccator) after the last addition of resin, ensuring all the pore spaces are filled. Samples are then cured for 24 hours each at 30°C, 40°C, 50°C and 60°C, allowing the resin to cool between each stage. Thin sections are prepared from the resin blocks using a standard method, and using oil- rather than water-based lubricant.

Hints

Label the container with a way-up arrow and sample number. Do this in a manner that will be permanent in an acetone environment. The wetter the sample, the more effective the embedding procedure, so keep sediment as wet as possible. Add acetone and resin slowly (using a syringe) down the sides of the containers. This prevents sample disturbance and allows fluids to soak in from the base first. It is preferable to make up a fresh batch of resin before each replacement, although large batches may be made up and stored anhydrous, in a freezer. Add the VCD, DER and NSA first and mix before adding the DMAE. VCD and DMAE may undergo an exothermic reaction.

References

ALGEO, T. J. & WOODS, A. D. 1994. Microstratigraphy of the Lower Mississippian Sunbury Shale: A record of solar-modulated climatic cyclicity. *Geology*, **22**, 795–798.

APLIN, A. C., BISHOP, A. N., CLAYTON, C. J., KEARSLEY, A. T., MOSSMANN, J.-R., PATIENCE, R. L., REES, A. W. G. & ROWLAND, S. J. 1992. A lamina-scale geochemical and sedimentological study of sediments from the Peru Margin (Site 680, ODP Leg 112). *In*: SUMMERHAYES, C. P., PRELL, W. L. & EMEIS, K. C. (eds) *Upwelling Systems: Evolution Since the Early Miocene.* Geological Society, London, Special Publication, **64**, 131–149.

BELIN, S. & KENIG, F. 1994. Petrographic analyses of organo-mineral relationships: depositional conditions of the Oxford Clay Formation (Jurassic), UK. *Journal of the Geological Society, London*, **151**, 153–160.

BENNETT, R. H., BRYANT, W. R. & KELLER, G. H. 1981. Clay fabric of selected submarine sediments: Fundamental properties and models. *Journal of Sedimentary Petrology*, **51**, 217–232.

BOND, G., BROECKER, W., LOTTI, R. & MCMANUS, J. 1992. Abrupt color change in isotope stage 5 in North Atlantic deep sea cores: implications of rapid change of climate driven events. *In*: KUKLA, G. & WEMT, E. (eds) *Start of a Glacial*. Springer, Berlin. 185–205.

BOUMA, A. G. 1969. *Methods for the Study of Sedimentary Structures*. Wiley, New York.

BRODIE, I. & KEMP, A. E. S. 1994. Variation in biogenic and detrital fluxes and formation of laminae in late Quaternary sediments from the Peruvian coastal upwelling zone. *Marine Geology*, **116**, 385–398.

—— & ——1995. Pelletal structures in Peruvian upwelling sediments. *Journal of the Geological Society, London*, **152**, 141–150.

BULL, D. & KEMP, A. E. S. 1996. Composition and origins of laminae in late Quaternary and Holocene sediments from the Santa Barbara Basin. *This volume*.

BYRNE, J. V. & EMERY, K. O. 1960. Sediments of the Gulf of California. *Bulletin of the Geological Society of America*, **71**, 983–1010.

CALVERT, S. E. 1966. Origin of diatom-rich varved sediments from the Gulf of California. *Journal of Geology*, **74**, 546–565.

—— & VEEVERS, J. J. 1962. Minor structures of unconsolidated marine sediments revealed by X-radiography. *Sedimentology*, **1**, 296–301.

CREVELLO, P. D., RINE, J. M. & LANESKY, D. E. 1981. A method for impregnating unconsolidated cores and slabs of calcareous and terrigenous muds. *Journal of Sedimentary Petrology*, **51**, 658–660.

CRUSIUS, J. & ANDERSON, R. F. 1992. Inconsistencies in accumulation rates of Black Sea sediments inferred from records of laminae and [210]Pb. *Paleoceanography*, **7**, 215–228.

DE GREER, G. 1912. A geochronology of the last 12,000 years. *Cong. Geol., Internat. 11th, Stockholm 1910, Comptes Rendus.*, 241–253.

DOMACK, E. W. 1984. Rhythmically bedded glacio-marine sediments on Whidbey Island, Washington. *Journal of Sedimentary Petrology*, **54**, 589–602.

DONEGAN, D. & SCHRADER, H. 1982. Biogenic and abiogenic components of laminated hemipelagic sediments in the central Gulf of California. *Marine Geology*, **48**, 215–237.

FISCHER, A. G. & ROBERTS, L. T. 1991. Cyclicity in the Green River Formation (lacustrine Eocene) of Wyoming. *Journal of Sedimentary Petrology*, **61**, 1146–1154.

GARRISON, R. E. & DOUGLAS, R. G. (eds) 1981. *The Monterey Formation and Related Siliceous Rocks of California*. Special Publication of the Society of Economic Paleontologists and Mineralogists, Pacific Section, 15. Society of Economic Paleontologists and Mineralogists, Pacific Section.

GOLDSTEIN, J. I., NEWBURY, D. E., ECHLIN, P., JOY, D. C., FIORI, C. & LIFSHIN, E. 1981. *Scanning Electron Microscopy and X-ray Microanalysis*. Plenum, New York.

GRIMM, K. A. 1992a. High-resolution imaging of laminated biosiliceous sediments and their paleoceanographic significance (Quaternary, site 798, Oki Ridge, Japan Sea). *In*: PISCIOTTO, K. A., INGLE JR, J. C., VON BREYMANN, M. T. & BARRON, J. (eds) *Proceedings of ODP, Scientific Results, 127/128*. Ocean Drilling Program, College Station, TX, 547–557.

——1992b. Preparation of weakly consolidated, laminated hemipelagic sediment for high-resolution visual microanalysis: an analytical method. *In*: PISCIOTTO, K. A., INGLE JR, J. C., VON BREYMANN, M. T. & BARRON, J. (eds) *Proceedings of ODP, Scientific Results, 127/128*. Ocean Drilling Program, College Station, TX, 57–62.

HAMBLIN, W. K. 1962. X-ray radiography in the study of structures in homogeneous sediments. *Journal of Sedimentary Petrology*, **32**, 201–210.

INGLE JR, J. C. 1981. Origin of Neogene diatomites around the North Pacific rim. *In*: GARRISON, R. E. & DOUGLAS, R. G. (eds) *The Monterey Formation and Related Siliceous Rocks of California*. Special Publication of the Society of Economic Paleontologists and Mineralogists, Pacific Section, 15. Society of Economic Paleontologists and Mineralogists, Pacific Section, 159–159.

JIM, C. Y. 1985. Impregnation of moist and dry unconsolidated clay samples using Spurr resin for microstructural studies. *Journal of Sedimentary Petrology*, **55**, 597–599.

KEMP, A. E. S. 1990. Sedimentary fabrics and variation in lamination style in Peru continental margin upwelling sediments. *In*: SUESS, E., VON HEUNE, R. *et al.* (eds) *Proceedings of ODP, Scientific Results, 112*. Ocean Drilling Program, College Station, TX, 43–58.

——1991. Mid-Silurian pelagic and hemipelagic sedimentation and palaeoceanography. *Special Papers in Palaeontology*, **44**, 261–299, pls. 1–3.

—— & BALDAUF, J. G. 1993. Vast Neogene laminated diatom mat deposits from the eastern equatorial Pacific Ocean. *Nature*, **362**, 141–144.

——, BALDAUF, J. G. & PEARCE, R. B. 1995. Origins and paleoceanographic significance of laminated diatom ooze from the eastern equatorial Pacific Ocean (ODP Leg 138). *In*: PISIAS, N. G., MAYER, L. A., JANECEK, T. R. *et al.* (eds) *Proceedings of ODP Scientific Results, 138*. College Station, TX (Ocean Drilling Program), 641–645.

KRINSLEY, D, PYE, K. & KEARSLEY, A. T. 1983. Application of backscattered electron microscopy in shale petrology. *Geological Magazine*, **120**, 109–114.

KUEHL, S. A., NITTROUER, C. A. & DEMASTER, D. J. 1988. Microfabric study of fine-grained sediments: observations from the Amazon subaqueous delta. *Journal of Sedimentary Petrology*, **58**, 12–23.

LANGE, C. B. & SCHIMMELMANN, A. 1994. Seasonal resolution of laminated sediments in Santa Barbara Basin: Its significance in paleoclimatic studies. *In*: REDMOND, K. T. & THARP, V. L. (eds) *Proceedings of the Tenth Annual Pacific Climate (PACLIM) Workshop, April 4–7, 1993*. California Department of Water Resources, Interagency Ecological Studies Program, Technical Report 36, 83–92.

MACQUAKER, J. H. S. 1994. A lithofacies study of the Peterborough Member, Oxford Clay Formation (Jurassic), UK: an example of sediment bypass in a mudstone sucession. *Journal of the Geological Society, London*, **151**, 161–172.

—— & GAWTHORPE, R. L. 1993. Mudstone lithofacies in the Kimmeridge Clay Formation, Wessex Basin, southern England: implications for the origin and controls of the distribution of mudstones. *Journal of Sedimentary Petrology*, **63**, 1129–1143.

MURPHY, C. P. 1986. *Thin Section Preparation of Soils and Sediments*. A.B. Academic, Berkhampsted.

PATIENCE, R. L., CLAYTON, C. J., KEARSLEY, A. T., ROWLANDS, S. J., BISHOP, A. N., REES, A. W. G., BIBBY, K. G. & HOPPER, A. C. 1990. An integrated biochemical, geochemical, and sedimentological study of organic diagenesis in sediments from Leg 112. *In*: SUESS, E., VON HUENE, R. *et al.* (eds) *Proceedings of ODP, Scientific Results, 112*. College Station, TX (Ocean Drilling Program), 135–153.

PEARCE, R. B., KEMP, A. E. S., BALDAUF, J. G. & KING, S. C. 1995. High resolution sedimentology and micropaleontology of laminated diatomaceous sediments from the eastern equatorial Pacific Ocean (ODP Leg 138). *In*: PISIAS, N. G., MAYER, L. A., JANECEK, T. R. *et al.* (eds) *Proceedings ODP Scientific Results, 138*. College Station, TX (Ocean Drilling Program), 647–663.

PHILLIPS, A. C., SMITH, N. D. & POWELL, R. D. 1991. Laminated sediments in prodeltaic deposits, Glacier Bay, Alaska. *In*: ANDERSON, J. B. & ASHLEY, G. M. (eds) *The Last Interglacial-Glacial Transition in North America*. Geological Society of America Special Paper, **261**, 51–60.

PIKE, J. & KEMP, A. E. S. 1996a. Preparation and analysis techniques for studies of laminated sediments. *This volume*.

——1996b. Silt aggregates in laminated marine sediments produced by agglutinating foraminifera. *Journal of Sedimentary Research*, **66**, 625–631.

POLYSCIENCES INC. 1986. Spurr Low-Viscosity Embedding Media. *Polysciences Data Sheet 127*.

PRIMMER, T. J. & SHAW, H. F. 1987. Diagenesis in shales: evidence from backscattered electron microscopy and electron microprobe analyses. *In*: SCHULTZ, L. G., VAN OLPHEN, H. & MUMPTON, F. A. (eds) *Proceedings of the International Clay Conference, Denver 1985*. The Clay Minerals Society, Denver, 135–143.

PYE, K. & KRINSLEY, D. H. 1984. Petrographic examination of sedimentary rocks in the SEM using backscattered electron detectors. *Journal of Sedimentary Petrology*, **54**, 877–888.

REYNOLDS, S. & GORSLINE, D. S. 1992. Clay microfabric of deep-sea, detrital mud(stone)s, California continental borderland. *Journal of Sedimentary Petrology*, **62**, 41–53.

RIPEPE, M., ROBERTS, L. T. & FISCHER, A. G. 1991. ENSO and sunspot cycles in varved Eocene oil shales from image analysis. *Journal of Sedimentary Petrology*, **61**, 1155–1163.

SANCETTA, C. 1995. Diatoms in the Gulf of California: Seasonal flux patterns and the sediment record for the past 15,000 years. *Paleoceanography*, **10**, 67–84.

SCHAAF, M. & THUROW, J. 1994. A fast and easy method to derive highest-resolution time-series datasets from drillcores and rock samples. *Sedimentary Geology*, **94**, 1–10.

SCHIMMELMANN, A., LANGE, C. B. & BERGER, W. H. 1990. Climatically controlled marker layers in Santa Barbara Basin sediments and fine-scale core-to-core correlation. *Limnology and Oceanography*, **35**, 165–173.

SOUTAR, A. & CRILL, P. A. 1977. Sedimentation and climatic patterns in the Santa Barbara Basin during the 19th and 20th centuries. *Geological Society of America Bulletin*, **88**, 1161–1172.

SPURR, A. R. 1969. A low-viscosity epoxy resin embedding medium for electron microscopy. *Journal of Ultrastructure Research*, **26**, 31–43.

TADA, R. 1991. Origin of rhythmical bedding in middle Miocene siliceous rocks of the Onnagawa Formation, northern Japan. *Journal of Sedimentary Petrology*, **61**, 1123–1145.

WHITE, S. H., SHAW, H. F. & HUGGETT, J. M. 1984. The use of back-scattered electron imaging for the petrographic study of sandstones and shales. *Journal of Sedimentary Petrology*, **54**, 487–494.

Image analysis and microscopic investigation of annually laminated lake sediments from Fayetteville Green Lake (NY, USA) Lake C2 (NWT, Canada) and Holzmaar (Germany): a comparison

BERND ZOLITSCHKA

Fachbereich Vl, Geologie, Universität Trier, D-54286 Trier, Germany

Abstract: Counts and measurements of annually laminated sediments are often time consuming and subjective. An automated and computerized method to accelerate the generation of varve data and to make these data more comprehensible is required. The approach taken here is to digitize thin sections of varved lacustrine sediments of different composition and mean lamination thickness. Density variations within an area scan of the digitized image are then measured. The data obtained provide peaks and troughs of light transmission corresponding to the pale and dark alternating laminae. After comparing these results with microscopic varve measurements it is clear that light transmission data alone do not allow the counting of varves. The composite nature of laminations causes high-frequency noise which is hard to distinguish from thin varves without additional microscopic information.

During the last few years, reconstruction of palaeoenvironmental conditions from high-resolution lacustrine and marine sediments has added much to the understanding of past global changes. Annually laminated or varved lake sediments provide particular information on a number of different parameters on a yearly to seasonal basis. There are, however, many problems related to this type of sediment.

First, varve formation and preservation is restricted to certain physico-chemical, limnological and hydrological conditions which are responsible for the scarcity of such records today and in the geological past.

Second, demonstrating that laminations are formed during one year of deposition is difficult. Although organic varves may contain many remains of organisms indicating a specific season of deposition, it is too time consuming to assure the presence of such seasonality for every single lamina. Furthermore, classic minerogenic varves lack biogenic material and therefore provide little means to test for annual deposition. Conversely, the seasonal cycle is the only process capable of forming rhythmic laminations in the scale of millimetres or thinner within a lake.

Third, even when laminated sediments have been recovered, and their laminations have been shown to be of an annual nature, the laborious process of counting and measuring hundreds and sometimes more than ten thousand laminae still has to be carried out. This process is generally based on subjective decisions of the investigator which renders it open to criticism. Both time-consuming measurements and subjective results demand an automated, standardized and computerized method in order to speed up investigations and to make them more easily comprehensible.

Methods of image analysis seem to be most suitable for this third problem. Rather than using an automated system, this study attempts to evaluate the potential of such a method under optimal conditions. Laminated sections from well-studied sediments of different composition and mean lamination thickness in combination with results obtained by microscopic investigation of the same part of the sediment are employed.

Methods

Thin sections were used as a source for images. Although their preparation often introduces artificial fissures and gaps, they provide the best structural information on thin laminae from soft lake sediments. Two different methods were applied for embedding unfrozen lake sediments. Shock-freezing with liquid nitrogen, freeze drying and impregnation with Araldite epoxy resin (Merkt 1971) was used for the sediments from Holzmaar. Thus the structure is well preserved although minor disturbances occur as the result of the freezing and drying processes. Such fine cracks can be avoided if the specimen is kept wet all the time. Therefore, sediments from the other two sites were embedded according to the method first described by Clark (1988) and later modified by Lamoureux (1994). First the water is replaced by acetone, then the acetone is replaced by Spurr epoxy resin. This method reduced the presence of artificial fissures considerably.

From Kemp, A. E. S. (ed.), 1996, *Palaeoclimatology and Palaeoceanography from Laminated Sediments*, Geological Society Special Publication No. 116, pp. 49–55.

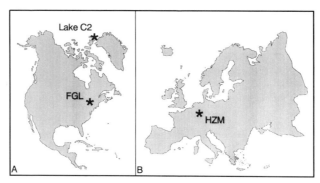

Fig. 1. Location of studied sites in North America (**A**) and in Europe (**B**); FGL, Fayetteville Green Lake; HZM, Holzmaar.

The thin sections were digitized with a laser scanner at a maximum setting of 300 DPI (dots per inch) corresponding to a resolution of slightly less than 0.1 mm. Before capturing the image, contrast and brightness were optimized in order to give best results on the monitor. After digitizing the whole thin section (100 × 15 mm) a commercially available Macintosh-based image analysing system (IMAGE 1.23) was used to magnify the image electronically. Sections with the best preserved laminations and without any major disruptions were selected for analysis. Disturbed or bent, non-parallel laminations were avoided. Therefore, no rectification of the image was necessary.

Data acquisition is based on densitometric measurements (along a line scan) or within a window (surface scan or area scan). Results are directly related to light transmission, which is determined by the composition of the sediment. Using the line scan option, one measurement is carried out every 0.085 mm along a predefined line perpendicular to laminations. Surface and area scans possess the same vertical resolution as line scans. They differ in that both types measure also parallel to the laminations within a user-defined window. Horizontally, 350 single points were always determined. Thus, a narrow window pro-

vides a more detailed and spiked record, while a wide window provides smoothed results. The principle of measurements is the same both with surface and area scans, but data processing is different. Surface plots display results in a three-dimensional form, which allow one to recognize the spatial distribution of light and dark areas not only along the vertical axis (perpendicular to laminations) but also along the horizontal axis (parallel to laminations). Area plots are calculated by stacking data along the horizontal axis into a single value forming a line-plot. For this investigation no filtering procedures were applied to manipulate the obtained data set.

Microscopic thin section examination was carried out using a petrographic microscope. This method enables one to decipher very fine structures like submillimetre-scaled laminations invisible to the naked eye.

Sediments

Three different types of laminated sediments from the Canadian Arctic, the United States and from Germany (Fig. 1) were used for this comparative study. All were recovered from the central basin of relatively small and deep

Table 1. *Morphometric data of Lake C2, Fayetteville Green Lake and Holzmaar*

Investigated site	Lake C2	Fayetteville Green Lake	Holzmaar
Location	northern Ellesmere Island, Canada 82°58′ N, 78°02′ W	north-central New York, United States 43°03′ N, 75°58′ W	Westeifel, Germany 50°07′ N, 6°53′ E
Elevation a.s.l.	1.5 m		425 m
Max. depth	84 m	52.5 m	20 m
Lake surface (L)	1.77 km²	0.258 km²	0.058 km²
Drainage basin (D)	26.6 km²	34.5 km²	2 km²
D/L	15.0	133.7	34.5

lakes (Table 1) which favours the formation and preservation of varved sediments. Previous studies (Ludlam 1969, 1981, 1984; Zolitschka 1990, 1991, 1996; Lamoureux & Bradley 1996) have shown that these deposits consist of annually laminated sediments, e.g. true varves.

Laminated sediments from nonglacial Arctic Lake C2 (north coast of Ellesmere Island, N.W.T., Canada) are minerogenic in composition. The pale summer layer is formed of coarse-grained sand and silt, while the dark late summer/fall layer is composed of fine silt and clay. Lateral persistence, lakeward thinning, variation in grain size and colour of these laminations, as well as present-day processes of highly seasonal sediment transfer into the lake basin, suggest that these are siliciclastic varves (Hardy 1996; Retelle & Child 1996; Zolitschka 1996). This is supported by measurements of ^{137}Cs and ^{210}Pb (Zolitschka 1996). Two samples from different locations within Lake C2 were analyzed: one from proximal site C2-16 and another from distal site C2-20.

The second investigated site is Fayetteville Green Lake, New York State, USA. Deposition within this lake is characterized by strongly seasonal calcite precipitation forming a pale carbonate late spring/early summer layer, followed by dark organic detritus (Ludlam 1969, 1981, 1984). This cyclicity is sometimes interrupted by massive or graded layers (Ludlam 1974).

Another sample was analysed from Holzmaar, Germany. The organogenic sediments consist of a pale diatomaceous spring/summer layer, and a dark layer of minerogenic and organic detritus that formed during the remaining part of the year (Zolitschka 1991). Evidence for the annual nature of laminations is given by sedimentology and palaeontology (diatoms and pollen). In addition, curves of magnetic palaeosecular variation (Haverkamp & Beuker 1993), AMS radiocarbon dates (Hajdas et al. 1995) and demonstration of solar forcing of varve thickness (Zolitschka 1992), corroborate the seasonal mode of deposition.

Results

Microscopic thin section investigation provides the basic information of varve thickness measurements (Fig. 2) and varve composition, which is necessary to evaluate the results of image analysis. Data on lamination thickness vary considerably between the four sections from the three sites. The lowest mean value of lamination thickness (0.29 mm) occurred in the distal sediment from Lake C2 (C2-20). In the proximal

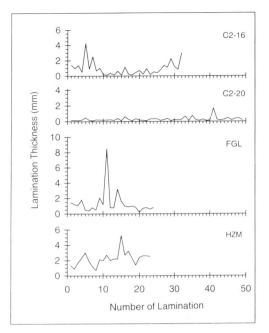

Fig. 2. Lamination thickness variations of varved sediments from Lake C2 (C2-16 and C2-20), Fayetteville Green Lake (FGL) and Holzmaar (HZM) as measured on a petrographic microscope.

site C2-16 mean lamination thickness was greater at 1.25 mm, and similar to that of Fayetteville Green Lake (1.29 mm). Lamination thickness is highest in the sediments from Holzmaar (2.17 mm). At Fayetteville Green Lake (layer 11) and Holzmaar (layer 15) finely laminated sediments are interrupted by both massive and slightly graded sections, where lamination thickness is more than doubled. These layers are probably caused by non-erosive low density turbidity currents or homogenities.

Application of image analysis to counting these laminae was begun with the line scan option for light transmission measurements. This option, however, produced signals with a high amount of noise, which is raised both by single pale sand grains within dark layers, or by black pieces of wood within light layers and was therefore rejected. Three-dimensional surface scans (Fig. 3) demonstrate this variability. Pale layers (low values) possess a high noise level, whereas with the dark layers (high values) the scan is consistent across the entire window.

The observed noise can be eliminated by using the area scan option and stacking all data points of the horizontal scan lines (Figs 4–8). The results then exhibit consistent variations in transmitted light. As all varves analysed are composed of pale

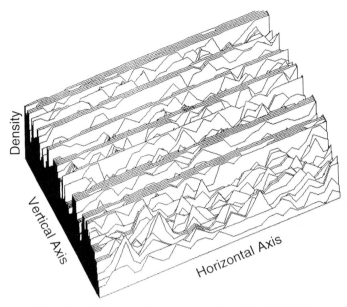

Fig. 3. Three-dimensional plot of a surface scan with seven varves from Lake C2 (C2-16, at around 2.5 cm sediment depth). Horizontal axis is parallel and vertical axis is perpendicular to laminations. Density in arbitrary units.

and dark layers, one successive minimum and maximum of these curves should represent one year.

From sediment core C2-16 a 4 cm long section was scanned with a window width of 1 mm (Fig. 4). In general, it is possible to arrive at a number of varves close to that obtained by microscopic thin section measurements, by counting the maxima of the densitometric scan. A conspicuous characteristic of this plot is the trend in the density data related to wedge-shaped thickness of the thin section. A similar

trend may be observed in the Holzmaar area scan (Fig. 8). Another problem caused by thin section preparation is the occurrence of gaps which appear as very translucent layers (Figs 4 & 5). Thinning at this part of the C2-16 section (0–15 mm) leads to low contrast between pale and dark layers, which makes them difficult to distinguish. Also the composite nature of laminations may cause high frequency noise, as in layers 16, 18, 19, 20, 24, 27 and 31.

Finally, there is a problem with the whole method because varves with a thickness of

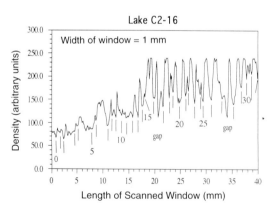

Fig. 4. Area scan of sediments from Lake C2 (C2-16, 0–4 cm); numbered lines indicate the varves as determined by thin section examination.

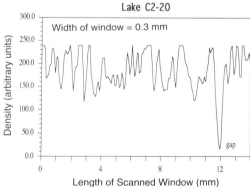

Fig. 5. Area scan of sediments from Lake C2 (C2-20, 10.3–11.7 cm).

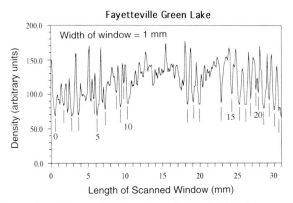

Fig. 6. Area scan of sediments from Fayetteville Green Lake (FGL, 14-67) with a window width of 1 mm; numbered lines indicate the varves as determined by thin section examination.

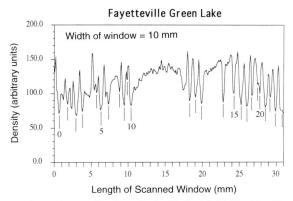

Fig. 7. Area scan of sediments from Fayetteville Green Lake (FGL, 14-67) like Fig. 6 but with a window width of 10 mm.

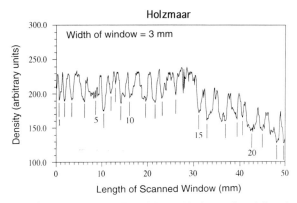

Fig. 8. Area scan of sediments from Holzmaar (HZM-C4, 55–60 m); numbered lines indicate the varves as determined by thin section examination.

0.1 mm are close to the resolution limit of the laser scanner, and therefore can not be identified from the area scan. This is most clearly observed in the area scan of the distal sediments from Lake C2 (C2-20) with most of the layers being very thin (Fig. 5). For this scan it is impossible

to arrive at the number of 48 varves as measured and counted with the microscope (Fig. 2).

With minimum lamination thickness of more than 0.2 mm, the sediments from Fayetteville Green Lake and Holzmaar are much more suitable for resolution by laser scanner. A scan with a window width of 1 mm from Fayetteville Green Lake (Fig. 6) provides an example of the influence of pale sand grains and of black pieces of wood within the turbiditic layers 11, 14 and 15 . These cause a typical high-frequency noise which may be reduced by widening the window to 10 mm (Fig. 7). Even when these turbidites have been clearly identified, it is still impossible, by counting the maxima of the plot, to produce the 24 varves observed microscopically. The reason is the composite nature of the thicker layers 1, 4 and 7.

The best results were obtained from sediments of Holzmaar. This 5 cm long section with a mean lamination thickness of 2.17 mm was scanned with a window width of 3 mm (Fig. 8). Except for a slight indistinctness of layers 4, 5 and 14, the identified 23 varves are recognizable in the area scan. However, the composite nature of laminations is again reflected by high frequency noise occurring within nearly every peak and trough of the light transmission data.

Conclusions

High variability of varve thickness, even within a very short record of less than 5 cm, in combination with the composite structure of laminations, makes it difficult to separate thin varves from high-frequency noise, without additional information from microscopic investigation. Furthermore, the density or light transmission data obtained do not improve on information gained by microscopy.

Problems with resolution of the image may be overcome either by using a scanner with higher resolution, or by employing a more sophisticated array such as a microscope with an attached video camera for producing the initial image (Ripepe et al. 1991). However, a very high-resolution process, adapted to very thin laminations, will probably cause new noise problems. Application of some kind of filtering to remove this noise will be extremely difficult, because this will also be capable of removing some of the very thin laminations. These results demonstrate that it is not appropriate to use stacked (and perhaps filtered and rectified) variations in light transmission alone as a source for establishing a varve timescale, simply by assuming that successive peaks and troughs relate to individual years and that the distance between peaks represents the thickness of varves as suggested by Ripepe et al. (1991). However, Petterson et al. (1993) demonstrated with more regularly varved Swedish sediments that thickness measurements with a microscope and with image analysis may give similar results.

In conclusion, problems with time consuming microscopic measurements and counting of varved sediments are not yet overcome. Semi-automated methods based on interactive thickness measurements of the microscopic sediment image are still very time consuming and do not exclude subjective decisions of the investigator. They do, however, give the advantage of slightly more comprehensibility and reproducibility. One of the major disadvantages of all image analysing systems is the need for sediments with well preserved laminations. For many reasons this is not always the reality. Therefore, microscopic investigation remains the most reliable method for sediments with varve thicknesses of a few millimetres or less, which is typical for nonglacial environments.

This study was carried out during a one-year post-doctoral appointment at the Department of Geology and Geography (now Department of Geosciences), University of Massachusetts. Sediment samples from Fayetteville Green Lake were kindly provided by Stuart Ludlam. Financial support was given by the Nato Science Fellowships Programme through the German Academic Exchange Service (DAAD) and by the German Science Foundation (DFG). Field work was supported by the National Science Foundation (NSF), Division of Polar Programs, through a grant to the University of Massachusetts, by Polar Continental Shelf Project (Ottawa), and by the German Science Foundation (DFG). The constructive comments made by Ray Bradley and Paddy O'Sullivan on an earlier version of this paper are highly appreciated. This is PALE contribution number 37.

References

CLARK, J. S. 1988. Stratigraphic charcoal analysis on petrographic thin sections: Application to fire history in north-western Minnesota. *Quaternary Research*, **30**, 81–91.

HAJDAS, I., ZOLITSCHKA, B., IVY, S. D., BEER, J., BONANI, G., LEROY, S. A. G., NEGENDANK, J. F. W., RAMRATH, M. & SUTER, M. 1995. AMS radiocarbon dating of annually laminated sediments from Lake Holzmaar, Germany. *Quaternary Science Reviews*, **14**, 137–143.

HARDY, D. R. 1996. Variability of stream flow and sediment transfer from a mountainous High Arctic watershed in response to climate. *Journal of Paleolimnology*, in press.

HAVERKAMP, B. & BEUKER, T. 1993. A palaeomagnetic study of maar lake sediments from the Westeifel. *In*: NEGENDANK, J. F. W. & ZOLITSCHKA, B. (eds) *Paleolimnology of European Maar Lakes*. Springer Lecture Notes in Earth Sciences, **49**, 349–365.

LAMOUREUX, S. F. 1994. Embedding unfrozen lake sediments for thin section preparation. *Journal of Paleolimnology*, **10**, 141–146.

—— & BRADLEY, R. S. 1996. A late Holocene varved sediment record of environmental change from northern Ellesmere Island, Canada. *Journal of Paleolimnology*, in press.

LUDLAM, S. D. 1969. Fayetteville Green Lake, New York. 3. The laminated sediments. *Limnology and Oceanography*, **14**, 848–857.

——1974. Fayetteville Green Lake, New York. 6. The role of turbidity currents in lake sedimentation. *Limnology and Oceanography*, **19**, 656–664.

——1981. Sedimentation rates in Fayetteville Green Lake, New York, USA. *Sedimentology*, **28**, 85–96.

——1984. Fayetteville Green Lake, New York. Vll. Varve chronology and sediment focusing. *Chemical Geology*, **44**, 85–100.

MERKT, J. 1971. Zuverlässige Auszählung von Jahresschichten in Seesedimenten mit Hilfe von Gross-Dünnschliffen. *Archiv für Hydrobiologie*, **69**, 145–154.

PETTERSON, G., RENBERG, I., GELADI, P., LINDBERG, A. & LINDGREN, F. 1993. Spatial uniformity of sediment accumulation in varved lake sediments in northern Sweden. *Journal of Paleolimnology*, **9**, 195–208.

RETELLE, M. & CHILD, J. 1996. Suspended sediment transport and deposition in a High Arctic meromictic lake, northern Ellesmere Island, Canada. *Journal of Paleolimnology*, in press.

RIPEPE, M., ROBERTS, L. T. & FISCHER, A. G. 1991. ENSO and sunspot cycles in varved Eocene oil shales from image analysis. *Journal of Sedimentary Petrology*, **61**, 1155–1163.

ZOLITSCHKA, B. 1990. Spätquartare jahreszeitlich geschichtete Seesedimente ausgewählter Eifelmaare. *Documenta naturae*, **60**, 1–226.

——1991. Absolute dating of late Quaternary lacustrine sediments by high resolution varve chronology. *Hydrobiologia*, **214**, 59–61.

——1992. Climatic change evidence and lacustrine vanes from maar lakes, Germany. *Climate Dynamics*, **6**, 229–232.

——1996. Recent of sedimentation in a High Arctic lake, northern Ellesmere Island, Canada. *Journal of Paleolimnology*, in press.

The potential for palaeoclimate records from varved Arctic lake sediments: Baffin Island, Eastern Canadian Arctic

KONRAD A. HUGHEN[1], JONATHAN T. OVERPECK[1,2],
ROBERT F. ANDERSON[3] & KERSTIN M. WILLIAMS[1]

[1] *Institute of Arctic and Alpine Research and Department of Geological Sciences,
University of Colorado, Boulder, Colorado 80309, USA*
[2] *NOAA Paleoclimatology Program, NGDC, Boulder, Colorado 80303, USA*
[3] *Lamont-Doherty Earth Observatory, Columbia University, Palisades,
New York 10964, USA*

Abstract: Tidewater lakes on Baffin Island in the eastern Canadian Arctic offer an excellent opportunity to study interannual to century-scale Arctic climatic change. Freeze-cores were analysed from three lakes in southeastern Baffin Island: Upper Soper Lake, Ogac Lake and Winton Bay Lake. The sediment record in each lake consists of massive sediments overlain by an organic-rich, finely laminated section which continues to the surface. The laminae in Ogac Lake were studied in detail and consist of two types. The light layers are composed almost entirely of intact diatom frustules, primarily *Chaetoceros* spp. The darker layers are dominated by clay and silt-sized terrigenous mineral grains, including abundant quartz and feldspars. These couplets are probably deposited as the result of diatom blooms in the late spring/summer growing season followed by settling of grains introduced by summer runoff. Sedimentation rates based on ^{210}Pb dates agree well with rates based on laminae counts in both Ogac and Winton Bay Lakes, indicating that the laminae couplets are annually deposited varves. Our experience suggests that shallow-silled tidewater lakes with varved sediments may be relatively common along the coast of Baffin Island. It should thus be possible to create a network of sites with annually dated palaeoclimate records.

High-resolution (annual to century-scale) palaeoclimate records from Arctic regions are important for understanding the global climate system. The patterns and causes of climate change at these time scales are poorly understood, even though the magnitude of change encompassed may be quite large (Overpeck 1991). Annual to decadal records are necessary because future climate change will take place on these timescales, and we must know the baseline natural variability against which future change will be measured. Several initiatives, including the IGBP-PAGES (International Geosphere-Biosphere Programme-Past Global Changes) and NSF-PALE (Paleoclimate of Arctic lakes and Estuaries), list obtaining high-resolution records of climate change for the past 1000 to 2000 years as a priority (IGBP 1992; Andrews & Brubaker 1991). With the exception of cores from a limited number of ice sheets, palaeoclimate records from the Arctic with this resolution and duration are rare.

Background

The Arctic is an important region for palaeoclimate research. Model results have shown that the

Arctic is particularly sensitive to global-scale perturbations, and that future greenhouse gas-induced temperature changes are likely to be amplified in Arctic regions (Houghton *et al.* 1990). This is primarily due to feedback effects associated with the albedo of snow and ice, as well as the stability of the Arctic troposphere. In addition to its sensitivity, the Arctic also exerts a strong influence on hemispheric climate. For example, the equator-pole temperature gradient, directly influenced by Arctic conditions, is likely to affect the patterns and intensity of atmospheric circulation over the entire Northern Hemisphere.

Baffin Island, in the eastern Canadian Arctic, is ideally located to monitor changes in Arctic climate (Fig. 1). Baffin Island is situated beneath one of the most prominent features of Northern Hemisphere circulation – a semipermanent trough in the upper westerlies. Studies have shown that longitudinal changes in the location of the trough axis affect surface climate conditions in the Baffin region (Keen 1980) as well as the adjacent Canadian mainland (Brinkmann & Barry 1972). Eastern Baffin Island is a high-latitude mountainous area with a near-by water source, conditions well suited for extended glacier growth. Studies have indicated that climatic conditions over Baffin Island are precariously close to

From Kemp, A. E. S. (ed.), 1996, *Palaeoclimatology and Palaeoceanography from Laminated Sediments*,
Geological Society Special Publication No. 116, pp. 57–71.

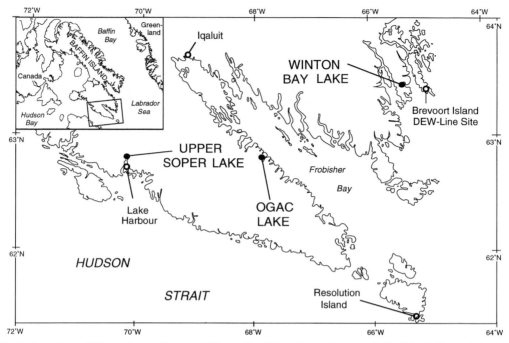

Fig. 1. Locations of Upper Soper, Ogac and Winton Bay Lakes in southeastern Baffin Island. Open circles indicate locations of stations collecting meteorological data, to be used in calibrating paleoenvironmental records from the lakes. Inset shows the location of Baffin Island in the eastern Canadian Arctic with the study site shown outlined.

those which would produce full glacierization over much of the island (Andrews *et al.* 1972). Additionally, Bradley & Miller (1972) showed that small glaciers and permanent snow banks responded to a changing climate on a timescale of less than a decade. In general, these results underscore the sensitivity of Baffin Island surface conditions to shifts in hemispheric or global climate, as well as the importance of palaeoclimate research in this region.

Some of the most widespread repositories of palaeoclimatic information on Baffin Island are lake sediments. Arctic climate, and its resultant influence on sediment production and deposition in Arctic lakes, displays strong seasonal fluctuations throughout the year. During the winter, lakes are sealed beneath one or more metres of ice, preventing sediments from entering the water column. Diatom growth is effectively prevented by the low incident angle of sunlight and the opacity of thick, partially snow-covered ice. In spring and early summer, strengthening sunlight, ice-free lake conditions and nutrients released from melting ice support diatom blooms (McLaren 1969), which are deposited as a distinct sediment layer. Terrigenous silts and clays introduced to the lakes by runoff from rainfall and

melting snow throughout the summer, settle out of suspension in late summer through winter and form another distinct sediment layer. Late-season settling of silts and clays forming a distinct sediment layer has been documented in other Arctic lakes (Retelle 1993; Zolitschka & Bradley 1993). This seasonal alternation of depositional regimes results in the annual production of a lamina couplet, which in most lakes is homogenized by benthic fauna. Tidewater lakes, however, are especially valuable for palaeoclimatic research due to their potential to preserve these annually layered sediments. The isostatic rebound history of Baffin Island has resulted in the uplift of fjords along the coasts, creating numerous deep, sheltered tidal lakes (Andrews 1991). Uplift of the fjord sills to sea level, or slightly above, isolates saltwater beneath a freshwater surface. Periodic incursions of seawater from high tides replenishes the saltwater and prevents it from being eventually flushed out by freshwater. The dense salt-water creates a strong pycnocline with the freshwater above, preventing deep mixing. Oxidation of organic matter uses up available oxygen and renders the bottom waters anoxic (McLaren 1967). Anoxic bottom water prevents benthic organisms from living in and disturbing

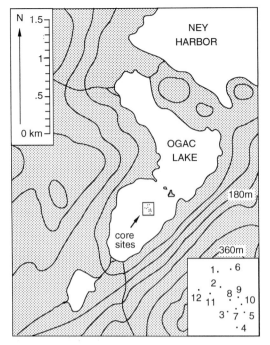

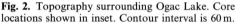

Fig. 2. Topography surrounding Ogac Lake. Core locations shown in inset. Contour interval is 60 m.

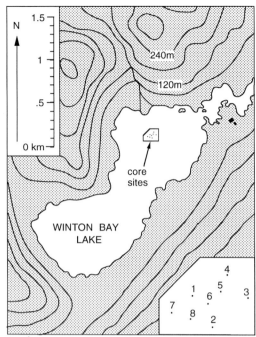

Fig. 3. Topography surrounding Winton Bay Lake. Core locations shown in inset. Contour interval is 60 m.

the sediments, and small-scale primary depositional structures such as annual laminations are preserved.

Ogac Lake and Winton Bay Lake lie in steep-walled valleys and were formed from uplifted fjords (Figs 2 & 3 Table 1). Upper Soper Lake, on the other hand, lies in a region of more subdued topography and does not appear to have been originally a fjord (Fig. 4 Table 1). All of these are tidewater lakes, with incursions of seawater occurring either monthly at high tide, or at least annually at extreme high tides. In contrast to typical Baffin Island freshwater lakes, Ogac, Winton Bay and Upper Soper Lake are deep, reaching approximately 50, 75 and 30 metres, respectively. The drainage basins vary widely in areal extent, although the lakes are of comparable size.

Coring expeditions prior to 1993 determined that these tidewater lakes all contain finely laminated sediments. The purpose of this research is to establish that the laminae couplets in these sediments are varves and to evaluate the potential usefulness of physical parameters such as varve thickness as palaeoclimate proxies. This paper presents preliminary results. The sedimentology and composition of the different laminae types are described in detail, and possible climatological interpretations are discussed. Evidence will also be presented that the laminae couplets in two of the three lakes studied so far are annually deposited varves. Finally, we will discuss the potential for creating a network of annually dated palaeoclimate records from around the Baffin Island region.

Table 1. *Baffin Island Lake data: location, elevation, size and depth. The elevation is measured against sea level. The maximum tidal range in Frobisher Bay is 14 m. All lakes receive tidal inflow. Ogac Lake only receives maximum spring tides*

Lake name	Latitude	Longitude	Elevation (m)	Size (ha)	Max. depth (m)
Upper Soper Lake	62°53′ N	69°52′ W	3	82	30.1
Winton Bay Lake	63°24′ N	64°21′ W	1	182	72.8
Ogac Lake	62°51′ N	67°20′ W	5	148	48.8

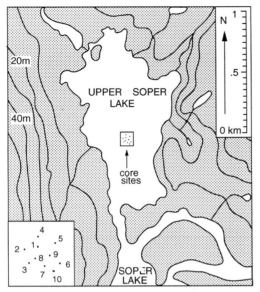

Fig. 4. Topography surrounding Upper Soper Lake. Core locations shown in inset. Contour interval is 10 m.

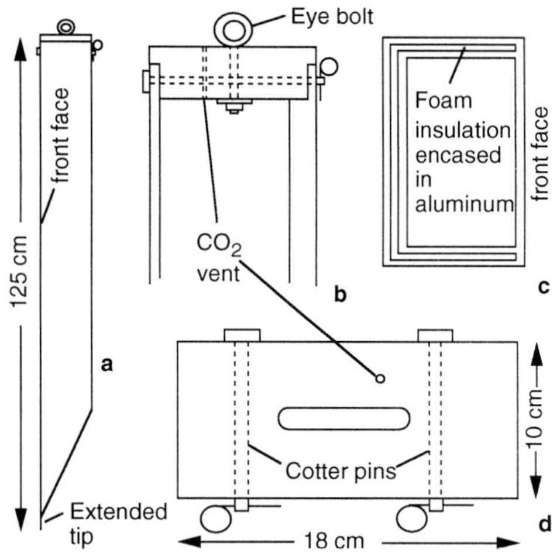

Fig. 5. Design of box-type freeze-corer. (**a**) longitudinal view showing assymmetrical point and sheer front face. (**b**) side view of top showing positions of cotter pins and CO_2 vent hole. (**c**) top view showing insulated sides and back. (**d**) top view showing positions of eye bolt, cotter pins and CO_2 vent hole.

Methods

Field work

A successful lake coring expedition to Baffin Island took place during the spring of 1993. April and May are the optimum months for lake coring in this region, because the danger of extreme winter storms is reduced, while it is still early enough before break-up for the ice to be a thick and stable platform. Based on previous knowledge of finely laminated sediments in Ogac, Upper Soper and Winton Bay Lakes, a concerted effort was made to recover 8 to 12 freeze-cores from each of these lakes. Complex logistics are involved in accessing remote lakes, obtaining undisturbed cores which preserve the fragile sediment–water interface, and then transporting these cores intact to the lab. Preserving the sediment–water interface is extremely important for studies involving annually laminated sediments, because it provides a modern baseline for both varve chronologies and isotopic (^{210}Pb, Pu) dating methods. Unfortunately, piston coring techniques which have proved useful in recovering long cores from deep lakes rarely, if ever, recover this interface. Freeze-coring techniques, developed to preserve fine laminae in soupy sediments (Shapiro 1958; Wright 1980), can retrieve even the most delicate structures of the uppermost sediments and interface. Freeze-coring, however, has its own

set of logistical requirements, including large amounts of dry ice (solid-phase CO_2) necessary for acquiring and transporting cores, that place constraints on the amount one can achieve during a field season.

The freeze-corer is a hollow aluminium cylinder or box with a pointed bottom and removable top where a rope can be attached (Figs 5 & 6). Immediately prior to being lowered into the

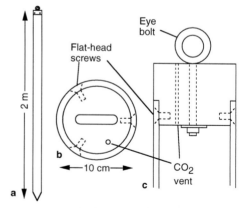

Fig. 6. Design of tubular-type freeze corer. (**a**) longitudinal view showing narrow profile and symmetrical point. (**b**) top view showing positions of eye bolt, CO_2 vent hole and cap screws. (**c**) side view of top showing positions of eye bolt, CO_2 vent hole and cap screws.

water, the core tube was filled with a slurry of finely crushed dry ice and methyl alcohol. This slurry was packed into the tube to eliminate pockets formed by the CO_2 released from the sublimating dry ice. The corer was lowered to a position approximately 5 m above the sediment surface, then allowed to drop freely into the sediments. The corer was then tied off snugly with a static rope to stabilize and prevent the tube from shifting or tipping over while within the sediments. The amount of time a freeze-corer should be left in place may vary with water depth and temperature. Experience showed that the optimum amount of time to leave our corer in place before pulling it back up was about 20 minutes. This allowed enough time for a thick slab of sediments to freeze on, but not so much time that the dry ice completely sublimated and allowed the frozen sediments to melt off. Approximately 27 litres (one cubic foot) of commercially compressed dry ice was required to obtain a single freeze-core, not including the amount needed for later transport. Cores were transported back to the lab intact in custom-built insulated boxes.

The exact design of the corer depends on the type of freeze-core desired. For the highest quality sediment–water interface, we used a rectangular design (Fig. 5). The longitudinal cross section shows the asymmetrical shape of the corer. The longest face is flat all the way to the blade at the base, where the corer penetrates the sediments. The widening from the point, which results in the greatest amount of sediment deformation upon impact, is limited to the rear surface. The transverse cross section (Fig. 5) shows that all of the surfaces but the front face are insulated. This focuses the freezing sediment onto the desired face and minimizes dealing with unwanted frozen sediment. For cores longer than one metre, we used a cylindrical corer (Fig. 6). The narrower profile and increased length allow this corer to penetrate much farther into the sediments, obtaining cores up to two metres in length. However, because of the lack of a sheer, continuous surface rising from the point, and the increased speed with which the corer is released into the sediments, the sediment–water interface is often deformed with a cylindrical corer. The best method is to use both types in any given lake, the box corer released slowly to obtain the interface, and the tubular corer released with more speed to obtain the longer record.

Laminae analysis

Once the cores arrive in the lab, they are photographed and sub-samples are embedded in epoxy resin to be made into petrographic thin sections, as well as sampling individual laminae for sedimentological and diatom analysis. To study sub-millimetre laminae between cores, clear, high-resolution photographs of frozen sediments were obtained. We did our photography in a freezer room kept below the temperature at which the cores are stored ($-20°C$). This prevented condensation from the surrounding air from collecting on the core surface and significantly blurring the image. Prior to photography each core surface was scraped with a plane or hand scraper, to remove superficial mud and expose the laminae. Then the surface was sealed with a wet sponge, producing a thin glaze of ice, which fills pits and surrounds individual crystals, rendering them invisible. The photos were assembled into full-length mosaics for comparing and correlating laminae between cores.

Embedding core samples in resin is an involved process and can be done in several ways. We chose to remove the interstitial water (ice) using either acetone replacement (Clark 1988) or freeze-drying. Removing water without destroying fragile structure required care, particularly in the uppermost sediments which were up to 99% water. Techniques which allow the ice to melt, such as oven drying or acetone exchanges, tended to reduce the more delicate sediments to dust. Freeze-drying, however, when done properly, preserved the structure of even the most fragile upper sediments. It is important to know the limitations of the freeze-dryer being used. For example, trying to dry too many samples at once resulted in the samples partially melting during the process and forming desiccation cracks or other undesirable features. Freeze-dried samples could also be embedded with a single application of resin, contrasted with at least eight applications necessary to replace acetone. Dyes were sometimes added to the resin to help enhance the contrast between laminae in some samples.

Thin sections allow high magnification studies of sediments and were used for measurement and counting of individual light and dark laminae, as well as for determining internal laminae composition and structure. The thin sections were studied using a light microscope and video camera attached to a computer to digitize images and analyze them using image analysis software (Thetford et al. 1991). Analysis consisted of counting and measuring varves and individual light and dark laminae, as well as identifying disturbances due to slumping or microturbidites. Backscattered electron imaging (BSEI; Kemp 1990) was used when extremely detailed analysis of a specific section was required. The high-resolution images produced by this technique were useful for identifying fine details including composition and structure within individual laminae. Backscattered electron images were produced on a JEOL 8600 electron microprobe.

Diatom analysis

Samples were taken of individual, sub-millimetre laminae from Ogac Lake cores for diatom analysis. Nine smear slides were made from a total of six light and three dark laminae. The samples were treated with hydrogen peroxide to digest organic material. A mounting medium of high refractive index (CargilleUs Meltmount) was used. The slides were examined with a Vanox Research Microscope and were counted for total diatom abundance and species diversity.

Lead-210 analysis

To prove the annual nature of the laminae couplets, we dated the sediments using ^{210}Pb (Appleby & Oldfield 1978). ^{210}Pb dating was carried out through isotope-dilution alpha spectrometric determination of the shortlived decay product ^{210}Po. Aliquots of dried sediment were dissolved in the presence of ^{208}Po yield monitor in Teflon beakers using a sequence of mineral acids (HNO_3, $HClO_4$, HF, and HCl). Following dissolution, samples were taken up in a dilute solution of HCl, and then adjusted to pH 2 with NH_4OH. Iron was reduced by adding $NH_2OH.HCl$ and Na-citrate, after which Po isotopes (natural ^{210}Po and tracer ^{208}Po) were spontaneously plated at 955°C onto silver disks to provide thin sources suitable for alpha spectrometry. Po isotopes were counted on low-background silicon surface-barrier detectors coupled to a multichannel analyzer with 512 channels per detector. Studies of recent sediments have shown ^{210}Po to be in radioactive equilibrium with ^{210}Pb, and thus it is a valid assumption to take the ^{210}Po activities as representative of ^{210}Pb activities.

Unsupported ^{210}Pb activities (that portion of the ^{210}Pb used for dating) were calculated by subtracting an estimate of the supported ^{210}Pb activity from the measured total ^{210}Pb activity of each sample. The average ^{210}Pb activity of samples from deep in the cores, at depths below the level where all of the unsupported ^{210}Pb has decayed, was taken to represent the supported ^{210}Pb activity.

Results

Core descriptions

During the 1993 coring expedition, 12 cores were taken from Ogac Lake, nine from Upper Soper Lake, and eight from Winton Bay Lake. We obtained a combination of shorter cores, including the sediment–water interface, and longer cores to secure the entire laminated record from each lake. The sediment sequences for the three lakes show the same general pattern. In each case a homogenous sediment unit is overlain by dark brown, organic-rich laminations which continue to the surface. This transition was probably caused by the isolation of deep saltwater and the consequent development of anoxic conditions as the lake basins were isostatically raised. The change from massive to laminated structure is typically abrupt, although some cores, especially from Winton Bay Lake, show a zone of alternating massive and laminated sediments where conditions seem to have fluctuated for a period before settling into anoxia. Within the laminated sections, disturbed layers and turbidites do occur, but are usually not pervasive throughout all of the cores recovered from a lake. Dropstones are common, probably originating from avalanches onto the ice, and range in size from 2 to 15 cm.

The quality of laminae preserved in the freeze-cores is excellent. At least three pristine sediment–water interfaces were recovered from each of Ogac and Upper Soper Lakes, and one was recovered from Winton Bay Lake; most of the cores contained at least part of the sediment–water interface. The lengths of the laminated sections vary between lakes. Ogac Lake possesses the longest laminated sequence; one box-type freeze-core is 127 cm long, and is laminated all the way to the base. Tubular cores recovered from the lake penetrated to a basal diamict, but they have not yet been photographed and analyzed, so it is unknown exactly how much overlap there is between these sediments and those in the box cores. Observations made in the field suggest that the laminated section in Ogac Lake is at least 150 cm long. Box cores from Upper Soper and Winton Bay Lakes contain the entire laminated sequences. Although intermittent laminae appear to greater depths in both lakes, undisturbed continuous records are approximately 50 cm and 40 cm in length, for Upper Soper Lake and Winton Bay Lake, respectively.

Laminae descriptions

The laminae in each lake appear to be biogenic, or diatom-rich, following the terminologies of

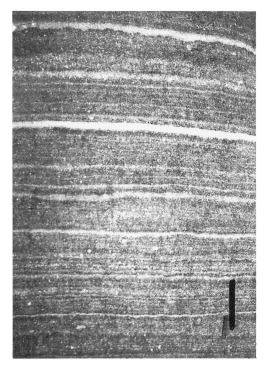

Fig. 7. Photograph of surface of freeze-core Soper 93-3A, depths 5 to 15 cm. Note the distinct separation of laminae. Scale bar = 1 cm.

Fig. 8. Photograph of surface of freeze-core Ogac 93-7B, depths 45 to 55 cm. Note the widely varying laminae thickness. Scale bar = 1 cm.

O'Sullivan (1983) and Saarnisto (1986). The laminae consist of alternating light tan and dark brown layers that are submillimetre to millimetre-scale in thickness. The appearance and colour of the laminae from lake to lake are generally similar. There are, however, subtle differences between lakes in the size, visual clarity, sharpness of contacts, and variance in thickness of the laminae. Upper Soper Lake has the most clearly defined laminae with sharp contacts and intermediate size (Fig. 7). Ogac Lake has laminae of intermediate definition, sharp contacts, and the greatest variance in thickness, ranging from submillimetre on average to an occasional layer over 1 cm (Fig. 8). Winton Bay Lake laminae are the thickest on average (>1 mm), but also the least clearly defined, with comparatively indistinct, gradational contacts (Fig. 9).

Marker laminae allow correlations between the various cores from each lake. All of the cores taken from Ogac and Upper Soper Lakes, within their respective basins, can be correlated to the individual layer along their entire lengths. Winton Bay Lake appears to have a more complex lake basin bathymetry, and thus although five of the eight cores can be correlated among themselves, the remaining three are difficult to correlate with any confidence. This is probably due as much to the lack of definition in the laminae as it is to true differences in deposition. Nevertheless, there are some sections within Winton Bay Lake cores which appear to be different from their neighbours. These differences in Winton Bay Lake cores, and even Ogac and Upper Soper Lake cores which show the same events but significantly different sedimentation rates, are surprising. In each lake, the cores were taken from the deepest areas, well away from shore, as only these areas contained laminated sediments. The cores were all taken within about ten metres of another core. The differences in sedimentation rate, and even possibly depositional regime in the case of Winton Bay, seem unusual for such close proximity, especially away from shore.

To date, we have only analyzed in detail laminated sediments from Ogac and Winton Bay Lakes, and we will restrict the following descriptions to these lakes. The sediments are

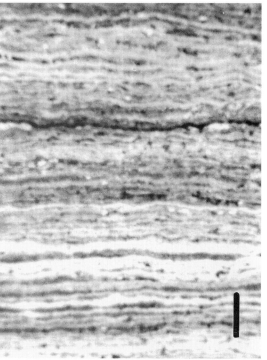

Fig. 9. Photograph of surface of freeze-core Win 93-lA, depths 25 to 35 cm. Note that the laminae are less distinct than Ogac or Upper Soper Lake laminae. Scale bar = 1 cm.

Fig. 10. Photomicrograph of laminae in thin section from piston core Ogac 91-2A, depths 27.5 to 28.5 cm. Laminae consist of alternating light, diatom-rich and dark, mineral-rich layers. Note the sub-millimeter thickness of laminae. Scale bar = 1 mm.

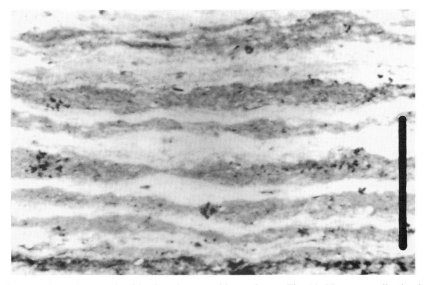

Fig. 11. Close-up photomicrograph of laminae in same thin section as Fig. 10. Note generally simple internal laminae structure, with single layers of mostly homogeneous composition. At the top of the figure is an example of an occasional layer with sub-laminae and more heterogeneous composition. Scale bar = 1 mm.

composed of light and dark laminae couplets (Fig. 10). The laminae vary in structural complexity, from single, homogenous layers to layers comprised of several sublaminae (Fig. 11). Sub-laminae, however, are not common and the individual laminae are generally of simple construction. Additionally, the repeating couplet sequence does not appear to be complex; it is consistently composed of two laminae, without obvious patterns of three or four different lamina types.

The internal composition of the two laminae types from Ogac and Winton Bay Lakes have been determined by high-magnification, high resolution BSEI, in addition to sampling individual laminae for diatom and sedimentological analysis. In general, the light-coloured laminae are composed entirely of intact diatom frustules, whereas the darker laminae contain silt and clay-sized terrigenous mineral grains (Fig. 12). The mineral grains within the dark laminae include abundant quartz and feldspars, which match the metamorphosed gneissic bedrock underlying

much of southeastern Baffin Island. A limited amount of carbonate bedrock outcrops near Upper Soper Lake, but these sediments did not respond to treatment with HCl and appear to contain little, if any, carbonate. Electron backscatter images show the distinct separation of internally homogenous diatom-rich and mineral-rich compositional elements (Fig. 12), with very little contamination of mineral grains into diatom layers and *vice versa*. Also apparent is the low morphological diversity of the diatom assemblages, an observation confirmed by taxonomic diatom analysis.

Comparisons between these laminated sediments and others from around the world show that these Arctic biogenic laminae seem to be among the least structurally and compositionally complex. Temperate lakes containing biogenic laminated sediments, including Lake Lovojarvi, southern Finland (Simola 1979; Saarnisto *et al.* 1977); Lake Soppensee, Switzerland (Hajdas *et al.* 1993; Lotter 1991); Lake Gosciaz, Poland (Rozanski *et al.* 1992; Goslar 1993);

Fig. 12. Electron backscatter image of laminae from the same thin section as Figs 10 & 11. Electron backscatter shows material density. In this image, dense mineral grains appear white, whereas less-dense organic matter appears darker, reversed from light microscope and core-surface images. Note distinct separation of diatom layers and mineral grain layers. Diatoms are identified as mature cells and resting spores of *Chaetoceros* spp. Mineral grains include quartz and feldspars. Note image has been rotated from horizontal. Scale bar = 100 μm.

Lake Holzmaar, Germany (Zolitschka *et al.* 1992; Leroy *et al.* 1993); Elk Lake, Minnesota (Dean *et al.* 1984; Anderson 1993); Crawford Lake, Ontario (Boyko-Diakonow 1979); and other North American lakes (Anderson *et al.* 1985) often have more than two layers deposited in one year, due to the bimodal nature of primary production at middle latitudes. Additionally, laminae in some of these lakes, such as Crawford Lake; Fayetteville Green Lake, New York (Ludlam 1979); and Cayuga Lake, New York (Ludlam 1979) include seasonal layers that are calcareous or chemical in origin, as well as diatoms.

These observations also apply to most marine varved sediments. Biogenic varves from the Gulf of California (Pike & Kemp 1996; Baumgartner *et al.* 1985; Donegan & Schrader 1982); Santa Barbara Basin, (Schimmelmann & Lange 1996; Soutar & Crill 1977); Saanich Inlet, British Columbia (Sancetta 1989; Sancetta & Calvert 1988; Gucluer & Gross 1964) and Cariaco Basin, Venezuela (Hughen *et al.* 1996; Peterson *et al.* 1991) are all structurally and compositionally complex, with multiple biogenic (plankton bloom) and/or minerogenic (runoff or wind) sub-laminae per year.

Other types of laminated sediments have simple structural and compositional makeup. Lakes and marine basins with chemical (ferrogenic or calcareous) or clastic laminated sediments often have remarkably well defined, two-part laminae couplets. Examples include Lake of the Clouds, Minnesota (Stuiver 1971; Anthony 1977); Lake C2, Ellesmere Island (Retelle 1993; Zolitschka & Bradley 1993; Zolitschka 1992); Rita Blanca lake beds (Anderson *et al.* 1985); Lake Rudetjarn, Sweden (Renberg 1981) and the Swedish and Finnish glacial varves (Stromberg 1990; Björck *et al.* 1992).

Diatom results

A comparison between the average diatom compositions of light and dark laminae is shown in Table 2. Within the dark laminae the species diversity is high with much fewer numbers overall, probably reflecting successive diatom blooms throughout the summer finally settling out during the winter. Considering the very rare diatoms seen within the dark laminae using BSEI, the assemblage may also reflect contamination from adjacent light layers during sampling. Within the light laminae the diatom species diversity is low, with the assemblage dominated by *Chaetoceros* spp. and *Nitzchia cylindra* (averages of 80% and 12%, respectively). The total diatom abundance in the light layers is almost an order of magnitude higher than in the dark laminae (Table 2). No mineral grains were analysed from the mineral-poor light laminae.

Lead-210 results

^{210}Pb concentrations were measured on sediments from Ogac and Winton Bay Lakes in order to confirm that the laminae couplets are annual varves (Figs 13 & 14). Somewhat different approaches were used to calculate sediment accumulation rates from ^{210}Pb values for the two lakes. In most lakes, linear sediment accumulation rates (cm/year) typically decrease with depth due to sediment compaction. In the Ogac Lake core used for ^{210}Pb dating, the effects of compaction seemed to be significant only in the uppermost 1 to 2 cm. Therefore, the sediment accumulation rate for this core was determined by fitting the downcore profile of unsupported ^{210}Pb activity to the equation:

$$^{210}Pb(z) = {}^{210}Pb(o)[\exp - (\lambda z/S)] \qquad (1)$$

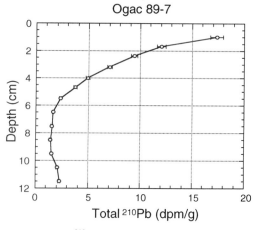

Fig. 13. Total ^{210}Pb concentration versus depth for Ogac Lake core Ogac 89-7. Plateau of measured values at depth indicates a supported ^{210}Pb value of about 1.5 dpm/g.

Table 2. *Diatom analysis of 6 light and 3 dark laminae in Ogac Lake varves*

	Light	Dark
Average diatom abundance (#/gram dry wt.)	528×10^6	68×10^6
Average number of species	6.5	19
% *Chaetoceros* spp.	80.0%	30.5%
% *Nitzchia cylindra*	11.5%	6.0%
Mineral grain abundance	low	high

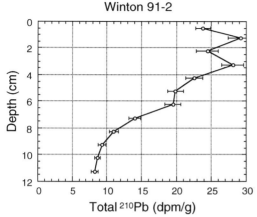

Fig. 14. Total ^{210}Pb concentration versus depth for Winton Bay Lake core Winton 91-2. High measured values for deepest samples and absence of clear plateau indicates that samples did not reach depth of supported ^{210}Pb value. High and low estimates of 6 and 1.5 dpm/g were used for supported ^{210}Pb values for the purposes of calculating sedimentation rates.

where ^{210}Pb is the activity of unsupported ^{210}Pb at depth (z) and at the core top (o), λ is the radioactive decay constant of [210]Pb (year^{-1}), and S is the sediment accumulation rate (cm/year). A least-squares fit to the Ogac Lake unsupported ^{210}Pb data produced a linear accumulation rate of 0.059 cm/year, with an overall uncertainty (including all components of analytical uncertainty plus the uncertainty associated with the value used for supported ^{210}Pb activity) conservatively estimated to be 0.010 cm/year (Fig. 15).

A more elaborate approach was required for Winton Bay Lake, where sediment compaction caused the linear accumulation rate to vary significantly throughout the entire depth interval containing unsupported ^{210}Pb activity. Under these conditions, it becomes necessary to regress unsupported ^{210}Pb activity against cumulative mass (g cm^{-2}; the mass accumulation rate [g cm^{-2} year^{-1}] may well remain constant despite large variability with depth in the linear accumulation rate). Cumulative mass was calculated using measured porosity values assuming a dry mineral density of 2.3 g cm^{-3}. The average mass accumulation rate was computed by fitting the unsupported ^{210}Pb activity profile to the equation:

$$^{210}\text{Pb}(z^*) = {}^{210}\text{Pb}(o)[\exp - (\lambda z^*/\text{MAR})] \quad (2)$$

where z^* is the cumulative mass (g cm^{-2}), MAR is the mass accumulation rate (g cm^{-2} year^{-1}), and the other parameters are as defined above.

To transpose MAR into linear accumulation rate, the porosity data was first smoothed by fitting it with a third-order polynomial, after which the linear accumulation rate (S, cm year^{-1}) was calculated as a function of depth using the equation:

$$S = \text{MAR}/([1 - \Phi)\rho_s] \quad (3)$$

where Φ is the porosity and ρ_s is the dry mineral density (assumed to be 2.3 g cm^{-3}). Conservative upper and lower limits for MAR were computed by incorporating both analytical uncertainties and an estimated uncertainty for the supported ^{210}Pb activity. Linear accumulation rates corresponding to these upper and lower limits are shown in Fig 15.

Although these results are preliminary, the good agreement between sedimentation rates in

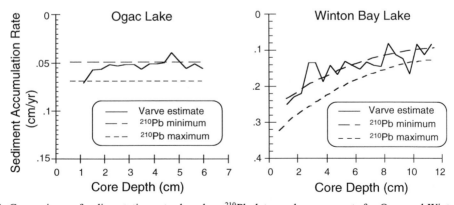

Fig. 15. Comparisons of sedimentation rates based on ^{210}Pb dates and varve counts for Ogac and Winton Bay Lakes. [210]Pb data are presented as maximum and minimum estimates (see text for explanation). Good overlap indicates that laminae couplets are annually deposited varves.

each lake indicate that the laminations are varves and will be usable for constructing chronologies. Additional ^{210}Pb and bomb nuclide geochronology is planned for Ogac and Winton Bay Lake cores, as well as for Upper Soper Lake cores. Sediments from Upper Soper Lake were not yet available in time for the preliminary ^{210}Pb analyses.

Discussion

The preliminary work presented here suggests that the construction of annually resolved chronologies will be possible for each of these lakes. The ^{210}Pb dating of Winton Bay and Ogac Lake sediments, combined with a plausible mechanism for seasonally variable sediment supply which matches the different laminae types, gives us reason to believe that the sediments are varved. In addition, the excellent correlations between cores within each lake will allow us to overcome localized gaps in the record due to disturbances, slumping, or erosion events. The quality and number of cores recovered will provide us with enough material for embedding and varve counting, as well as additional radiometric and isotopic dating, and diatom and pollen analyses. Initial varve counts and the length of the laminated sequence give us estimates of the temporal length of record expected from the lakes. Ogac Lake will provide the longest record by far, which will probably be about 3000 years long. Upper Soper Lake will probably yield a continuous record back about 500 to 600 years. Winton Bay Lake's record, due to the shortness of its laminated sequence, combined with its high sedimentation rate, will probably only reach back about 200 to 300 years.

The evaluation of varve and laminae thickness records as palaeoclimate proxy indicators is continuing. The techniques being used in this study allow thickness records to be constructed not only of complete varves, but of individual light and dark laminae as well. Although direct comparisons of thickness records to instrumental data have not yet been made, there are theoretical considerations which would link lamina thickness to a particular climate parameter for a given year. The light laminae are composed of an almost monospecific diatom assemblage, primarily *Chaetoceros* spp. This is a spring-blooming, marine species which prefers well-stratified water column conditions. In a study of Ogac Lake, thermal stratification occurred just below the halocline, where *Chaetoceros* spp. live (McLaren 1967). Although Ogac Lake is nutrient limited at middle depths

(McLaren 1969), high production rates were sustained throughout the summer in deeper waters, about 30 m and below the halocline, at 2–3 m (McLaren 1969). This was attributed to apparent nutrient replenishment, possibly by a tidal influx of zooplankton leading to decay and nitrification (McLaren 1969). Because nutrient replenishment occurs, at least in Ogac Lake, nutrient depletion may not play as important a role in varying diatom growth as availability of sunlight, ice-free lake conditions, or degree of thermal stratification. The extent of a diatom bloom (thickness of a light lamina), therefore, may be related to summer temperature, which controls lake ice extent and thickness, and thermal stratification. 'Summer temperature', as used here, may refer to the number of days above a certain temperature, irrespective of peak values, such as melting degree days, or the temperature integrated over the course of a summer, with peak values playing an important role, or some combination.

Dark laminae are formed by post summer settling of terrestrial mineral grains washed into the lakes by runoff. Runoff, and its capacity for carrying suspended grains, is influenced by several factors (Retelle 1993). Snow melt and precipitated water are the major sources of runoff which may carry grains. If snow melt is an important factor, spring and early summer temperature will contribute to the thickness of the dark laminae as well as light laminae. This is complicated by the fact that rapidity of onset of spring/early summer melting will probably be more important than the number of warm days or peak temperatures reached in mid- or late summer. However, approximately half of the annual precipitation in southeastern Baffin Island falls in July, August and September, and contributes to runoff directly as rainfall, rather than as snow melt. This would serve to make seasonal runoff a consequence of late summer/fall precipitation as well as of spring/early summer temperature. The importance of precipitated water to runoff capacity has been emphasized by workers in the high Arctic (Retelle 1993). The potential independence of depositional regimes controlled by different aspects of the climate system would reduce the amount of noise masking the climate signals contained in either the light or dark laminae thickness record. The separation of sediment–climate links would simplify the identification of climatic parameters controlling the thickness of both laminae types. The ability to measure laminae separately will aid greatly in the task of identifying proxy relationships and constructing palaeoclimate records.

Conclusions

This research demonstrates the potential for palaeoclimatic reconstruction held in anoxic Arctic lakes. Freeze-coring technology has proven to be an effective tool for consistently recovering easily disturbed Arctic lake sediments. Methods of freeze-drying and embedding are also useful even for the most uncompacted sediments near the sediment–water interface. These techniques will allow us to count and measure varves and individual light and dark laminae right to the sediment surface, giving us a present-day baseline for annual chronologies. Extended down core, our work will provide the accurate age models needed for geochronological, palynological, palaeolimnological and other studies. These varve chronologies will also provide the dating accuracy necessary for comparing instrumental climate data with varve and laminae thickness records on an annual basis. Independent evidence that the laminae couplets are annually deposited varves is provided by ^{210}Pb dating of sediments from Ogac and Winton Bay Lakes in addition to a mechanism linking laminae composition with seasonal sediment production.

Varve and laminae thickness records may be useful as palaeoclimate proxies. Some theoretical bases for climate–sediment flux relationships exist, but no thickness records have yet been completed to test against instrumental data. The ability to measure light and dark laminae separately increases the potential for identifying meaningful climate proxy relationships. If the sediment records in these lakes prove to be related to changing climates, the rewards will be substantial. An annually dated Ogac Lake record of palaeoenvironmental change back to 3000 years BP will be a valuable addition to other Holocene records such as those from Greenland. Even the shorter records of Upper Soper and Winton Bay Lakes will shed light on the temporal and spatial patterns of climatic changes associated with events such as the Little Ice Age.

Future work on this project will address several issues. First of all, sediments from all three lakes will be sampled at higher resolution for ^{210}Pb and bomb-nuclide analyses. These independent isotopic and radiometric dating techniques will confirm that the laminae couplets in each lake are indeed varves. Second, electron backscatter imagery combined with single-lamina diatom analyses of laminae from the three lakes will be compared to see if the same sedimentary processes are operating at each site. This will help interpret the varve and laminae thickness records as well as any differences which may be found between lakes. Finally, averaged multiple-core thickness records for light laminae, dark laminae and whole varves must be constructed to produce a record that is representative of the lake as a whole. Once constructed, these records will be calibrated to instrumental climate data from around the region in an effort to define and quantify the palaeoenvironmental information contained in the varves. Ultimately, it is hoped that a network of annual palaeoclimate records will be produced from around the Baffin Bay region. The conditions responsible for producing deep anoxic lake waters – isostatic uplift of coastal fjords – appear to be widespread, and the potential for discovering additional varved sediment records is high.

We wish to thank our colleagues at INSTAAR and the members of the Iqaluit Research Station (Science Institute of the Northwest Territories), Bradley Air, and the Canadian Arctic DEW-line sites for their logistical support on Baffin Island. J. Crusius and J. Sindt played key roles in the evolution of our freeze corer designs. We thank specifically L. Doner, G. Miller, L. Miller, B. Mode, A. Nesje and R. S. Webb for help in the field and J. T. Andrews for our first introduction to Ogac Lake. Funding was provided by the US National Science Foundation and US National Oceanic and Atmospheric Administration. This paper is PALE contribution number 13.

References

ANDERSON, R. Y. 1993. The varve chronometer in Elk Lake: Record of climatic variability and evidence for solar-geomagnetic-^{14}C-climate connection. *In:* BRABURY, J. P. & DEAN, W. E. (eds) *Elk Lake. Minnesota: Evidence for Rapid Climate Change in the North-Central United States*, Geological Society of America Special Paper, **276**, 45–67.

——, DEAN, W. E., BRADBURY, J. P. & LOVE, D. 1985. Meromictic Lakes and Varved Lake Sediments in North America. *United States Geological Survey Bulletin*, 1607.

ANDREWS, J. T. 1991. Relative sea levels, northeastern margin of the Laurentide Ice Sheet, on timescales of 10^3 and 10^7 a. *In:* SABADINI, R., LAMBECK, K. & BOSCHI, E. (eds) *Glacial Isostasy, Sea-level and Mantle Rheology.* Kluwer, Dordrecht, 143–163.

—— & BRUBAKER, L. B. 1991. *Paleoclimate of Arctic Lakes and Estuaries: Science and Implementation Plan.* Unpublished Proceedings of a Steering Committee Meeting, University of Colorado, Boulder.

——, BARRY, R., MILLER, G. H. & WILLIAMS, L. 1972. Past and present glaciological responses to climate in eastern Baffin Island. *Quaternary Research*, **2**, 303–314.

ANTHONY, R. S. 1977. Iron-rich rhythmically laminated sediments in Lake of the Clouds, northeastern Minnesota, *Limnology and Oceanography*, **22**, 45–54.

APPLEBY, P. G. & OLDFIELD, F. 1978. The calculation of lead-210 dates assuming a constant rate of supply of unsupported ^{210}Pb to the sediment. *Catena*, **5**, 1–8.

BAUMGARTNER, T., FERREIRA-BARTRINA, V., SCHRADER, H. & SOUTAR, A. 1985. A 20-year varve record of siliceous phytoplankton variability in the central Gulf of California. *Marine Geology*, **64**, 113–129.

BJÖRCK, S., CATO, I., BRUNNBERG, L. & STROMBERG, B. 1992. The clay varve based Swedish time scale and its relation to the late Weichselian radiocarbon chronology. *In*: BARD, E. & BROECKER, W. S. (eds) *The Last Deglaciation: Absolute and Radiocarbon Chronologies*, NATO ASI Series 1, 2, Springer, Berlin, 25–44.

BOYKO-DIAKONOW, M. 1979. The laminated sediments of Crawford Lake, southern Ontario, Canada. *In*: SCHLUCHTER, Ch. (ed.) *Moraines and Varves*. Balkema, Rotterdam, 303–307.

BRADLEY, R. S. & MILLER, G. H. 1972. Recent climatic change a··d increased glacierization in the eastern Canadian Arctic. *Nature*, **237**, 385–387.

BRINKMANN, W. A. R. & BARRY, R. 1972. Palaeoclimatological aspects of the synoptic climatology of Keewatin, Northwest Territories, Canada. *Palaeogeography Palaeoclimatology Palaeoecology*, **1**, 77–91.

CLARK, J. S. 1988. Stratigraphic Charcoal Analysis on Petrographic Thin Sections: Applications to Fire History in Northwestern Minnesota. *Quaternary Research*, **30**, 81–91.

DEAN, W. E., BRADBURY, J. P., ANDERSON, R. Y. & BARNOSKY, C. J. 1984. Variability of Holocene climatic change: Evidence from varved lake sediments. *Science*, **226**, 1191–1194.

DONEGAN, D. & SCHRADER, H. 1982. Biogenic and abiogenic components of laminated hemipelagic sediments in the central Gulf of California. *Marine Geology*, **48**, 215–237.

GOSLAR, T. 1993. Revision of the varve time scale of the Late Glacial laminated sediments of Lake Gosciaz (Central Poland). *Palaeoclimatology and Palaeoceanography from Laminated Sediments* (Abstracts), Geological Society of London.

GUCLUER, S. M. & GROSS, M. G. 1964. Recent marine sediments in Saanich Inlet, a stagnant marine basin. *Limnology and Oceanography*, **9**, 359–376.

HAJDAS, I., IVY, S. D., BEER, J., BONANI, G., IMBODEN, D., LOTTER, A. F., STURM, M. & SUTER, M. 1993. AMS radiocarbon dating and varve chronology of Lake Soppensee: 6000 to 12000 ^{14}C years BP. *Climate Dynamics*, **9**, 107–116.

HOUGHTON, J. T., JENKINS, G. J. & EPHRAUMS, J. J. (eds) 1990. *Climate Change The IPCC Assessment*. Cambridge University Press.

HUGHEN, K. A., OVERPECK, J. T., PETERSON, L. C. & ANDERSON, R. F. 1996. The nature of varved sedimentation in the Cariaco Basin, Venezuela, and its palaeoclimatic significance. *This volume*.

IGBP 1992. *PAGES Past Global Changes Project Report no. 19. Proposed Implementation Plans for Research Activity*, Stockholm.

KEEN, R. A. 1980. Temperature and circulation anomalies in the eastern Canadian Arctic summer 1946–76. *INSTAAR Occasional Paper*, **34**, University of Colorado, Boulder.

KEMP, A. E. S. 1990. Sedimentary Fabrics and Variation in Lamination Style in Peru Continental Margin Upwelling Sediments. *In*: SUESS, E. & VON HUENE, R. (eds) *Proceedings of the Ocean Drilling Program Scientific Results*, **112**, 43–58.

LEROY, S., SUBIRES, J., ZOLITSCHKA, B. & CORNET, C. 1993. The late Glacial by pollen and diatoms in the annually laminated sediments of Holzmaar, (Eifel, Germany). *Palaeoclimatology and Palaeoceanography from Laminated Sediments* (Abstracts), Geological Society of London.

LOTTER, A. F. 1991. Absolute dating of the late-Glacial period in Switzerland using annually-laminated sediments. *Quaternary Research*, **35**, 321–330.

LUDLAM, S. D. 1979. Rhythmite deposition in lakes of the northeastern United States. *In*: SCHLUCHTER, Ch. (ed.) *Moraines and Varves*. Balkema, Rotterdam, 295–302.

MCLAREN, I. A. 1967. Physical and Chemical Characteristics of Ogac Lake, a Landlocked Fiord on Baffin Island. *Journal of the Fisheries Research Board of Canada*, **24**, 981–1015.

——1969. Primary Production and Nutrients in Ogac Lake, a Landlocked Fiord on Baffin Island. *Journal of the Fisheries Research Board of Canada*, **26**, 1561–1576.

O'SULLIVAN, P. E. 1983. Annually-Laminated Lake Sediments and the Study of Quaternary Environmental Changes – A Review. *Quaternary Science Reviews*, **1**, 245–313.

OVERPECK, J. T. 1991. Century- to Millenium-Scale Climatic Variability during the Late Quaternary. *In*: BRADLEY, R. (ed.) *Global Changes of the Past*. UCAR/Office for Interdisciplinary Earth Studies, Boulder, CO, 139–172.

PETERSON, L. C., J. T. OVERPECK, N. G. KIPP, & IMBRIE, J. 1991. A high resolution late Quaternary upwelling record from the anoxic Cariaco Basin, Venezuela. *Paleoceanography*, **6**, 99–119.

PIKE, J. & KEMP, A. E. S. 1996. Preparation and analysis techniques for studies of laminated sediments. *This volume*.

RENBERG, I. 1981. Improved methods for sampling, photographing and varve-counting of varved lake sediments. *Boreas*, **10**, 255–258.

RETELLE, M. J. 1993. Interpretation of Laminated Lacustrine Sediments from Northern Ellesmere Island, Canada, Based on Seasonal and Annual Sediment Trap Yields and Process Studies. *The 23rd Annual Arctic Workshop Abstracts*, Byrd Polar Research Center, The Ohio State University, Columbus, Ohio, 67.

ROZANSKI, K., GOSLAR, T., DULINSKI, M., KUC, T., PAZDUR, M. F. & WALANUS, A. 1992. The Late Glacial-Holocene transition in central Europe derived from isotope studies of laminated sediments from Lake Gosciaz (Poland). *In*: BARD, E.

& BROECKER, W. S. (eds) *The Last Deglaciation: Absolute and Radiocarbon Chronologies*, NATO ASI Series, 1, 2, Springer, Berlin, 69–80.

SAARNISTO, M. 1986. Annually laminated lake sediments. *In*: BERGLUND, B. E. (ed.) *Handbook of Holocene Palaeoecology and Palaeohydrology*, 343–370.

——, HUTTUNEN, P. & TOLONEN, K. 1977. Annual lamination of sediments in Lake Lovojarvi, southern Finland, during the past 600 years. *Annales Botanici Fennici*, **14**, 35–45.

SANCETTA, C. 1989. Processes controlling the accumulation of diatoms in sediments: A model derived from British Columbian fjords. *Paleoceanography*, **4**, 235–251.

—— & CALVERT, S. E. 1988. The annual cycle of sedimentation in Saanich Inlet, British Columbia: implications for the interpretation of diatom fossil assemblages. *Deep-Sea Research*, **35**, 71–90.

SHAPIRO, J. 1958. The core-freezer-a new sampler for lake sediments, *Ecology*, **39**, 748.

SCHMIMMELMANN, A. & LANGE, C. B. Tales of 1001 varves: a review of Santa Barbara Basin sediment studies. *This volume*.

SIMOLA, H. 1979. Microstratigraphy of sediment laminations deposited in a chemically-stratifying eutrophic lake during the years 1913–1976. *Holarctic Ecology*, **2**, 160–168.

SOUTAR, A. & CRILL, P. A. 1977. Sedimentation and climatic patterns in the Santa Barbara Basin during the l9th and 20th centuries. *Geological Society of America Bulletin*, **88**, 1161–1172.

STROMBERG, B. 1990. A connection between the clay varve chronologies in Sweden and Finland, *Annales Acadamiae Scientiarum Fennicae* AIII, 154.

STUIVER, M. 1971. Evidence for the variation of atmospheric ^{14}C content in the late Quaternary. *In*: Turekian, K. K. (ed.) *The Late Cenozoic Glacial Ages*. Yale University Press, 57–70.

THETFORD, R. D., D'ARRIGO, R. D. & JACOBY, G. C. 1991. An Image Analysis System for Determining Densitometric and Ring-Width Time Series. *Canadian Journal of Forest Research*, **21**, 1544–1549.

WRIGHT, H. E. 1980. Cores of soft lake sediments. *Boreas*, **9**, 107–114.

ZOLITSCHKA, B. 1992. Sedimentation in a High Arctic lake – First Results. *The 22nd Arctic Workshop Program and Abstracts*, Institute of Arctic and Alpine Research, University of Colorado, Boulder, Colorado, 150.

—— & BRADLEY, R. 1993. Multiple core study of laminated sediments from the High Arctic, (Lake C2, Ellesmere Island, Northwest Territories, Canada) and their palaeoclimatic implications. *Palaeoclimatology and Palaeoceanography from Laminated Sediments*. (Abstracts), Geological Society of London.

——, HAVERCAMP, B. & NEGENDANK, J. F. W. 1992. Younger Dryas oscillation-varve dated microstratigraphic, palynological, and paleomagnetic records from Lake Holzmaar, Germany. *In*: BARD, E. & BROECKER, W. S. (eds) *The Last Deglaciation: Absolute and Radiocarbon Chronologies*. NATO ASI Series, 1, 2, Springer, Berlin, 81–101.

Varved sediments in Sweden: a brief review

GUNILLA PETTERSON

Department of Environmental Health, Umeå University, S-901 87 Umeå, Sweden

Abstract: This paper gives an introduction to studies made on varved (annually laminated) sediments in Sweden. Varve studies began in the late 19th century when Gerard de Geer correlated glacial clay-varves in order to establish a geochronology over the latest deglaciation. Both the chronology and its connection to present sediment deposition has continuously been revised during this century. As well as for glacial clay-varve deposits, annual formation of sediment laminae has also been found in estuarine, marine and lacustrine environments. Varved sediments, of different origins, first mainly used for dating purposes, have today also become an important tool for detailed reconstructions of past environmental changes.

Studies of varved sediments in Sweden began around 1884 when Gerard de Geer, in a lecture at the Geological Society in Stockholm, proposed that glacial clay-varves were formed annually and suggested that they could be used to establish a timescale for the latest deglaciation (de Geer 1884). Since then, varved sediments have been found in several other sedimentary environments. This paper gives an introduction to research in Sweden on glacial clay-varves, post-glacial varves deposited in river estuaries, marine varved sediments in the Baltic Sea and in fjords on the Swedish west coast (Fig. 1a), and varved sediments in lakes (Fig. 1b).

Glacial clay-varves

During the deglaciation, more than 10 000 years ago, melt water from the retreating ice sheet transported mineral material to the ice margin. Where the ice margin was in contact with fresh water, e.g. the Baltic Ice Lake, clay particles were deposited as varves. Seasonal changes in melt water discharge were reflected as changes in grain size and grain colour. De Geer and twenty students began in 1905 to search for deposits of clay-varves from Scania in the south of Sweden to the river valley of Ångermanälven in the north (Fig. 1a). Varve series from neighbouring

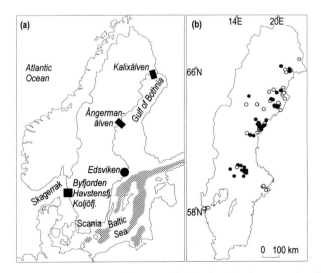

Fig. 1. Sites for varved sediments in different environments in Sweden; (**a**) river estuaries (Cato 1987; Widerlund & Roos 1994), the Baltic Sea (Jonsson *et al.* 1990), fjords on the west coast (Cato, pers. comm., Nordberg, pers. comm.) and (**b**) lakes (Petterson *et al.* 1993). In lakes marked with open circles between 10 and 100 recent varves have been formed, often as a result of cultural eutrophication. In lakes marked with filled circles varves have formed throughout the entire history of the lake. The longest recorded varve series covers about 9000 years.

From Kemp, A. E. S. (ed.), 1996, *Palaeoclimatology and Palaeoceanography from Laminated Sediments*, Geological Society Special Publication No. 116, pp. 73–77.

sites were correlated through varve thickness, i.e. the formation of the uppermost varves at one site and the lowest varves at another site, slightly closer to the ice margin, occurred at the same time giving the same pattern in varve thickness. These glacial clay-varves were connected to recent deposits of post-glacial varves and by 1910 a clay-varve chronology covering 12 000 years could be presented (de Geer 1912). De Geer and his collaborators continued their work both in Sweden and other glaciated areas, e.g. North America (Antevs 1922), Himalayas (Norin 1927) and East Africa (Nilsson 1931) and then tried to link these varve series to the Swedish clay-varve chronology (de Geer 1940). Since then, complementary studies have been made (e.g. Borell & Offerberg 1955; Fromm 1991; Strömberg 1989) and according to preliminary results the Swedish Time Scale now covers 13 257 varve years (Wohlfarth et al. 1994). Recent research on clay-varves also deals with environmental history during the deglaciation (Björck & Möller 1987), calibration between the Late Weichselian ^{14}C chronology and varve years (Björck et al. 1987, 1992; Wohlfarth et al. 1993), deglaciation and shore displacement (Risberg et al. 1991) and climate change during deglaciation (Ringberg 1984).

Post-glacial varves deposited in river estuaries

Post-glacial varves are common in the river beds of the large rivers in northern Sweden. Varved sediments were formed by mineral material transported by the rivers and deposited in their estuaries. Owing to subsequent land uplift (due to isostatic recovery, >200 m), these deposits were later eroded by the rivers and thus exposed in steep sandy river banks. Lidén (1913) found post-glacial varves in the valley of Ångermanälven. The connection between the period covered by the clay-varve chronology and the present was made through shore displacement data, post-glacial varves from several river valley and post-glacial varves found above glacial clay-varves in the lake Ragundasjön (Ahlmann 1924; de Geer 1912, 1940; Lidén 1938). There have, however, been uncertainties in the connection, and it was not until the 1980s, when Cato (1987) cored in the estuary of Ångermanälven (Fig. 1a) that a precise connection between the clay-varve chronology period and the present could be established (Boygle 1993).

Recent varve formation has also been discovered in the estuary of Kalixälven (Fig. 1a). This estuary is only 10–15 m deep and the varves

are formed and preserved because of density separation at the permanent halocline, few burrowing animals present in the low productivity waters of the Gulf of Bothnia, and ice cover on the estuary from December to May (Widerlund & Roos 1994). Cato (1987) demonstrated a correlation between varve thickness and water discharge during the 20th century in the estuary of Ångermanälven. There are also indications of such a relationship in the varved sediments of the estuary of Kalixälven.

Marine varved sediments in the Baltic Sea and in fjords on the Swedish west coast

Isostatic uplift and eustatic sea-level rise opened and closed the connection between the Baltic Sea and the Atlantic Ocean several times during and after the deglaciation. Alternating sequences with varved and non-varved sediments were deposited during these changes (Ignatius et al. 1981). Stratigraphic studies of the glacial clay-varves have been performed in order to connect them to the Swedish clay-varve chronology (Andrén & Risberg 1994; Björck et al. 1990). Today, brackish conditions prevail in the Baltic which means that varves are formed in the deep bottom areas. During the 1940s varves also began to form outside these areas, owing to eutrophication caused by an increased influx of phosphorus and nitrogen, and because of a stagnation in the inflow of oxygenated saline waters via the Skagerrak. Since then, the area in the south of the Baltic Sea in which varves are being formed has increased from about 20 000 km^2 to 70 000 km^2 (Jonsson et al. 1990) (Fig. 1a). Both varved and non-varved sediments have been investigated for changes in the deposition of organic material during this century (Jonsson & Carman 1994), and the sedimentation rate in the bay of Edsviken (Fig. 1a) has been studied with X-ray radiography and ^{210}Pb-dating (Axelsson & El-Daoushy 1989).

Glacial clay-varves representing the years 13 500–14 000 BP have been found near the Swedish west coast (Bergsten & Nordberg 1992). Further north, recent varves are being formed in the fjord Byfjorden (Cato, pers. comm.) and possibly also in Koljöfjorden and Havstensfjorden (Nordberg, pers. comm.) depending on anoxic conditions in the bottom waters (Fig. 1a).

Varved sediments in lakes

Swedish laminated lake sediments were first briefly described by Stålberg (1923) and Granlund (1931). However, varves were not studied in detail until the 1970s when Digerfeldt *et al.* (1975) investigated Järlasjön, and Renberg (1976) Rudetjärn. Although X-ray techniques (Axelsson 1983; Axelsson & Händel 1972) have been very useful for rendering sedimentary structures in cores of recent sediments more visible, the introduction of the freeze-coring technique has been of major importance for studies of recent varved sediments. The *in situ* freezing concept was brought to Sweden during the 1970s from North America via Finland (Saarnisto 1975, 1986). *In situ* freezing has made it possible for detailed year-by-year analyses of unconsolidated sediments (Renberg *et al.* 1984), and for counting of recent sediment varves (Renberg 1981*a*).

About one hundred lakes with varved sediments have been found in Sweden (Fig. 1b). They form and preserve varves even though they are dimictic, unlike many North American, Canadian and European sites which are meromictic (Anderson *et al.* 1985, Larsen & MacDonald 1993, O'Sullivan 1983). Lakes with varved sediments seem to be more common in the northern than the southern part of Sweden, which is very likely a climatic effect. In northern Sweden, the change between summer and winter occurs over only a few weeks, and the periods in spring and autumn without thermal stratification in lakes are very short. This stratification preserves the anoxic conditions within the bottom waters during summer and winter. When repeated each year, bottom fauna are reduced in numbers, and bioturbation declines. The best varves to study are found at the very deepest point of a lake since anoxic conditions begin there, but varves are also distinguishable outside this deepest area (Anderson *et al.* 1994). Investigation of lakes in Sweden indicates that they must be at least seven metres deep in order to develop thermal stratification, and in order to produce anoxic conditions they must also posses a relatively high biological production. Varves become more distinguishable and more consolidated if there is a regular inflow of minerogenic material (Renberg 1982).

The climate in northern Sweden, with marked seasonal changes, favours the formation of varves, since alternating layers of different material are deposited over the course of a year (Fig. 2). During the short and intense snow melt period (May), a layer of mineral grains from the catchment is deposited. The period of biological production (June–September) is marked by deposition of mostly organic material. This is sometimes followed by more mineral material from the catchment, depending on the

Fig. 2. Photograph of varves from the lake Kassjön, (about 2000 years BP). One varve is about 0.5 mm thick and consists of a light coloured minerogenic spring layer, a darker organic summer layer and a thin dark organic winter layer. The amount of deposited minerogenic and organic material changes between the years depending on environmental changes in the lake and catchment area.

amount of rain during October–November. When lakes are covered with ice (November–April), fine-grained organic material is deposited. The annual cycle of deposition has been studied with SEM (Renberg 1976) and thin-sections (Renberg 1981*b*). Comparison between surface sediment cores from one particular lake, over consecutive years, demonstrate that a new varve is deposited each year and that although they are compacted further down the core, their appearance is preserved (Renberg 1986*a*, Petterson *et al.* 1993).

In Sweden, varved lake sediments have been mainly used as a dating tool for palaeoenvironmental studies, such as vegetation history (Segerström 1990), heavy metal pollution (Renberg 1986*b*), deposition of carbonaceous particles (Renberg & Wik 1985), and diatom stratigraphy (Anderson *et al.* 1994). Research more related to the varves themselves has dealt with e.g. varve thickness and climate (Renberg *et al.* 1984), X-ray radiography of sediment cores (Bodbacka 1986), changes in sediment accumulation depending on land-use (Segerström *et al.* 1984), and methodological development of the image analysis technique for studying changes in the appearance of varves (Petterson *et al.* 1993). Freeze coring equipment has been developed by Renberg (1981*a*) and Renberg & Hansson (1993).

References

AHLMANN, H. W:SON. 1924. Ragundasjöns geomorfologi. *Geological Survey of Sweden.* Series Ca 12 [in Swedish].

ANDERSON, N. J., KORSMAN, T. & RENBERG, I. 1994. Spatial heterogeneity of diatom stratigraphy in varved and non-varved sediments of a small, boreal-forest lake. *Aquatic Sciences,* **56,** 40–58.

ANDERSON, R. Y., DEAN, W. E., BRADBURY, J. P. & LOVE, D. 1985. Meromictic lakes and varved lake sediments in North America. *U.S. Geological Survey Bulletin,* 1607.

ANDRÉN, T. & RISBERG, J. 1994. Late Quaternary development in the northwestern Baltic Sea. – An introduction. *In*: HICKS, S., MILLER, U. & SAARNISTO, M. (eds) *Laminated sediments.* European Symposium/Ravello – June 1991. PACT 41. Council of Europe, Rixensart, Belgium. 35–44.

ANTEVS, E. 1922. The recession of the last ice sheet in New England. *American Geographical Society, Research Series* **11.**

AXELSSON, V. 1983. The use of X-ray radiographic methods in studying sedimentary properties and rates of sediment accumulation. *Hydrobiologia,* **103,** 65–69.

—— & EL-DAOUSHY, F. 1989. Sedimentation in the Edsviken Bay studied by the X-ray radiographic and the Pb-210 methods. *Geografiska Annaler,* **71A,** 87–93.

—— & HÄNDEL, S. K. 1972. X-radiography of unextruded sediment cores. *Geografiska Annaler,* **54A,** 34–37.

BERGSTEN, H. & NORDBERG, K. 1992. Late Weichselian marine stratigraphy of the southern Kattegatt, Scandinavia: evidence for drainage of the Baltic Ice Lake between 12,700 and 10,300 years BP. *Boreas,* **21,** 223–252.

BJÖRCK, S. & MÖLLER, P. 1987. Late Weichselian environmental history in southeastern Sweden during the deglaciation of the Scandinavian ice sheet. *Quaternary Research,* **28,** 1–37.

——, SANDGREN, P. & HOLMQUIST, B. 1987. A magnetostratigraphic comparison between ^{14}C years and varve years during the Late Weichselian, indicating significant differences between the time-scales. *Journal of Quaternary Science,* **2,** 133–140.

——, DENNEGARD, B. & SANDGREN, P. 1990. The marine stratigraphy of the Hano Bay, SE Sweden, based on different sediment stratigraphic methods. *Geologiska Föreningens i Stockholm Förhandlingar,* **112,** 265–280.

——, CATO, I., BRUNNBERG, L. & STRÖMBERG, B. 1992. The clay-varve based Swedish Time Scale and its relation to the Late Weichselian radiocarbon chronology. *In*: BARD, E. & BROECKER, W. S. (eds) *The Last Deglaciation: Absolute and Radiocarbon Chronologies.* NATO ASI Series 1, 2, Springer, Berlin, 25–44.

BODBACKA, L. 1986. Sediment accumulation in Lakes Lilla Ullfjärden and Stora Ullfjärden, Sweden. *Hydrobiologia,* **143,** 337–342.

BORELL, R. & OFFERBERG, J. 1955. Geokronologiska undersökningar inom Indalsälvens dalgång mellan Bergeforsen och Ragunda. *Geological Survey of Sweden.* Series Ca, **31** [in Swedish].

BOYGLE, J. 1993. The Swedish varve chronology – a review. *Progress in Physical Geography,* **17,** 1–19.

CATO, I. 1987. On the definitive connection of the Swedish Time Scale with the present. *Geological Survey of Sweden.* Series Ca, **68.**

DE GEER, G. 1884. Summary of a lecture. Om möjligheten av att infora en kronologi for istiden. *Geologiska Föreningens i Stockholm Förhandlingar,* **7,** 3 [in Swedish].

——1912. A geochronology of the last 12 000 years. *Compte rendu du XI Congrès Géologique International (Stockholm 1910),* 241–253.

——1940. Geochronologia Suecica Principles. *Kungliga Svenska Vetenskapsakademins Handlingar.* Bd 18, 6. Stockholm.

DIGERFELDT, G., BATTARBEE, R. W. & BENGTSSON, L. 1975. Report on annually laminated sediment in lake Järlasjön, Nacka, Stockholm. *Geologiska Föreningens i Stockholm Förhandlingar,* **97,** 29–40.

FROMM, E. 1991. Varve chronology and deglaciation in south-eastern Dalarna, central Sweden. *Geological Survey of Sweden.* Series Ca, **77.**

GRANLUND, E. 1931. Kungshamnsmossens utvecklingshistoria jämte pollenanalytiska åldersbestämningar i Uppland. *Geological Survey of Sweden.* Series C, **368** [in Swedish].

IGNATIUS, H., AXBERG, S., NIEMISTÖ, L. & WINTERHALTER, B. 1981. Quaternary geology of the Baltic Sea. *In*: VOIPIO, A. (ed.) *The Baltic Sea.* Elsevier Oceanography Series, **30,** 54–104.

JONSSON, P. & CARMAN, R. 1994. Changes in deposition of organic matter and nutrients in the Baltic Sea during the twentieth century. *Marine Pollution Bulletin,* **28,** 417–426.

——, CARMAN, R. & WULFF, F. 1990. Laminated sediments in the Baltic – A tool for evaluating nutrient mass balances. *Ambio,* **19,** 152–158.

LARSEN, C. P. S. & MACDONALD, G. M. 1993. Lake morphometry, sediment mixing and the selection of sites for fine resolution palaeoecological studies. *Quaternary Science Reviews,* **12,** 781–792.

LIDÉN, R. 1913. Geokronologiska studier över det finiglaciala skedet i Ångermanland. *Geological Survey of Sweden.* Series Ca, **9** [in Swedish].

——1938. Den senkvartära strandförskjutningens förlopp och kronologi i Ångermanland. *Geologiska Föreningens i Stockholm Förhandlingar,* **60,** 397–404 [in Swedish].

NILSSON, E. 1931. Quarternary glaciations and pluvial lakes in British East Africa. *Geografiska Annaler*, **13**, 249–349.

NORIN, E. 1927. Late glacial clay varves in Himalaya connected with the Swedish time-scale. *Geografiska Annaler*, **9** 157–161.

O'SULLIVAN, P. E. 1983. Annually-laminated lake sediments and the study of Quaternary environmental changes – a review. *Quaternary Science Reviews*, **1**, 245–313.

PETTERSON, G., RENBERG, I., GELADI, P., LINDBERG, A. & LINGREN, F. 1993. Spatial uniformity of sediment accumulation in varved lake sediments in northern Sweden. *Journal of Paleolimnology*, **9**, 195–208.

RENBERG, I. 1976. Annually laminated sediments in Lake Rudetjärn, Medelpad province, northern Sweden. *Geologiska Föreningens i Stockholm Förhandlingar*, **98**, 355–360.

——1981a. Improved methods for sampling, photographing and varve-counting of varved lake sediments. *Boreas*, **10**, 255–258.

——1981b. Formation, structure and visual appearance of iron-rich, varved lake sediments. *Verhandlungen Internationale Vereinigung für Theoretische und Angewandte Limnologie*, **21**, 94–101.

——1982. Varved lake sediments – geochronological records of the Holocene. *Geologiska Föreningens i Stockholm Förhandlingar*, **104**, 275–279.

——1986a. Photographic demonstration of the annual nature of a varve type common in N. Swedish lake sediments. *Hydrobiologia*, **140**, 93–95.

——1986b. Concentration and annual accumulation values of heavy metals in lake sediments: Their significance in studies of the history of heavy metal pollution. *Hydrobiologia*, **143**, 379–385.

—— & HANSSON, H. 1993. A pump freeze corer for recent sediments. *Limnology and Oceanography*, **38**, 1317–1321.

—— & WIK, M. 1985. Carbonaceous particles in lake sediments – pollutants from fossil fuel combustion. *Ambio*, **14**, 161–163.

——, SEGERSTRÖM, U. & WALLIN, J.-E. 1984. Climatic reflection in varved lake sediments. *In*: MÖRNER, N.-A. & KARLEN, W. (eds) *Climatic Changes on a Yearly to Millennial Basis*. Reidel, Dordrecht, 249–256.

RINGBERG, B. 1984. Cyclic lamination in proximal varves reflecting the length of summers during Late Weichsel in southernmost Sweden. *In*: MÖRNER, N.-A. & KARLEN, W. (eds) *Climatic Changes on a Yearly to Millennial Basis*. Reidel, Dordrecht, 57–62.

RISBERG, J., MILLER, U. & BRUNNBERG, L. 1991. Deglaciation, Holocene shore displacement and coastal settlements in eastern Svealand, Sweden. *Quaternary International*, **9**, 33–37.

SAARANISTO, M. 1975. Pehmeiden järvisedimenttien näytteenottoon soveltuva jäädytysmenetelmä. [A freezing method for sampling soft lake sediments]. *Geologi*, **26**, 37–39 [in Finnish, summary in English].

——1986. Annually laminated lake sediments. *In*: BERGLUND, B. E. (ed.) *Handbook of Holocene Palaeoecology and Palaeohydrology*. Wiley, Chichester, 343–370.

SEGERSTRÖM, U. 1990. *The natural Holocene vegetation development and the introduction of agriculture in northern Norrland, Sweden. Studies of soil, peat and especially varved lake sediments.* PhD Thesis, Umea University, Sweden.

——, RENBERG, I. & WALLIN, J.-E. 1984. Annual sediment accumulation and lake use history; investigations of varved lake sediments. *Verhandlungen Internationale Vereinigung fur Theoretische und Angewandte Limnologie*, **22**, 1396–1403.

STRÖMBERG, B. 1989. Late Weichselian deglaciation and clay varve chronology in eastcentral Sweden. *Geological Survey of Sweden. Series Ca*, **73**.

STÅLBERG, N. 1923. Några undersökningar over Vättergyttjans beskaffenhet. En preliminär översikt. *Skrifter utgivna av Södra Sveriges fiskeriförening*, 88–95 [in Swedish, summary in German].

WIDERLUND, A. & ROOS, P. 1994. Varved sediments in the Kalix river estuary, northern Sweden. *Aqua Fennica*, **24**, 163–169.

WOHLFARTH, B., BJÖRK, S., HOLMQVIST, B., LEMDAHL, G. & ISING, J. 1994. Ice recession and depositional environment in the Blekinge archipelago of the Baltic Ice Lake. *Geologiska Föreningens i Stockholm Förhandlingar*, **116**, 3–12.

——, BJÖRK, S., POSSNERT, G., LEMDAHL, G., BRUNNBERG, L., ISING, J., OLSSON, S. & SVENSSON, N.-O. 1993. AMS dating Swedish varved clays of the last glacial/interglacial transition and the potential/difficulties of calibrating Late Weichselian 'absolute' chronologies. *Boreas*, **22**, 113–128.

Conflicting indicators of palaeodepth during deposition of the Upper Permian Castile Formation, Texas and New Mexico

A. B. LESLIE[1,2], A. C. KENDALL[1], G. M. HARWOOD† & D. W. POWERS[3]

[1] *Earth Science Research, School of Environmental Sciences, University of East Anglia, Norwich NR4 7TJ, UK*
[2] *Present address: Scottish Lime Centre, PO Box 251, Edinburgh EH6 4DW, UK*
[3] *Department of Geological Sciences, University of Texas at El Paso, El Paso, Texas 79968, USA*

Abstract: The Upper Permian Castile Formation of the Delaware Basin, Texas and New Mexico has commonly been interpreted as a deep-basin, deep-water evaporite deposit, based on the lateral continuity of mm-thick, anhydrite–calcite laminations. The identification of pseudomorphs after bottom-growth gypsum crystals throughout the Castile Formation implies that during much of the time, deposition occurred within a partially desiccated basin from supersaturated brines at most 40 m deep. The pseudomorphs are commonly poorly preserved within nodular horizons which were previously interpreted as mesogenetic alterations of former laminae. The nodular horizons themselves contain pseudomorphs after displacive lenticular gypsum crystals that suggest a synsedimentary, shallow-water origin for both bottom-growth and nodular textures. In contrast, structureless beds of sulphate up to 0.5 m, thick which are most common in the eastern part of the basin, appear to have been deposited from density currents as turbidites, implying relatively deep water. The evidence provided by pseudomorphs after gypsum within the Castile Formation suggests that average water depths across the basin were at times less than 40 m and could not have exceeded 200 m. Synsedimentary fault control may have partly accommodated the turbiditic sedimentation in the eastern part of the basin, although the depositional environment would require a remarkable coincidence between deposition and subsidence. The evidence as a whole suggests a delicate balance between subsidence, sedimentation and water depth during deposition of the Castile Formation.

The Upper Permian (Ochoan) Castile Formation of the Delaware Basin, Texas and New Mexico, consists of calcite and anhydrite, interlaminated on a millimetre scale, that are themselves interbedded with halite on a scale of 30 to 100 m (Figs 1 & 2). The laminae, of which there are over 260 000 within the Castile Formation (Anderson 1982), can be correlated laterally in cores for up to 100 km (Anderson *et al.* 1972; Dean & Anderson 1974; Dean *et al.* 1975; Anderson 1966). This lateral correlation was established by counting individual laminae within several cores (Anderson *et al.* 1972), and the time equivalence of individual laminae within cores across the basin has been used in numerous studies of the Castile Formation (Anderson *et al.* 1972; Anderson & Dean 1996). The Castile Formation evaporites pass upwards into the Salado Formation which contains proportionally more halite (Fig. 2) and abundant pseudomorphs after bottom-growth gypsum. The Salado Formation was formed in a shallow saline lagoon environment

† Deceased

affected by repeated flooding events (Lowenstein 1988; Holt & Powers 1990).

The even laminations and trace elemental homogeneity of the anhydrite in the Castile Formation (Dean 1978) have been used to argue for a substantial water depth (600 m) at the beginning of Ochoan time, with a gradual upwards shallowing until shallow-water conditions, with regular exposure of the sediment surface, predominated during deposition of the Salado Formation. Schreiber (1986) compared the Castile Formation with the Late Pleistocene Lisan Formation of the Dead Sea Rift, interpreted to have been deposited in brine depths of 400 to 600 m (Katz *et al.* 1977), and observed that the rapid filling of the Delaware Basin by the Castile Formation (in less than 300 000 years), and the relative lack of dolomitization of the margins, argue against significant drawdown of water level and exposure of the marginal Capitanian carbonates.

This study was carried out to test a series of observations made in core which suggested a shallow-water origin for parts of the Castile Formation (Kendall & Harwood 1989). For this

From Kemp, A. E. S. (ed.), 1996, *Palaeoclimatology and Palaeoceanography from Laminated Sediments*, Geological Society Special Publication No. 116, pp. 79–92.

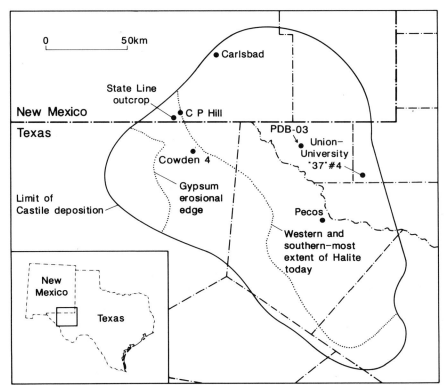

Fig. 1. Location map for the Delaware Basin, Texas and New Mexico showing the geographical distribution of cores sampled within this study. The nine cores from the CP Hill area are not individually marked.

reason the oldest unit, the Anhydrite 1 member, was examined since the potential difference between shallow-water (40 m) and deep-water (600 m) models are greatest when the basin has not been filled with evaporites. Anhydrite 1 member samples from a total of seventeen well cores were examined (Fig. 1), nine from the northwest Delaware Basin in the CP Hill area of New Mexico, and the other eight, three of which are shown on Fig. 1, in Texas.

Questioning the deep-water origin of the Castile Formation is of more than parochial interest. The Castile Formation has been instrumental in the formulation of many depositional models that have been applied to basin-filling evaporites worldwide. The discrepancy between the proportions of evaporites in the Castile Formation and those expected from evaporation of seawater led King (1947) to propose a reflux model wherein the most soluble salts escaped from the basin as an outflow of dense brine over a basin-confining sill. This model has been modified by Schmalz (1969) for other basins. The Castile Formation has been used to illustrate a number of features of deep-water

evaporative basins such as the size of inlets (Lucia 1972), hydrostatic conditions (Shaw 1977) and control of evaporation by atmospheric humidity (Kinsman 1976). Thus any conclusion about conditions during evaporite deposition will have implications for the interpretation of many other evaporite sequences.

Textural variations within the Castile Formation

The Castile Formation is characterized by cycles of upwards-increasing salinity, the smallest of which are the millimetre-scale (probably annual, Anderson 1982) laminae. The largest cycle incorporates both the Castile and Salado Formations, which form one upwards-increasing salinity trend associated with shallowing of the depositional environment (Fig. 2). Four intermediate-scale cycles define the anhydrite and halite members of the Castile Formation. The Anhydrite 1 member contains approximately 38 000 anhydrite laminae (Anderson 1982), ranges between 65 and 100 m in thickness and is thickest in the east of the basin (Fig. 3). This

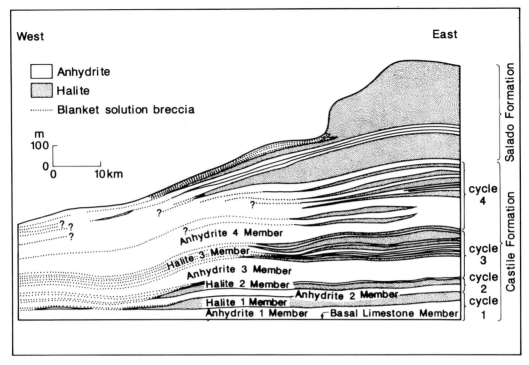

Fig. 2. Sketch cross section through the Castile and Salado formations showing the basal Anhydrite 1 member, from which the core samples were collected. The thickening from west to east is caused by a regional tilt and dissolution of the evaporites in the west of the basin. Taken from Anderson *et al.* (1972).

thickening of the Anhydrite 1 member has been interpreted by Anderson & Dean (1996) as a result of a predominant western supply of meteoric waters into the Delaware Basin (Fig. 3). Within the Anhydrite 1 member, a number of cycles are developed (Fig. 4; Dean & Anderson 1978). Carbonate is most abundant at the base of cycles, and laminae thicken from <1 mm to >4 mm and become progressively more anhydrite-rich upwards. In all of the cycles, basal laminae contain sulphate, indicating that brines were constantly supersaturated with respect to gypsum during deposition. The cycles are topped by thick-bedded anhydrite consisting of an amalgamation of numerous beds, 1 to 3 cm in thickness, that have a nodular or flaser structure. These were originally interpreted as recrystallized laminae (Anderson *et al.* 1972). Within the Anhydrite 1 member five major cycles between 5 and 17 m in thickness, topped by thick-bedded anhydrite, have been identified and correlated across the basin in all but one of the cores examined (Fig. 5).

The lateral continuity of the thick-bedded anhydrite at the top of each cycle (Fig. 5) in itself suggests a primary origin. In each of the cores studied, in addition to the well documented correlation of laminae (Anderson *et al.* 1972; and others), individual thick beds of sulphate are laterally consistent over much of the basin (Robinson & Powers 1987). This lateral continuity of beds would not be expected if they were a burial diagenetic feature (Dean *et al.* 1975).

Bottom-growth gypsum

Within the thick-bedded anhydrite horizons at the tops of cycles, pseudomorphs after bottom-growth gypsum have been identified. Holser (1976, his Fig. 23) includes a photograph of such a texture from the Union-University 37-4 well, but the pseudomorphs were not identified as such. The bottom-growth texture is identified by preserved crystal terminations (Fig. 6a) within thick or nodular beds, either anhydrite in core or rehydrated to gypsum at the State Line outcrop. The beds commonly have relatively flat bases and irregular tops, some of which resemble swallow-tail twinned gypsum crystal terminations. The terminations are outlined by a drape of calcite and organic matter in the overlying

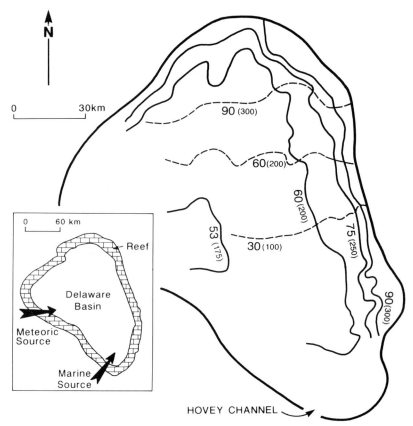

Fig. 3. Isopachyte map of the Anhydrite 1 and Halite 1 members showing thickening of anhydrite (solid lines) from west to east and thickening of halite (dashed lines) from south to north. Thicknesses in metres, approximate feet given in brackets. The Hovey Channel is assumed to be the site of entry of marine waters into the basin. Adapted from Anderson & Dean (1996).

laminae. The bottom-growth crystals are variably preserved, and commonly there is little indication of the original presence of the texture other than the carbonate drape overlying terminations. In some cores the detail of individual crystals can be discerned where a single bottom-growth horizon is overlain by laminated evaporites (Fig. 6a). Where well preserved, the horizons resemble beds of cm-high, bottom-growth gypsum crystals (Arakel 1980; Hardie & Eugster 1971; Richter-Bernburg 1973; Schreiber 1978; Schreiber *et al.* 1976, 1977; Truc 1978) which are known informally as 'gypsum grass', or anhydrite replacements of such a texture (Crawford & Dunham 1982; Lowenstein 1982; Nassichuck & Davies 1980; Nurmi & Freidman 1977).

Displacive gypsum

As well as bottom-growth textures, pseudomorphs after lenticular crystals up to 5 cm long

have been identified within many of the thick-bedded anhydrite horizons (Figs 6b & 6c). In some cases, these crystals have disrupted laminae; in others, laminae have been wedged apart and thickened by the addition of sulphate (Fig. 6b), forming elongate beds of fluctuating thickness. In extreme cases, such as the tops of major cycles, displacive sulphate forms a thick bed with little discernable structure (Fig. 6c) in which only a few pseudomorphs are visible. In these cases, the thick bed may originally have consisted of bottom-growth structures that have become obscured by displacive growth.

Resedimented sulphate within the Castile Formation

In the Anhydrite 1 member of the Union-University 37-4 core (Fig. 1) there are abundant beds of sulphate up to 10 m in thickness,

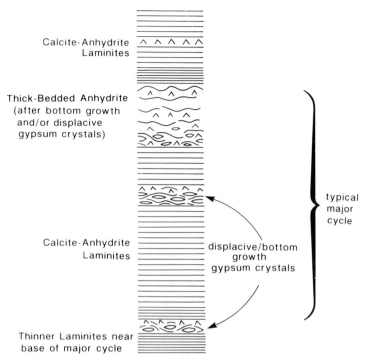

Calcite-Anhydrite
Laminites

Thick-Bedded Anhydrite
(after bottom growth
and/or displacive
gypsum crystals)

typical
major
cycle

Calcite-Anhydrite
Laminites

displacive/bottom
growth
gypsum crystals

Thinner Laminites near
base of major cycle

Fig. 4. Diagrammatic log showing a major cycle within the Anhydrite 1 member, between 5 and 17 m in thickness. The single cycle is further subdivided by the presence of less prominent beds of bottom-growth and displacive pseudomorphs which form the tops of sub-cycles.

interbedded with laminated evaporites, that have very flat boundaries and little or no internal structure (Fig. 6d). These thick beds, interpreted to be the product of turbidity

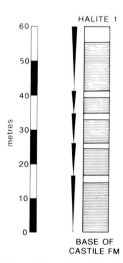

HALITE 1

metres

60
50
40
30
20
10
0

BASE OF
CASTILE FM

Fig. 5. Sketch log of the Anhydrite 1 member which has been subdivided into five major sub-cycles topped by thick-bedded anhydrite.

currents, form roughly 30% of the total thickness of the Anhydrite 1 member in the Union-University 37-4 well. In this well it is difficult to subdivide the Anhydrite 1 member into cycles as a result of both disruption of the succession and missing core, although the uppermost cycle of the Anhydrite 1 member can clearly be correlated between 37-4 and cores in the west of the basin, indicating that lateral correlations were basin-wide. In the PDB-03 research well in the centre of the basin (Fig. 1), thick turbiditic units are present but they are less common, and the five cycles in the Anhydrite 1 member are easily identifiable. The cores from around the CP Hill area have some thin units interpreted as clastic units (C. Latimer, pers. comm. 1992) but these are thin and different in character from the distinctive beds in the eastern and central basin. Robinson & Powers (1987) described a fan-shaped accumulation of clastic sulphate breccias overlying the Anhydrite 1 member and the Halite 1 member at CP Hill. They interpreted the breccia beds as a gravity driven deposit, although it is possible that this deposit represents the solution breccia that is the lateral equivalent of the Halite 1 member.

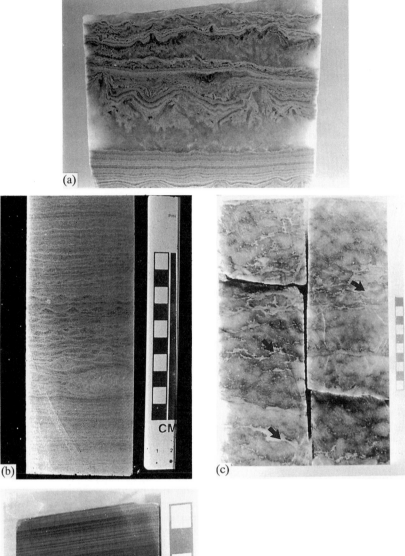

Fig. 6. Textures observed within the Anhydrite 1 member. (**a**) Pseudomorphs after bottom-growth gypsum interbedded with laminated Castile. Note the flat bases to pseudomorph beds and the drape of overlying sediments over terminations. (**b**) Laminated Castile disrupted by growth of displacive lenticular gypsum, now anhydrite. (**c**) Thick bed at top of salinity cycle composed of amalgamated displacive gypsum nodules. Few distinctive crystal shapes are still recognizable (arrowed). (**d**) Thick, relatively structureless bed of sulphate within laminated Castile, interpreted to be turbiditic in origin.

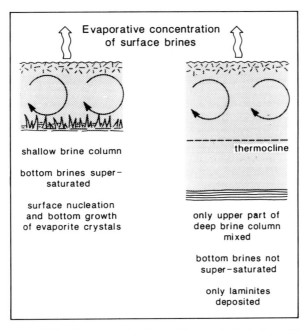

Fig. 7. The effects of changes in brine depth on style of deposition on the basin floor. The thermocline is unlikely to have been deeper than 40 m.

Discussion

Implications for basin water depths

Crusts composed of upwards-growing gypsum crystals have been described from other evaporite basins and have been interpreted as shallow-water features. In the Silurian Michigan Basin, crusts are restricted to the shallow marginal parts of the basin during deposition of the Anhydrite 1 member evaporite of the Salina Group (Nurmi & Freidman 1977). Crusts overlie massive gypsum that covers stromatolites in shallowing-upwards cycles in the Messinian of the Apennines (Vai & Ricci-Lucchi 1977). Schreiber (1978), in discussing gypsum crusts, states that 'comparatively little gypsum is formed...at depths below about 5 m'. Gypsum crusts from basinal locations in the Mississippian–Pennsylvanian Otto Fiord Formation of the Canadian Arctic (Nassichuck & Davies 1980) form stratigraphic units that climb at most a few tens of metres up the sides of the basin slopes.

The shallow-water origin of bottom-growth textures can be argued using other lines of evidence. Whereas the laminated evaporites appear to have formed by settling of surface-precipitated calcium sulphate (Anderson *et al.* 1972), bottom brines must be supersaturated with respect to gypsum for crusts to precipitate on the basin floor. Evaporative concentration and supersaturation of the brine can only occur at the surface. This can easily explain the precipitation of surface-nucleated gypsum crystals and their settling to form laminae. To form crusts, however, a shallow body of brine in which surface processes can fully mix the surface waters to create a homogeneous brine column (Fig. 7) is required. Warren (1982) records bottom-growth gypsum in Holocene saline lakes up to 10 m deep. Neev & Emery (1967) describe how a rope lowered into Dead Sea waters was coated with gypsum down to a depth of 40 m. The maximum storm wavelength, however, suggests a wave base of 5 m, indicating that whilst saturated waters can be related to surface processes, they are not simply constrained by mixing of the upper water column by wave action. Thus bottom-growth textures in the Castile Formation, while still indicative of shallow water, can be taken to have formed in waters up to at least 40 m deep.

Undercurrents of denser brine may have contributed to the supply of sulphate to the basin floor. In the uppermost cycle of the Anhydrite 1 member, however, pseudomorphs after bottom-growth crusts underlie halite and therefore record precipitation of gypsum within basinal brines approaching halite saturation. If the crusts were formed by density underflows, then they would be expected to occur at the base of salinity cycles, where the bottom brines would be less dense and more easily displaced by underflows. In a deep-water setting immediately prior to halite deposition, any brine capable of displacing the almost halite-saturated bottom brines must have been previously concentrated to such an extent that all the calcium sulphate should have been formed elsewhere. It is probable, therefore, that bottom-growth crusts indicate relatively shallow conditions across the Delaware Basin during their formation.

The pseudomorphs after displacive gypsum crystals associated with the thick-bedded horizons are further evidence for supersaturation of bottom brines. Displacive gypsum is unlikely to have been emplaced during deep burial, and is clearly not a late, uplift texture since the gypsum crystals are replaced by anhydrite. The most likely origin for the displacive gypsum crystals is during very early burial, while the sediment was flushed by supersaturated basin brines. Newly deposited sulphates have a porosity of 10–50% (Sonnenfeld & Perthusiot 1989), and this would aid both the penetration of brines into the sediment and the interstitial growth of crystals.

The absence of truncation or dissolution of the gypsum crusts suggests that freshening events, which may have been responsible for the deposition of calcite-rich laminae in the overlying cycle (Adams 1944; Briggs 1957), did not dissolve the upstanding gypsum crystals. It is possible that the freshening events primarily affected surface waters and that subsequent mixing reduced the salinity contrast. This, as well as the lateral consistency of the crusts and nodular sulphate horizons, suggests that waters may have been several tens of metres deep. The carbonate-rich laminae may, however, have formed from either seasonal organic (Richter-Bernburg 1958, 1986; Dean 1967) or temperature-controlled blooms (Neev & Emery 1967). Both these interpretations require no freshening event, but the concentration of both crusts and displacive crystals at the tops of salinity cycles suggests that both textures were the result of a progressive shallowing during a cycle, resulting in precipitation at and below the basin floor when the water column became sufficiently shallow.

Deposition of the Castile Formation laminae

Laminae are intimately associated with bottom-growth pseudomorphs (Fig. 6a) throughout the Anhydrite 1 member and in overlying anhydrite members in the Castile Formation. The laminae immediately underlying pseudomorphs must have been deposited in similar water depths, but are not different from most laminae in the Castile Formation, although laminae near the tops of salinity cycles are thicker (Fig. 4). Gypsum–carbonate laminites from the Messinian of Sicily ('balatino', Hardie & Eugster 1971) resemble Castile laminae but contain desiccation cracks and bird footprints, clear evidence of a shallow-water origin. In Solar Lake on the Sinai Peninsula, regular carbonate–gypsum laminae were evenly deposited in a brine pool less than 5 m deep (Gerdes & Krumbein 1987, their Fig. 27). Conversely, thin, even laminae are also typical of the Late Pleistocene Lisan Formation (Begin *et al.* 1974) and the Permian Zechstein (Richter-Bernburg 1958, 1986; Taylor 1980), which have been interpreted as being deposited in basins up to 600 m deep.

Relatively deep water can be inferred for the Bell Canyon Formation which underlies the Castile Formation (Cys 1978) and has been interpreted to have been deposited when the basin was filled (600 m) with marine waters. Bottom-growth pseudomorphs ten metres above the Bell Canyon contact indicate rapid drawdown of the waters in the basin. Underlying the laminated anhydrite is a unit of laminated limestone which contains 600 laminae (Anderson 1982). If these laminae are also annual, then there would be sufficient time to generate a gypsum-saturated brine and then to reduce the level sufficiently to generate bottom-growth gypsum on the basin floor near the base of the Anhydrite 1 member. At the base of the Anhydrite 1 member, 500 m of seawater would be required to fill the basin. Given that there is no dissolution of gypsum, it would require 2000 years (at an evaporation rate of 2 m/year and with an addition of a further 1500 m of seawater) to generate 500 m of gypsum-saturated brine.

The constant and abrupt changes between laminated and thick-bedded anhydrite suggest that actual changes in water depth were not as great as at the beginning of evaporite deposition. The tops of salinity cycles are synchronous, basin-wide events, and the beginning of a new cycle records the addition of water to the basin. The added water was not sufficient to reduce salinity to the point where sulphate was not produced, since laminae at the base of cycles still contain anhydrite. This places certain constraints on the volume (and composition) of water entering the basin. Within the constraints of the gypsum saturation field (80–90% by volume evaporation of seawater) the volume of brine in the basin could be doubled during dilution by marine water, and increased by even more if some of the incoming waters were already concentrated by passage through saline back-reef sediments.

The distribution and accommodation of turbiditic units

The concentration of turbiditic beds in the east of the basin is itself problematic regardless of the water depths proposed during Castile Formation deposition. There is unfortunately only one core, the Union-University 37-4 in which the Anhydrite 1 member is not completely recovered, in which the turbidites significantly thicken the succession, but thickening of the Anhydrite 1 member due to turbidites appears to be confined to the eastern margin of the basin (Fig. 3; Anderson & Dean in press, their Figs 3 & 4). The few turbidites that occur in the PDB-03 Research well in the centre of the basin do not disrupt the sequence of cycles and do not significantly increase the thickness of the Anhydrite 1 member. The reason for the concentration of turbidites in the east of the basin is not known, but it may be connected with the possible fault control on that side of the basin (Hills 1970; Holt & Powers 1990). It is possible that the predominant palaeowind direction was significant in controlling the distribution of turbiditic beds. A wind direction from the northeast and southeast might have caused greater instability of marginal sediments in the eastern side of the basin. This mechanism is not wholly satisfactory, however, since there appears to be no concentration of clastic beds in the east of the basin during the earlier Permian, when wind directions would have been the same (Walker et al. 1991).

The problem of accommodating the turbiditic material in the east of the basin is particularly pertinent to a shallow-water model. If it is accepted that water depths were at times approaching 40 m then there is little leeway for the input of a thick turbiditic succession (30 m in the Anhydrite 1 member in the Union-University 37-4 well) without causing shallowing and therefore exposure of the succession, or at least a greater proportion of bottom-growth textures. If localized subsidence is proposed as a means of balancing the additional input of sediments (Fig. 8), the coincidence of sediment supply and subsidence is surprising, but can be explained by slight fault activity along the basin-bounding

fault. The late Permian was a time of tectonic quiescence, although some movement of the West Platform Fault during or after the Ochoan has been demonstrated (Hills 1970; Holt & Powers 1990). This tectonic activity could account for both a source of material and accommodation of the turbidites within the basin.

Constraints on basin hydrology

Some theoretical constraints can be placed upon the level of drawdown of waters in the basin. Holser (1976) states that water level in a partially isolated basin will drop by only 30% of the total brine depth before influx through a permeable barrier is sufficient to balance evaporation and to prevent reflux of dense brines out of the basin. For the Castile Formation this would imply that the water level would, at most, drop 200 m at the start of the Ochoan leaving a brine column 300 to 400 m in depth. In a depression on the scale of the Delaware Basin, however, waters could enter through one part of the margin, allowing reflux at other locations (Fig. 3). The Hovey Channel (Fig. 3) in the south of the basin (Anderson & Dean in press), if taken as the source of influx of marine waters, would indicate that there is scope for reflux of brines from the basin in other parts of the basin margin. Furthermore, the strength of inflow is a function of the horizontal thickness of the barrier between marine waters and the basin, which controls the strength of the hydrological head driving waters into the basin. Palaeogeographic reconstructions suggest that the Hovey Channel formed a significant barrier to the incursion of marine waters into the Delaware Basin. This would have decreased the hydrological drive of waters into the basin, reducing the constraints on reflux and allowing greater drawdown of brine level.

Relative humidity of the atmosphere can also be considered as a constraint upon evaporation of the brines (Kinsman 1976). To evaporate a brine beyond the field of halite supersaturation and precipitate potash salts, the atmosphere within the basin must have a low relative humidity ($<67\%$). If the relative humidity is greater, as might be expected in the Delaware Basin (Kinsman 1976), then the brine will not become more saline and an equilibrium is achieved in which brines will precipitate halite, or gypsum if salinities are reduced by influx of less saline waters. This also constrains the amount of drawdown that can initially take place. With recharge from the basin margins as well as marine waters, drawdown could in theory continue,

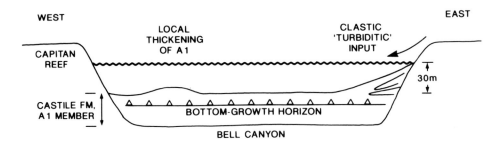

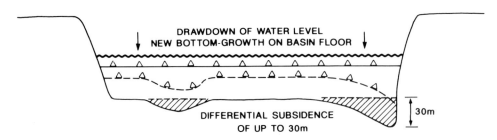

Fig. 8. Sketch section across the Delaware Basin showing the relationship between turbidite deposition and subsidence in a shallow basin. If water depth are less then 30 m then deposition must be exactly matched by subsidence if the bottom-growth crusts represent a depth-controlled texture. Slight fault movements along the eastern boundary of the basin (Hills 1970; Holt & Powers 1990) could have accommodated the additional clastic input, allowing bottom-growth textures to develop across the entire basin floor.

maintaining the equilibrium between atmospheric humidity and brine salinity, until the basin was predominantly empty.

Implications of shallow-water deposition for the basin margins

The shallow-water (40 m brine depth) model for the Castile has importance for the consideration of diagenetic processes on the basin margins. If several hundred metres of drawdown (Maiklem 1971) took place during or just before deposition of the evaporites, this implies that the basin was rapidly isolated from marine waters. The Delaware Basin was surrounded by exposed, narrow carbonate platforms that pass into extensive red bed–evaporite shelves. It is possible that much of the Castile Formation formed by the remobilization of older Permian sulphate salts on the basin margins (the Chalk Bluff facies of King 1948). If this was the case, then much of the water supplied to the basin must have come from movements of groundwaters through the surrounding margins (Fig. 9), as is the case with Lake MacLeod,

western Australia (Logan 1987). The extent of marine water input into the basin would then be a function of the topography and permeability of the carbonates at the Hovey Channel.

There is some evidence for evaporite recycling in the low and variable bromine content of the halite (Holser 1966), which may indicate reprecipitation from a second cycle of brine mobilization. This reprecipitation may have occurred *in situ*, or as a result of dissolution of salts elsewhere and precipitation in the Delaware Basin (Holser 1966).

Drawdown of basin waters during deposition also implies exposure of the marginal carbonates to vadose diagenesis. It is possible that the vadose cements reported by Given & Lohmann (1986) may have formed at this time, although the relatively short period of exposure (250 000 years) may not have given rise to cements of any volumetric importance.

Also of importance to reef diagenetic history is the inferred movement of brines into the basin through the carbonate rim (Figs 9 & 10). In the Guadalupe Mountain outcrops it appears that fluid flow through the margin was mostly concentrated in pre-existing fracture systems,

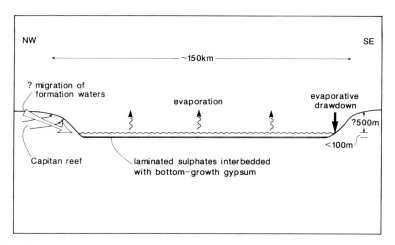

Fig. 9. Sketch section across the Delaware Basin during deposition of the Anhydrite 1 member. Drawdown of waters in the basin would lead to influx of formation waters through the marginal carbonates and into the basin.

without altering the main body of reef carbonate. In contrast, the outcrops in the Apache and Glass Mountains are dolomitized (Harris *et al.* 1988). Brine migration into the basin could have caused this partial dolomitization of the margin (Cercone 1988; Kendall 1992).

As evaporites progressively filled the basin, brines would reflux through the margins (Fig. 10), dolomitizing the carbonates and precipitating calcium sulphate. Preliminary outcrop studies have identified large volumes of pore-filling calcite, emplaced after dissolution of the original sulphate (Darke & Harwood 1990), associated with dolomitized fore-reef and reef

deposits. Scholle *et al.* (1992) have also suggested that the coarsely crystalline calcites are replacements of earlier void-filling sulphates. It is not possible to conclusively prove that these former sulphates were emplaced during the early Ochoan. Some occlusion of porosity, however, is required to prevent loss of basinal brines during deposition of the evaporites. As noted in the Recent Lake MacLeod basin, a basinal and lateral aquitard is necessary in evaporite basins in order for brines to persist and evaporites to form (Logan 1987). If the extensive porosity in the marginal carbonates was not occluded by sulphate during the Ochoan, then there would

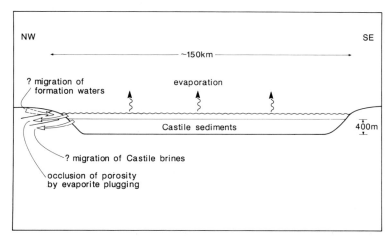

Fig. 10. As the basin is filled by the Castile Formation, dense brines will reflux out of the basin, increasing the potential for diagenetic alteration of the carbonate margins.

have been no lateral seal and evaporites would not have formed (Kendall 1992).

Conclusions

Pseudomorphs after bottom-growth and displacive lenticular gypsum crystals at the tops of sub-cycles within the Castile suggest that water depths were periodically sufficiently shallow (40 m) to permit mixing of the entire water column. Although crystal textures are commonly poorly preserved, the horizons containing the pseudomorphs can be identified and correlated across the Delaware Basin, implying that shallow water conditions were basin-wide. The cycles within the Anhydrite 1 member represent an upwards increase in salinity and decrease in water depth, with abrupt deepening of water at the beginning of the overlying cycle. The addition of water at the beginning of each cycle, however, was not sufficient to reduce salinity to the extent that sulphate was not precipitated, implying that even at the start of salinity cycles, water depths were not greater than 200 m. Laminated sediments are intimately associated with the pseudomorphs, suggesting that water depths did not exceed 200 m during deposition of laminae and that the basin was not filled with brine, thus questioning the original deep-water, deep-basin interpretation of the Castile Formation.

Substantial turbidite deposits in the eastern part of the basin suggest that local subsidence coincided with turbidite input, preserving lateral continuity of the evaporites and preventing exposure of parts of the basin floor.

If the basin was not filled during deposition of the Castile Formation, then some movement of groundwaters through the basin margins would be expected. During the deposition of the Anhydrite 1 member, when the basin was not filled with sediment, substantial inflow of groundwaters into the basin would take place, causing diagenesis of the carbonate basin margins. The inflow into the basin during deposition of the Anhydrite 1 and Halite 1 members, and reflux of brines out of the basin during later stages of basin fill would both be expected to have an influence on the diagenetic history of the margins. As yet it has not been possible to prove that any diagenetic phase was formed during the deposition of the Castile, but plugging of much of the basin margin porosity by sulphate, now replaced by coarse calcite, may have taken place during reflux of brines out of the basin.

This study was funded by NERC research grant NR3/7871. Thanks are due to Walter Dean, Mitch Harris and Roger Anderson for allowing access to cores and for helpful discussion. Diagrams were drawn by Philip Judge, photography by Sheila Davies. The comments of an anonymous reviewer greatly improved the manuscript

Note added in proof. Gillian Harwood died on 12 March 1996 after a long illness.

References

ADAMS, J. E. 1944. Upper Permian Ochoa series of Delaware Basin, west Texas and southeastern New Mexico. *AAPG Bulletin*, **28**, 1596–1625.

ANDERSON, R. Y. 1982. A long geoclimatic record from the Permian. *Journal of Geophysical Research*, **87**, 7285–7294.

—— 1996. Seasonal sedimentation: a framework for reconstructing climatic and enviromental change. *This volume.*

—— & DEAN, W. E. 1995. Filling the Delaware Basin: hydrologic and climatic controls on the Permian Castile Formation varved evaporite: *In*: SCHOLLE, P. A., PERYT, T. M. & ULMER-SCHOLLE, D. S. (eds) *The Permian of the Northern Pangea*. Vol. 2: Sedimentary Basins and Economic Resources. Springer, New York, 61–78.

——, ——, KIRKLAND, D. W. & SNIDER, H. I. 1972. Permian Castile varved evaporite sequence, west Texas and New Mexico. *Bulletin of the Geological Society of America*, **83**, 59–86.

ARAKEL, A. V. 1980. Genesis and diagenesis of Holocene evaporitic sediments in Hutt and Leemen Lagoons, Western Australia. *Journal of Sedimentary Petrology*, **50**, 1305–1326.

BEGIN, Z. B., EHRLICH, A. & NATHAN, Y. 1974. Lake Lisan, the Pleistocene precursor of the Dead Sea. *Bulletin of the Geological Survey of Israel*, **63**.

BRIGGS, L. I., JR 1957. Quantitative aspects of evaporite deposition. *Paper of the Michigan Academy of Sciences, Arts and Letters*, **42**, 115–123.

CERCONE, K. R. 1988. Evaporative sea-level drawdown in the Silurian Michigan Basin. *Geology*, **16**, 837–390.

CRAWFORD, G. A. & DUNHAM, J. B. 1982. Evaporite sedimentation in the Permian Yates Formation, Central Basin Platform, Andrews County, West Texas. *In*: HANDFORD, C. R., LOUCKS, R. G. & DAVIES, G. R. (eds) *Depositional and Diagenetic Spectra of Evaporites – a Core Workshop*. Society of Economic Paleontologists and Mineralogists Core Workshop, **3**, 238–275.

CYS, J. M. 1978. Transitional nature and significance of the Castile-Bell Canyon contact. *In*: AUSTIN, G. S. (ed.) *Geology and Mineral Deposits of Ochoan Rocks in Delaware Basin and Adjacent Areas*. New Mexico Bureau of Mines and Mineral Resources Circular, **159**, 53–56.

DARKE, G. & HARWOOD, G. M. 1990. Time constraints on sulphate-related diagenesis, Capitan Reef Complex, West Texas and New Mexico. *AAPG Bulletin*, **74**, 638.

DEAN, W. E. 1967. *Petrologic and Geochemical Variations in the Permian Castile Varved Anhydrite, Delaware Basin, Texas and New Mexico.* PhD dissertation, University of New Mexico.

—— 1978. Theoretical versus observed successions from evaporation of sea water. *In*: DEAN, W. E. & SCHREIBER, B. C. (eds) *Marine Evaporites.* Society of Economic Paleontologists and Mineralogists Short Course Notes, **4**, 74–85.

—— & ANDERSON, R. Y. 1974. Trace and minor element variations in the Permian Castile Formation, Delaware Basin, Texas and New Mexico, revealed by varve calibration. *4th International Symposium on Salt.* Northern Ohio Geological Society, **1**, 15–20.

—— & ——1978. Salinity cycles: evidence for subaqueous deposition of Castile Formation and lower part of Salado Formation, Delaware Basin, Texas and New Mexico. *In*: AUSTIN, G. S. (ed.) *Geology and Mineral Deposits of Ochoan Rocks in Delaware Basin and Adjacent Areas.* New Mexico Bureau of Mines and Mineral Resources Circular, **159**, 15–20.

——, DAVIES, G. R. & ANDERSON, R. Y. 1975. Sedimentological significance of nodular and laminated anhydrite. *Geology*, **3**, 367–372.

GERDES, G. & KRUMBEIN, W. E. 1987. *Biolaminated Deposits.* Lecture Notes in Earth Sciences, **9**, Springer, New York.

GIVEN, R. K. & LOHMANN, K. C. 1986. Isotopic evidence for the early meteoric diagenesis of the Reef facies, Permian Reef Complex of West Texas and New Mexico. *Journal of Sedimentary Petrology*, **56**, 1–18.

HARDIE, L. A. & EUGSTER, H. P. 1971. The depositional environment of marine evaporites: a case for shallow, clastic accumulation. *Sedimentology*, **16**, 187–220.

HARRIS, P. M., GARBER, R. A. & GROVER, G. A. 1988. Upper Permian Capitan Reef: Revision of outcrop model (abstract). *AAPG Bulletin,* **72**, 194.

HILLS, J. H. 1970. Late Paleozoic structural directions in southern Permian Basin, west Texas and southern New Mexico. *AAPG Bulletin,* **54**, 1809–1827.

HOLSER, W. T. 1966. Bromide geochemistry of salt rocks. *2nd Symposium on Salt.* Northern Ohio Geological Society, **2**, 248–275.

—— 1976. Mineralogy of evaporites. *In*: BURNS, R. G. (ed.) *Marine Minerals.* Mineralogical Society of America Short Course Notes, **6**, 211–294.

HOLT, R. M. & POWERS, D. W. 1990. *Geologic mapping of the air intake shaft at the Waste Isolation Pilot Plant.* U.S. Department of Energy Report DOE-WIPP **90-051**.

KATZ, A., KOLODNY, Y. & NISSENBAUM, A. 1977. Geochemical evolution of the Pleistocene Lake Lisan – Dead Sea system. *Geochimica et Cosmochimica Acta*, **41**, 1609–1626.

KENDALL, A. C. 1992. Evaporites. *In*: WALKER, R. G. & JAMES, N. P. (eds) *Facies Models: Response to Sea Level Change.* Geological Association of Canada, 375–409.

—— & HARWOOD, G. M. 1989. Shallow-water gypsum in the Castile Formation – significance and implications. *Society of Economic Paleontologists and Mineralogists Core Workshop*, **13**, 451–457.

KING, P. B. 1948. Geology of the southern Guadalupe Mountains, Texas. *U.S. Geological Survey Professional Paper*, **215**.

KING, R. H. 1947. Sedimentation in Permian Castile Sea. *AAPG Bulletin*, **311**, 470–477.

KINSMAN, D. J. J. 1976. Evaporites: relative humidity control of primary mineral facies. *Journal of Sedimentary Petrology*, **46**, 273–279.

LOGAN, B. W. 1987. *The MacLeod Evaporite Basin, Western Australia.* American Association of Petroleum Geologists Memoir, **81**.

LOWENSTEIN, T. 1982. Primary features in a primary evaporite deposit, the Permian Salado Formation of West Texas and New Mexico. *In*: HANDFORD, C. R., LOUCKS, R. G. & DAVIES, G. R. (eds) *Depositional and Diagenetic Spectra of Evaporites – a core workshop.* Society of Economic Palentologists and Mineralogists Core Workshop 3, 276–304.

LOWENSTEIN, T. K. 1988. Origin of depositional cycles in a Permian "saline giant": The Salado (McNutt Zone) evaporites of New Mexico and Texas. *Geo-logical Society of America Bulletin*, **100**, 592–608.

LUCIA, F. J. 1972. Recognition of evaporite–carbonate shoreline sedimentation. *In*: RIGBY, J. K. & HAMBLIN, W. K. (eds) *Recognition of Ancient Sedimentary Environments.* Society of Economic Paleontologists and Mineralogists Special Publication, **16**, 160–191.

MAIKLEM, W. R. 1971. Evaporative drawdown – a mechanism for water-level lowering and diagenesis in the Elk-Point Basin. *Bulletin of Canadian Petroleum Geologists* **19**, 467–503.

NASSICHUCK, W. W. & DAVIES, G. R. 1980. Stratigraphy and sedimentation of the Otto Fiord Formation – a major Mississippian–Pennsylvanian evaporite of subaqueous origin in the Canadian Arctic Archipelago. *Bulletin of the Geological Survey of Canada*, **286**.

NEEV, D. & EMERY, K. O. 1967. *The Dead Sea – Depositional Processes and Environment of Evaporites.* Bulletin of the Geological Survey of Israel, **41**.

NURMI, R. D. & FREIDMAN, G. M. 1977. Sedimentology and depositional environments of basin-center evaporites, Lower Salina Group (Upper Silurian), Michigan Basin. *In*: FISHER, J. H. (ed.) *Reefs and Evaporites – Concepts and Depositional Models.* American Association of Petroleum Geologists Studies in Geology 5, 23–52.

RICHTER-BERNBURG, G. 1958. Die Korrelierung isochroner Warven im Anhydrit des Zechstein 2 (Zweiter Beitrag). *Geologische Jahrbuch*, **75**, 629–646.

—— 1973. Facies and Paleogeography of the Messinian evaporites in Sicily. *In*: DROOGER, C. W. (ed.) *Messinian Events in the Mediterranean.* North-Holland Publishing Company, Amsterdam, 124–141.

——1986. Zechstein 1 and 2 Anhydrites: facts and problems of sedimentation. *In*: HARWOOD, G. M. & SMITH, D. B. (eds) *The English Zechstein and Related Topics.* Geological Society, London, Special Publication, **22**, 157–163.

ROBINSON, J. Q. & POWERS, D. W. 1987. A clastic deposit within the lower Castile Formation, western Delaware Basin. *In*: POWERS, D. W. & JAMES, W. C. (eds) *Geology of the Western Delaware Basin, West Texas and Southeastern New Mexico.* El Paso Geological Society Guidebook, **18**, 69–79.

SCHMALZ, R. F. 1969. Deepwater evaporite deposition – a genetic model. *AAPG Bulletin*, **53**, 798–823.

SCHOLLE, P. A., ULMER, D. S. & MELIM, L. A. 1992. Late-stage calcites in the Permian Capitan Formation and its equivalents, Delaware Basin margin, west Texas and New Mexico: evidence for replacement of precursor evaporites. *Sedimentology*, **39**, 207–234.

SCHREIBER, B. C. 1978. Environments of subaqueous gypsum deposition. *In*: DEAN, W. E. & SCHREIBER, B. C. (eds) *Marine Evaporites.* Society of Economic Paleontologists and Mineralogists Short Course Notes, **4**, 43–73.

——1986. Arid shorelines and evaporites. *In*: READING, H. G. (ed.) *Sedimentary Environments and Facies.* Blackwell, Oxford, 189–228.

——, FREIDMAN, G. M, DECIMA, A. & SCHREIBER, E. 1976. Depositional environments of Upper Miocene (Messinian) evaporite deposits of the Sicilian Basin. *Sedimentology*, **23**, 729–760.

——, CATALANO, R. & SCHREIBER, E. 1977. An evaporitic lithofacies continuum: Latest Miocene (Messinian) deposits of Salemi Basin (Sicily) and a modern analog. *In*: FISHER, J. H. (ed.) *Reefs and Evaporites – Concepts and Depositional Models.*

American Association of Petroleum Geologists Studies in Geology, **5**, 169–180.

SHAW, A. B. 1977. A review of some aspects of evaporite deposition. *Mountain Geology*, **14**, 1–16.

SONNENFELD, P. & PERTHUSIOT, J.-P. 1989. *Brines and Evaporites.* American Geophysical Union Short Course in Geology, **3**.

TAYLOR, J. C. M. 1980. Origin of the Werraanhydrit in the U.K. southern North Sea – a reappraisal. *In*: FUCHTBAUER, H. & PERYT, T. M. (eds) *The Zechstein Basin with Emphasis on Carbonate Sequences.* Contributions to Sedimentology, **9**, 91–113.

TRUC, G. 1978. Lacustrine sedimentation in an evaporitic environment: the Ludian (Palaeogene) of the Mormoiron Basin, southeastern France. *In*: MATTER, A. & TUCKER, M. E. (eds) *Modern and Ancient Lake Sediments.* International Association of Sedimentologists Special Publication, **2**, 187–203.

VAI, G. B. & RICCI-LUCCHI, 1977. Algal crusts, autochthonous and clastic gypsum in a cannibalistic evaporite basin: a case history from the Messinian of Northern Appenines. *Sedimentology*, **24**, 211–244.

WALKER, D., GOLONKA, J., REID, A. M. & TOMLINSON REID, J. 1991. The effects of Late Paleozoic paleolatitude and paleogeography on carbonate sedimentation in the Midland Basin, Texas. *In*: CANDELARIA, M. (ed.) *Permian Basin Plays-Tomorrows Technology Today.* West Texas Geological Society Symposium Publication, **91–89**, 139–163.

WARREN, J. K. 1982. The hydrologic setting, occurrence and significance of gypsum in late Quaternary salt lakes in South Australia. *Sedimentology*, **29**, 609–637.

A survey of occurrences of Holocene laminated sediments in California Borderland Basins: products of a variety of depositional processes

D. S. GORSLINE, ENRIQUE NAVA-SANCHEZ &
JANETTE MURILLO DE NAVA

*Department of Geological Sciences, University of Southern California,
Los Angeles, California 90089–0740, USA*

Abstract: Laminated sediments are found in several contemporary depositional settings in basins of the California Continental Borderland. These primary structures are preserved either as a result of anoxia (or near anoxia) in the bottom water when oxygen demand is greater than the oxygenation rate, or as a result of sedimentation rates in excess of bioturbation rates. The controlling processes in the anoxic settings include cyclic seasonal variation in the composition of hemipelagic deposition (hemipelagic laminae), and/or formation and destruction of bacterial mats as a result of periodic bottom water flushing of normally low-oxygen bottom waters (cyclic organic-rich laminae). Non-bioturbated hemipelagic laminations occur in San Pedro, Santa Monica and Santa Barbara Basins. Low-oxygen bottom waters and exclusion of macrobenthos have been characteristic of San Pedro Basin for a few decades. These conditions have existed in Santa Monica Basin for a few centuries and in Santa Barbara Basin for most of the Holocene. Physical sedimentation processes include rapid deposition from turbidity currents. Fine grained contemporary distal turbidites occur in Santa Monica, San Pedro, San Diego, Santa Cruz and San Nicolas Basins. Turbidity current flows into the deep basin floor occur at century scale in Santa Monica Basin and at somewhat longer frequencies in San Pedro Basin. Frequency of such flows is at the kilo-year level in San Diego Trough, Santa Cruz Basin and San Nicolas Basin. Such events are of interest because they show that submarine canyons will remain active through a high sea-level period. In narrow-shelf, active margins, canyons may not be shut off as generally occurs on passive margins when sea level rises rapidly, as is typical of climatically driven sea level fluctuations.

Primary, thin (mm to cm) laminations in sedimentary formations have been studied in many localities from a variety of geological ages (e.g. Antevs 1925; Calvert 1966; Hallam 1967; Duff *et al.* 1967; Fischer & Roberts 1991). This interest stems from the possibility that such records may reveal seasonal to multi-year climatic cycles that may have influenced the sedimentary conditions in those deposition sites. This study presents a comprehensive review of radiographs illustrating late Holocene small-scale sedimentary structures in the entire California Borderland off the United States southwest coast. It identifies a number of sites for study of active lamination-forming processes.

The primary laminations described in this paper are the product of: (1) fine-grained hemipelagic deposition from multiple sources which have different seasonal maxima (e.g. Emery & Hulsemann 1962; Soutar & Crill 1977); (2) by processes of bacterial mat formation and destruction affected by variations in bottom water conditions as a result of periodic basin water flushing on a seasonal or multi-year frequency (e.g. Sholkovitz & Gieskes 1971; Reimers *et al.* 1990; Grant 1991; Christensen 1991; Christensen *et al.* 1994; Schimmelmann &

Lange 1996); or, (3) by deposition from fine distal turbidity currents, at rates fast enough to prevent bioturbational mixing, or in low-oxygen content bottom water (Schwalbach *et al.* 1996). I am limiting inclusion of these latter products of waning flows to those of mm to cm scale (cf. O'Brien 1996). Larger turbidity currents in their proximal parts will typically contain bedding units much thicker than the mm to cm scale.

Definition

Definitions of bedding units of various scales have been discussed and suggested throughout most of the history of sedimentary research (e.g. Otto 1938; Campbell 1967). For the purposes of this paper the following simple definition has been used: 'Laminated sediments are preserved primary sedimentary structures deposited by a variety of events or processes, the individual units of which are of mm to cm thicknesses.'

The term *lamina* as used in sequence stratigraphy is of similar scale (Van Wagoner *et al.* 1990), but is strictly a scale-defined term. For

From Kemp, A. E. S. (ed.), 1996, *Palaeoclimatology and Palaeoceanography from Laminated Sediments*, Geological Society Special Publication No. 116, pp. 93–110.

Table 1. *Summary of basin floor hemipelagic sedimentological data for Santa Barbara, Santa Monica, San Pedro, San Diego, Santa Cruz and San Nicolas Basins. Data for surface 2 cm of sediment*

Basin name	Mean diam. phi[†]	Std. dev. phi	Skewness phi	Kurtosis phi	% CaCO$_3$	% TOC*
Santa Barbara	7.2	2.0	0.2	2.1	5.8	3.0
Santa Monica	6.2	2.1	0.4	5.5	7.4	2.8
San Pedro	5.9	1.8	0.6	4.3	7.6	2.3
San Diego	6.8	2.0	0.3	3.2	8.2	2.9
Santa Cruz	6.1	2.0	0.2	2.8	12.1	4.2
San Nicolas	5.9	2.1	0.4	3.4	24.3	4.4

Basin	% Gr (note)	% Sa[‡]	% Si	% Cl	Bulk density	No. of samples
Santa Barbara	0	3	61	36	1.20	110
Santa Monica	2	11	54	31	1.27	185
San Pedro	0	24	50	26	1.19	85
San Diego	0	8	60	30	1.22	183
Santa Cruz	1	24	41	33	1.35	173
San Nicolas	1	31	38	29	1.30	106

* Total organic carbon.
† Phi notation, where phi = $-\log_2$ diameter in mm.
‡ Gr = gravel, Sa, = sand, Si = silt, Cl = clay.

example, a section of a turbidite may be laminated, or a given environment may produce sequences of laminae. An older term, *sedimentation unit* (Otto 1938), is scale equivalent but stresses the idea of the lamina (thin bed) as the product of a single depositional event.

The event may be a turbidity current, or be a seasonal (or longer period) response to such effects as the cyclic occurrence of plankton blooms, or as a result of a major flood in the neighbouring source area. They may be the result of periodic, climatically driven ocean water intrusions. These structures are preserved by the reduction or elimination of bioturbating faunas in low-oxygen environments, where oxygen demand exceeds oxygenation rate. They may be preserved when deposition of sediments creates beds that are too thick for penetration by burrowers, or are textural barriers to local infaunal burrowing. Destruction of such primary structures may also be by physical processes such as post-depositional fluidization or slumping.

Data source

We have collected more than 2400 box cores distributed more or less uniformly over the region, principally during the period from 1975 to 1986. All have been collected and curated with care using a uniform procedure to ensure best possible recovery, and virtually all have been x-radiographed. The top 1 to 2 cm of all of the cores have been analysed for standard sedimentological parameters including textural descriptions, carbonate content, total organic carbon content, some bulk density measurements for later samples, and down-core analyses for some selected cores (Table 1).

Radiometric ^{210}Pb dates have been determined for about 20 cores, mostly in Santa Monica Basin (Chistensen 1990; Christensen *et al.* 1994), and a few in Santa Barbara Basin (also see in Schimmelmann & Lange 1996). In addition, some older radiocarbon dates were obtained by Emery & Bray (1962) and these, plus high resolution acoustic profiles, have been used to obtain total Holocene sedimentation rates and basin sediment mass (Schwalbach 1982; Table 2). These tabulated data give a

Table 2. *Holocene total basin sedimentation rates and sediment accumulation masses for Santa Barbara, Santa Monica, San Pedro, San Diego, Santa Cruz and San Nicolas Basins (after Schwalbach & Gorsline 1985)*

Basin name	Holocene mass ($\times 10^{10}$ tonnes)	Sedimentation rate (mg cm^{-2} a^{-1})
Santa Barbara	2.35	80–130
Santa Monica	1.40	30–60
San Pedro	0.36	20–50
Santa Cruz	0.46	20–40
San Nicolas	0.33	10–20
San Diego	0.18	*c.* 10

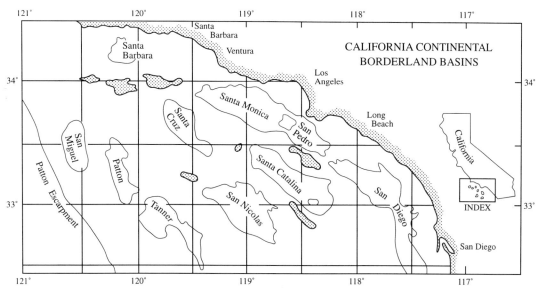

Fig. 1. Borderland basin location map. This paper discusses fine laminations of several origins from Santa Barbara, Santa Monica, San Pedro, San Diego, Santa Cruz and San Nicolas Basins.

general picture of the physical character and the accumulation rates of the sediments exhibiting laminations.

Until late 1993, the entire set of radiographs had not been systematically viewed for sedimentary structures. Savrda *et al.* (1984) examined bottom photographs and some of the radiographs of cores in San Pedro, Santa Monica and Santa Barbara basins, but the great majority had not previously been examined.

Setting

The setting for the study is the Continental Borderland off southern California (Shepard & Emery 1941; Teng 1985; Gorsline 1992). This area includes 2 or 3 parallel rows of deep, active-margin basins (Fig. 1), with basin floor depths ranging from 600 m to over 3000 m and sill depths from 250 m to over 2200 m (Emery 1960). In all basins discussed here, the box cores studied were essentially limited to those from the basin floors. The preliminary scan of the radiographs showed that, with only three exceptions in cores from the slopes of Santa Cruz Basin, the laminated structures were confined to the basin floor environment.

Non-bioturbated hemipelagic laminations

Examples of hemipelagic laminated sediments are shown in Fig. 2 (Santa Monica Basin and

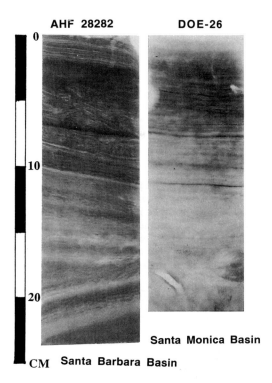

Fig 2. The print of the x-radiograph of core AHF 28282 shows typical laminations from the central anoxic area of nonbioturbation in Santa Barbara Basin. Core DOE-26 is from the central part of the nonbioturbated anoxic zone in Santa Monica Basin. Locations are shown in Figs 7 & 9 respectively.

Santa Barbara Basin). These also show comparative examples of annual (?) hemipelagic laminae (Santa Barbara Basin: Soutar & Crill 1977) and multi-year laminae (Santa Monica Basin: Christensen 1990; Christensen *et al.* 1994). Typically turbidites appear as discrete lamina; hemipelagic laminations appear as laminasets (Campbell 1967). The majority as seen in the radiographs are even parallel laminae, either continuous or discontinuous, to use Campbell's nomenclature (1967). In a few instances some wavy parallel and nonparallel types are seen in the thin turbidites.

Only San Pedro, Santa Monica and Santa Barbara Basins contain low-oxygen basin floor environments at the present time. Piston core studies by Prensky (1973) and Fleischer (1970) have reported laminated zones at depths of pre-Holocene age in San Clemente and Santa Catalina Basins. The occurrences are noted here since they indicate that periods of anoxic bottom water conditions have been more widespread in late Pleistocene time. Those piston cores have not been reviewed here since this paper is specifically about the Holocene locations.

A recent Ocean Drilling Program hydraulic piston core in Santa Barbara Basin illustrates the variation between glacial and interglacial times and the response of the sediments (Kennett *et al.* 1995). The core confirms that the zones of laminations occur during the Holocene and become infrequent in Late Pleistocene time.

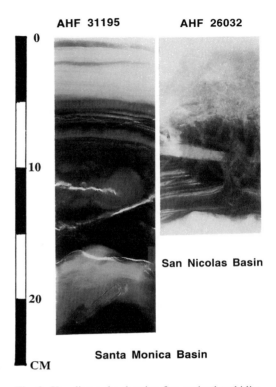

Fig. 3. X-radiographs showing fine-grained turbidity current laminations from Santa Monica Basin (core AHF 31195) and San Nicolas Basin (core AHF 26032). Locations are shown in Figs 9 & 16 respectively. San Nicolas Basin is aerated and bioturbation has partially destroyed the laminated structures. In Santa Monica Basin the core location is in the nonbioturbated zone.

Fine laminated distal turbidites

Examples of thin, fine grained turbidites are shown in Fig. 3 (Santa Monica basin, San Nicolas Basin). Contemporary (late Holocene) distal turbidite laminae are present in several basins where active extended channel/levee-systems deliver turbidity currents to the basin floor. These features, and the associated events that occur at frequencies from decadal (Santa Monica outer fan) to multi-century (e.g. Santa Cruz basin floor, San Nicolas Basin floor), have been described in detail by Gorsline (1996), Schwalbach *et al.* (1996).

General sedimentological characteristics

Table 1 summarizes the average surficial (0–2 cm) sedimentological characteristics of the basin floor sediments of each of the six basins discussed. Several trends are evident that have been generally described by Emery (1960).

Carbonate bulk content (weight percent) generally increases offshore and is primarily a result of dilution of the biogenic carbonate in the inner basins that are open to continental detrital influx. Contents range from about 6% in Santa Barbara Basin to 24% in San Nicolas Basin.

Total organic carbon (TOC) bulk content shows a similar trend that reflects both decreasing dilution by terrigenous sediment, and greater mineralization due to increasing travel and fall distances of organic debris to the basin floors (Table 1). This pattern is overprinted in the inner basins by the effect of increasing distance from the northern areas of upwelling. Thus Santa Barbara Basin sediments average 3 % TOC, decreasing south in the inner basins, to Santa Monica at 2.8%, San Pedro at 2.3%, and then progressively increasing to 2.9% in San Diego Trough, 4.2% in Santa Cruz Basin, and 4.4% in San Nicolas Basin, as detrital input and dilution decline away from the northern inner borderland sources.

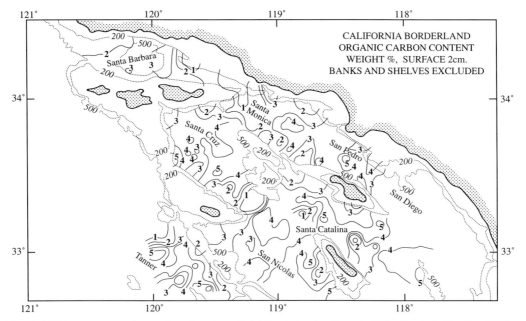

Fig. 4. Weight percent total organic carbon in surficial Borderland sediments (Schwalbach & Gorsline 1985).

Note again that the more distant basins contain more refractory organic material in the sediments indicating long exposure to bacterial decay, and that the patterns of TOC and carbonate *fluxes* are the reverse of the *bulk content* patterns (Schwalbach & Gorsline 1985) (Figs 4 & 5).

The textural parameters in all basins are typically in the hemipelagic range with mean diameters in the fine silt grade, moderately sorted, slightly positively skewed, and slightly platykurtic (Table 1). Thin turbidites within the top 30–50 cm in all basins are typically coarser silts, strongly positively skewed, and leptokurtic.

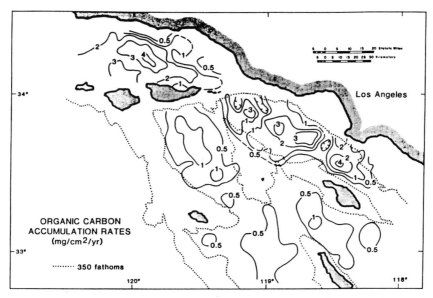

Fig. 5. Total organic carbon accumulation rates in surficial bottom sediments of the Borderland (Schwalbach & Gorsline 1985).

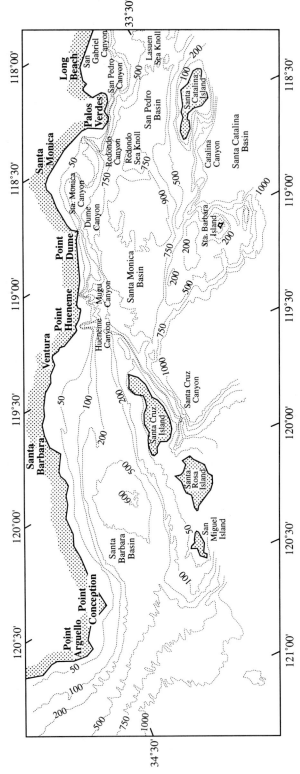

Fig. 6. Bathymetry of Santa Barbara, Santa Monica and San Pedro Basins.

These are all distal turbidites and are usually less than 5 cm in thickness.

These laminated sediments are the contemporary equivalent of laminated siltstones and silty shales in the stratigraphic record. In the example of the hemipelagic laminations, the lamina sets can be traced over much of the basin floors, for distances of tens of km. The turbidites are often also laid down basin-wide, or at least over distances of 10 km or more.

Basin floor laminated sediment occurrences

Santa Barbara Basin

The laminated hemipelagic sediments of Santa Barbara Basin have been described in considerable detail by Emery & Hulsemann (1962); Soutar & Crill (1977); Reimers et al. (1990); Grant (1991) and by Wolfgang Berger's group at Scripps Institution of Oceanography (e.g. Lange & Berger 1993; Schimmelmann & Lange 1996). Thornton (1981) has described the overall basin character and the depositional processes operating in the basin.

This basin has been used frequently as a modern analog of ancient anoxic basin floors such as the Miocene Monterey siliceous shales,

diatomites and phosphatic shales (e.g. Isaacs 1984; Isaacs et al. in review). The authors share Isaacs' opinion that the analogy is probably not exact, and has been overworked. The rates of accumulation are much higher than those cited for the laminated Monterey sections, the terrigenous component is a much larger fraction of the sediment, and the level of productivity is probably lower as evidenced by the lower biogenic silica content of the modern sediments.

Santa Barbara basin is directly influenced by seasonal variations in the California Current and associated spring flushing of that basin's deep waters. The west sill is at a depth of 450 m (Fig. 6); the east sill is at less than 250 m. Bacterial mats (Beggiotoa spp., Grant 1991; Reimers et al. 1990) flourish as oxygen drops below 5 mM, and then are destroyed or migrate into the sediments as the flushing renews the oxygen. The renewals are limited to periods of less than three months and subsequent oxygen content decline typically before oxygen levels return to anoxic condition (Sholkovitz & Geiskes 1971; Reimers et al. 1990; Grant 1991).

The Santa Barbara sub-sill basin floor encompasses an area of 830 km^2 below depths of 300 m (maximum basin depth 600 m). Within this area the anoxic zone (nonbioturbation zone) within

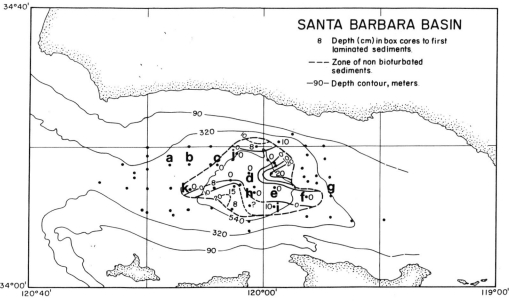

Fig. 7. Zone of laminated hemipelagic sediments within box core depth in Santa Barbara Basin. Note that the peripheral cores have been bioturbated to as much as 20 cm at some time within the past few decades. All cores beyond this zone are bioturbated to base-of-core (c. 50 cm). 'O' denotes cores laminated to the present sediment–water interface; '10' denotes thickness of core in which the surficial parts are disturbed by bioturbation. All other cores outside the zones are bioturbated from top to bottom. Locations of cores (see examples in Figs 8a–e) are as follows: (a) AHF 27147; (b) AHF 27152; (c AHF 27163; (d) AHF 28282 (shown in Fig. 2); (e) AHF 27173; (f) AHF 28275; (g) AHF 28267; (h) AHF 28283; (i) AHF 28278; (j) AHF 27164; (k) AHF 27151.

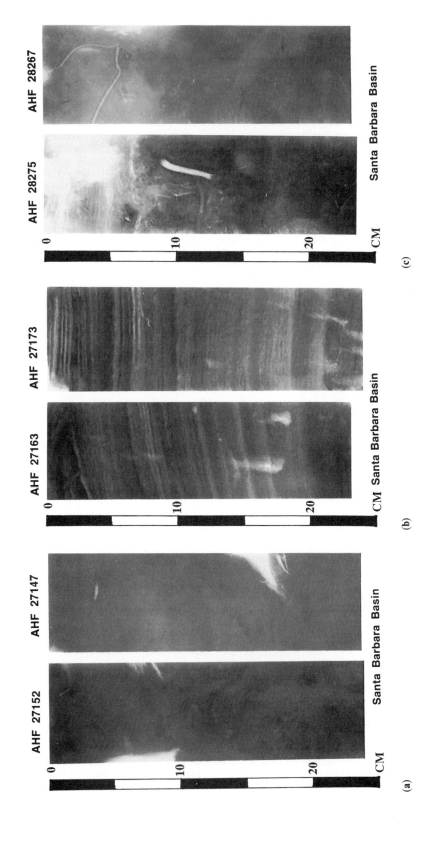

AHF 27152   AHF 27147   Santa Barbara Basin   (a)

AHF 27163   AHF 27173   CM Santa Barbara Basin   (b)

AHF 28275   AHF 28267   Santa Barbara Basin   (c)

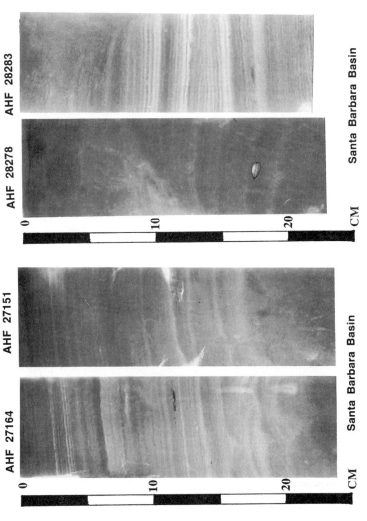

Fig. 8. (a) X-radiographs of cores AHF 27147 and AHF 27152 in the bioturbated area outside the central anoxic zone (Fig. 7, locations a, b). No laminations are preserved in these sediments. **(b)** X-radiographs of cores AHF 27163 and AHF 27173 showing a probable transported shell in 27163 at 12 cm. No bioturbation is present in these central anoxic zone cores (Fig. 7, locations c, e). Some deformation at the top of 27163 is probably due to sampler. **(c)** X-radiographs of cores AHF 28275 and AHF 28267. showing transition from laminated to completely bioturbated sediments at the eastern end of the central anoxic zone. AHF 28275 shows that lamination has been preserved for the past few decades but prior to that time, the anoxic zone had decreased in area to west of this core location (Fig. 7, locations f, g). **(d)** X-radiographs of cores AHF 27164 and AHF 27151. These cores show no bioturbation to below 20 cm in core. These are in the present central anoxic zone (Fig. 7, locations j, k). **(e)** X-radiographs of AHF 28278 and AHF 28283 showing bioturbation of the top 10+ cm in each core representing a shrinkage in the area of the anoxic zone (Fig. 7, locations h, i).

AHF 28283

AHF 28278

Santa Barbara Basin

CM 0 10 20

(e)

AHF 27164

AHF 27151

Santa Barbara Basin

CM 0 10 20

(d)

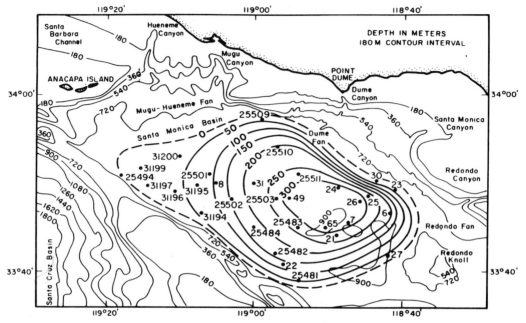

Fig. 9. Isochrons of the base of the nonbioturbated zone in Santa Monica Basin. The low oxygen condition has spread from the central basin floor to cover the entire basin floor at present. Note locations of Cores in Figs 2 and 3. Note location of DOE 26 (Fig. 2) on east side of oldest age isochron.

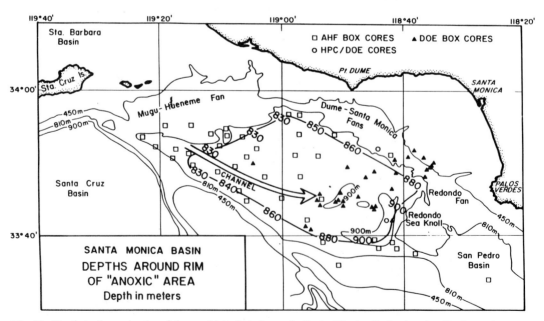

Fig. 10. Depth to the perimeter of the nonbioturbated zone in Santa Monica Basin showing the dynamic topography of that bottom water type. Relief is 70 metres and the surface slopes up to the northwest conforming to the clockwise bottom water circulation gyre. Exchange time is about 250 days (Hickey 1991).

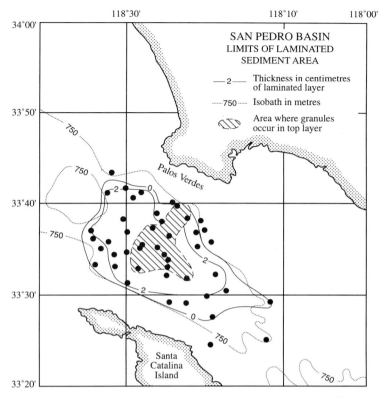

Fig. 11. San Pedro Basin nonbioturbated zone and area of dump site deposits of drilling muds and chips. The outline of that inner zone is the 5 cm isopach. In the center (see Fig. 12), the dump thickness is about 10 cm. Centre of the deposit exactly coincides with the coordinates of the historic dump site centre. Mass within that area approximates the rough total of dumped material. Dumping records are not totally reliable. Core AHF 24195 is at location 1; core AHF 24213 is at location 2.

the box core sampling depths of about 50 cm, includes about 400 km². Figure 7 shows the area of present non-bioturbation and the area of box cores with non-bioturbated zones. Figure 7 also shows the depth in cm to the first laminated zone occurrence in the box core surficial sediments. Within the 300 m contour 110 cores were examined to establish the boundaries shown in the figures.

Figures 8a–e show ten radiographs spanning the anoxic zone. Note that the cores near the margins of the anoxic zone typically show interruptions in the laminated zone cores due to episodes of bioturbation. Locations are shown in Fig. 7. The radiographs are a representative selection of the approximately 20 box cores in that zone.

Schimmelmann *et al.* (1992) noted that articulated mollusk shells (*Macoma leptonoidea*) forming a well defined layer occur in the central Santa Barbara Basin in box cores of 55–57 cm, corresponding to the period 1835–1840 AD. None of our box cores exceed 50 cm in length and thus probably missed the deeper layer.

Isolated articulated and disarticulated mollusk shells are seen in 9 of 20 cores in our collections that sample the zone encompassing laminated sediments within the 50 cm depth sampled by box coring. In those nine, there were two occurrences of shells at core tops (AHF 27151, 27156); in three cores (AHF 26202, 27160, 27742) isolated shells were seen at depths in the range of 12–18 cm; in five cores (AHF 26289, 27742, 27744, 28275, 28282) isolated shells were noted in a depth range of 28–40 m. If sedimentation rates are relatively uniform over the central deep basin floor, then these occurrences do not appear to cluster at a well defined depth horizon. The shells frequently occurred in bioturbated zones, but several were seen within laminated sections.

It is likely that these isolated shells represent either transported 'clasts' (disarticulated specimens) or individual 'pioneers' (articulated specimens) that entered the deep anoxic area during periods of minor oxygenation. It is interesting that they seem to fall into two rough levels as

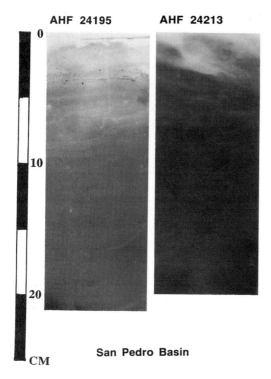

AHF 24195   AHF 24213

San Pedro Basin

CM

Fig. 12. X-radiographs of San Pedro box cores (see locations in Fig. 11) illustrating the laminae in the central basin floor (AHF 24195) and on the fringe of the non-bioturbated zone (AHF 24213). The coarse near-surface layer in AHF 24195 is drilling mud and cuttings deposited in an historic dump site closed in 1973.

noted above. Using Schimmelmann *et al.* (1990) dating curve for the basin the two levels would correspond very roughly to the decade centred on 1960 (12–18 cm) and to the decade centred on 1900 (28–40 m).

Santa Monica Basin

Malouta *et al.* (1981) first described the laminated hemipelagic sediments in Santa Monica Basin, and Christensen (1991), Flocks (1993) and Christensen *et al.* (1994) have described the cyclicity, geochemistry and extent of this zone. Reynolds has described the flow characteristics of one of the turbidites (1987).

In Santa Monica basin, bottom waters are apparently flushed only by the multi-year 'El Niño' events (Hickey 1991, 1992; Berelson 1991) due to the separation of that basin from the California Current system by three to four intervening basins (see deep circulation patterns in Emery 1960). Annual pulses are damped out (Christensen *et al.* 1994) by the long passage

over sills and through basin subsill water masses. Figure 6 shows the bathymetry; the controlling sills are all about 750 m deep, and are located at the eastern end of the basin.

Figure 9 shows the area of non-bioturbation and the isochrons that illustrate the rate of spreading of the zone. The thickest part of the zone, 26 cm, coincides with the maximum isochron area. This area encompasses 1300 km^2 which is most of the basin floor area below 700 m (an area of about 1700 km^2. Maximum basin depth is 910 m. This area was defined on the basis of study of 185 radiographs by Christensen (1991) and Flocks (1993). Since their studies were very well done, I reviewed only a random selection of radiographs from this basin for this paper.

Figure 10 shows the depths of the perimeter of the nonbioturbated zone and illustrates the dynamic influence of bottom water circulation. The zone actually slopes up to the northwest with a relief of about 70 metres and reflects the motion of the clockwise bottom water gyre (Hickey 1991).

San Pedro Basin

Savrda and others (1984) noted the probable nonbioturbated condition of the San Pedro Basin sediments, but the low oxygen condition has been known for several decades since the pioneering marine benthic studies of Hartman & Barnard (1958).

Figure 11 illustrates the extent of the zone of lamination in San Pedro Basin floor, and also shows the superimposed outline of an historic drilling mud and cuttings dump site. Maximum thickness at the centre of the area is about 10 cm. This dump site was abandoned in response to federal legislative regulations in the early 1970s. The larger, outer area of laminations which includes about 500 km^2, represents much of the basin floor area below 750 m (700 km^2). This is also the approximate depth of the sills (Fig. 6). This is a very thin laminated zone with a maximum thickness of 2.5 cm, and is very recent; probably no more than the last 20 years (before 1992, last coring date). The maximum basin depth is 904 m.

Figure 12 shows prints of radiographs from centre and margin of the nonbioturbated zone and also shows the dump material. Locations are shown in Fig. 11.

Santa Cruz Basin

This basin exhibits examples of thin laminations of distal turbidity current origin (Barnes 1970).

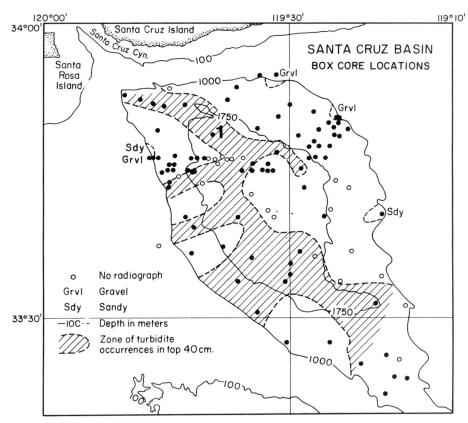

Fig. 13. Area of Holocene (top 0.5 to 1 m depths in cores) turbidity current deposits with fine-grained laminae. Main sources are from Santa Cruz Canyon to the north, the southern sill and small gullies between large slope slumps on the west (Barnes 1970). AHF 26046 is at location 1.

Only three box cores exhibit possible hemipelagic laminae in the top 1–2 cm of the cores.

Figure 13 shows the distribution of the zone of Holocene turbidite occurrences in the basin floor sediments. All of these are less than 5 cm in thickness, and contain finely laminated sequences with varying degrees of bioturbation (e.g. Fig. 14). As noted, this basin contains three core sites on the slopes (locations shown in Fig. 15) which contain some laminations that may be nonbioturbated primary hemipelagic structures. The zones are all less than 2 cm thick and occur at the tops of the cores. These cores are at depths greater than 1000 m and well below the depth of the core of the oxygen minimum zone (c. 500 m) in the adjacent eastern Pacific waters (Reid 1965).

Santa Cruz basin floor area is about 2200 km². Maximum basin floor depth is 1950 m. Within the sampling depth of box cores (about 40–50 cm), about 35% of the basin floor area has been affected by contemporary turbidity current deposition.

San Nicolas Basin

In San Nicolas Basin Reynolds & Gorsline (1987) have described the contemporary complex of immature submarine fans (Fig. 16) based on high-resolution acoustic profiles. Figure 17 shows the area of box core sites that contain fine turbidite laminations of less than 5 cm thickness (Fig. 3). Core location is shown in Fig. 17.

The preservation of these laminations is either due to rapid sedimentation to depths below active bioturbation, or low oxygen bottom water conditions that inhibit burrowers. Varying degrees of bioturbation of these laminated zones occur. Bottom waters are moderately oxygenated at present.

San Nicolas Basin basin floor encompasses over 3000 km². The basin has a maximum depth of about 1800 m. Within box core sampling depth (40–50 cm), only about 15% has been affected by modern turbidity current flows.

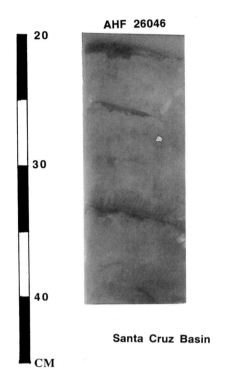

AHF 26046

20

30

40

Santa Cruz Basin

CM

Fig. 14. X-radiograph of Santa Cruz fine-grained turbidity current laminations (location in Fig. 13). Core AHF 26046 (2040 m) contains three thin partially bioturbated laminated silt layers deposited by distal turbidity currents. Another bioturbated thin layer is found at about 12 cm, but cannot be easily discerned in a radiograph print and is not shown here.

San Diego Trough

This basin has been filled to its sills (Emery 1960) and the basin floor is oxygenated and actively bioturbated. However, box cores contain thin distal turbidite laminations in parts of the basin floor as illustrated in Fig. 18. These have been partially bioturbated in some areas. They are additional illustrations of laminations due to thin distal fine turbidity current flows. Figure 19 shows a representative fine-grained turbidite lamination set.

No known anoxic areas occur in this basin whose basin floor depths of 1350 m are well below the oxygen minimum zone in the adjacent eastern Pacific Ocean. The basin is also distal from the major areas of upwelling in the northern borderland and Point Conception area, so organic infall rate is probably slow and there is much time in transit for mineralization of organics to occur.

It is likely that by-passing occurs when turbidity currents are actively flowing from the coastal canyon sources. Navy Fan has formed over the western sill of the trough where it abuts San Clemente Basin (Normark & Piper 1972). Flows also pass through to the southern end of Santa Catalina Basin (Gaal 1966).

San Diego Trough floor includes an area of about 4000 km². Sill depths are about 1350 m the same as the depth of the basin floor as noted above. At low sea levels when the basin floor channels are most active, much turbidity current material bypasses to Santa Catalina and San Clemente Basins.

Summary and conclusions

This paper reviews the locations of contemporary laminated basin floor sediments in the basins of the northern half of the California Continental Borderland. Its main objective is to provide a complete review of the radiographs of box cores collected over the entire area and to serve as a basis for further studies by interested workers of these intriguing regimes.

The locations of distal basin floor turbidites in Santa Cruz, San Diego Trough and San Nicolas Basins are interesting in that they record events of Holocene high-sea level time when such deposition processes should be much reduced and restricted to upper fan and canyon axis environments. Schwalbach *et al.* (1996) suggest that in small active margin basins, the submarine canyons may continue to be active through a sea level cycle. In addition, the increased vertical fall from shore to basin floor may emphasize erosion of channels on the fan surface and leveed channel extension into the basin plain. Note that these thin distal turbidites require a significant initial flow volume to permit them to reach the central basin floor. Such events appear to occur at century frequency to millennial frequency as one progresses from inner to central basins (Santa Monica to San Nicolas Basins).

For students of anoxic (nonbioturbated) basin floor depositional regimes, the basins of interest are obviously San Pedro, Santa Monica and Santa Barbara Basins in order of progressively longer term low oxygen conditions. If San Diego Trough is included as an oxygenated floor in the same bathymetric sequence, then conditions for a study of a spectrum of bottom water oxygen contents can be found in this region and could be an interesting geochemical study. The cores described in this paper are available as 1 to 2 cm

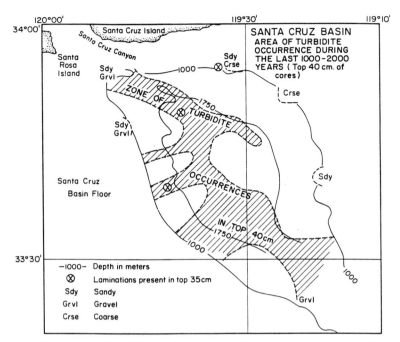

Fig. 15. Locations of box cores in Santa Cruz Basin that contain laminations in top 1–2 cm of cores. These are all on the basin slopes well below the depth of the east Pacific oxygen minimum. Locations are indicated by crossed circles.

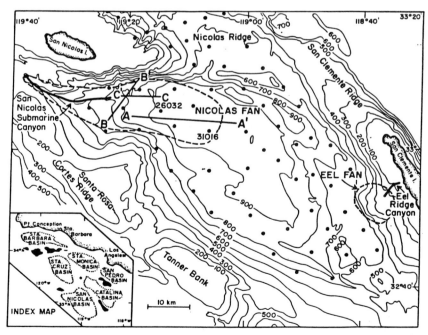

Fig. 16. Locations of submarine fans and fan complexes in San Nicolas Basin (figure from Reynolds & Gorsline (1987); profile lines are tracks of high-resolution profiles not shown here). Note location of core AHF 26032 (Fig. 3).

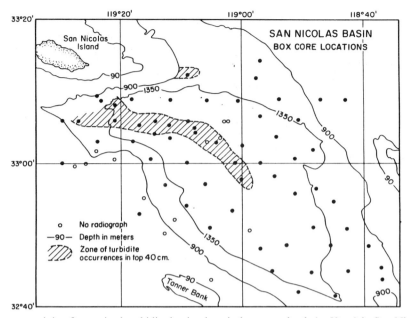

Fig. 17. Zone containing fine-grained turbidite laminations in box core depth (*c.* 50 cm) in San Nicolas Basin. Most turbidite laminae are partially bioturbated. See example in Fig. 3).

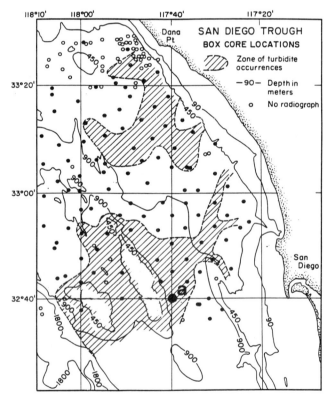

Fig. 18. Zones containing fine-grained turbidite laminations in San Diego Trough. Location of Core AHF 28893 (Fig. 19) is shown by large dot marked 'a'.

AHF 28893

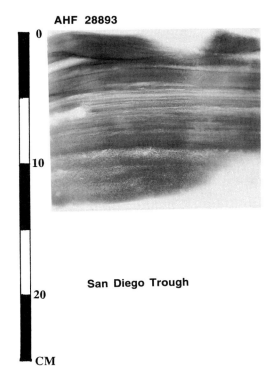

San Diego Trough

Fig. 19. Radiograph of fine-grained turbidite laminations in core AHF 28893 from San Diego Trough (see Fig. 18). Note multiple events and scour.

sections and as dry slabs of the original cores. Interested workers are invited to utilize the collection.

Enrique Nava-Sanchez did the major work of first-order review, and isolated those radiographs that I have then examined in greater detail. Janette Murillo de Nava studied and tabulated data for the radiographs of the Santa Barbara Basin cores. The paper was much improved by the reviews of Drs. Arndt Schimmelmann and Steven Lund. Support for field work and previous laboratory studies was provided by the National Science Foundation and the Department of Energy over the past 25 years.

References

ANTEVS, E. 1925. Retreat of the last ice sheet in eastern Canada. *Canada Geological Survey Memoir*, **146**.

BARNES, P. W. 1970. *Marine geology and oceanography of Santa Cruz Basin off southern California*. PhD Thesis, University of Southern California, Los Angeles.

BERELSON, W. M. 1991. The flushing of two deep sea basins, southern California Borderland. *Limnology and Oceanography*, **36**, 1150–1166.

CAMPBELL, C. V. 1967. Lamina, laminaset, bed and bedset. *Sedimentology*, **8**, 7–26.

CALVERT, S. E. 1966. Origin of diatom-rich, varved sediments from the Gulf of California. *Journal of Geology*, **74**, 546–565.

CHRISTENSEN, C. J. 1991. *An Analysis of Sedimentation Rates and Cyclicity in the Laminated Sediments of Santa Monica Basin, California Continental Borderland*. MS Thesis, University of Southern California, Los Angeles.

——, GORSLINE, D. S., HAMMOND, D. E. & LUND, S. P. 1994. Initiation and expansion of "anoxic" basin-floor conditions, Santa Monica Basin, California Borderland over the past four centuries. *Marine Geology*, **116**, 399–418.

DUFF, P. M. D., HALLAM, A. & WALTON, E. K. 1967. *Cyclic Sedimentation*. Elsevier, Amsterdam.

EMERY, K. O. 1960. *The Sea off Southern California*. Freeman, San Francisco.

—— & BRAY, E. E. 1962. Radiocarbon dating of California basin sediments. *American Association of Petroleum Geologists, Bulletin*, **46**, 1839–1856.

—— & HULSEMANN, J. 1962. The relationships of sediments, life and water in a marine basin. *Deep-Sea Research*, **8**, 165–180.

FISCHER, A. G. & ROBERTS, L. T. 1991. Cyclicity in the Green River Formation (Lacustrine Eocene) of Wyoming. *Journal of Sedimentary Petrology*, **61**, 1146–1155.

FLEISCHER, P. 1970. Mineralogy and sedimentation history, Santa Barbara Basin, California. *Journal of Sedimentary Petrology*, **42**, 49–58 .

FLOCKS, J. G. 1993. *Transport Mechanisms and Element Distribution in a Semi-enclosed, Anoxic Marine Basin, and the Influences of Natural and Anthropogenic Input, California Continental Borderland*. MS Thesis, University of Southern California, Los Angeles.

GAAL, R. A. P. 1966. *Marine Geology of the Santa Catalina Basin Area, California*. PhD Thesis, University of Southern California, Los Angeles.

GORSLINE, D. S. 1992. The geologic setting of Santa Monica and San Pedro Basins, California Continental Borderland. *Progress in Oceanography*, **30**, 1–36.

——1996. Depositional events in Santa Monica Basin, California Borderland, over the past five centuries. *Sedimentary Geology*, **103**, 46–61.

GRANT, C. W. 1991. *Distribution of bacterial mats (Beggiatoa spp.) in Santa Barbara Basin, California: a modern analog for organicrich facies of the Monterey Formation*. MS Thesis, California State University, Long Beach.

HALLAM, A. 1967. The depth significance of shales with bituminous laminae. *Marine Geology*, **5**, 481–493.

HARTMAN, O. & BARNARD, J. V. 1958. *The benthic fauna of the deep basins off southern California*. Allan Hancock Pacific Expeditions, University of Southern California, Los Angeles, **22**.

HICKEY, B. M. 1991. Variability in a deep coastal basin off southern California. *Journal of Geophysical Research*, **96**, 16689–16708.

——1992. Circulation over the Santa Monica-San Pedro Basins and shelf. *Progress in Oceanography*, **30**, 37–48.

ISAACS, C. M. 1984. Hemipelagic deposits in a Miocene basin, California: toward a model of lithologic variation and sequence. *In*: STOW, D. A. V. & PIPER, D. J. W. (eds) *Fine-grained Sediments: Deep Water Processes and Facies*. Geological Society, London, Special Publications, **15**, 481–496.

——, PIPER, D. Z., TENNYSON, M. E., INGLE, J. C., & BAUMGARTNER, T. R. in press. Depositional setting of the Monterey Formation revisited. *U.S. Geological Survey Professional Paper*.

KENNETT, J. P., BALDAUF, J. G. & LYLE, M. (eds) 1995. *Proceedings of the Ocean Drilling Program, Scientific Results*, **146** (Pt. 2) College Station, TX (Ocean Drilling Program).

LANGE, C. B. & BERGER, W. H. 1993. Paleoclimatic significance of Santa Barbara laminated sediments: a history of upwelling and El Nino events. EOS, *Transactions of the American Geophysical Union*, **74**, 372.

MALOUTA, D. N., GORSLINE, D. S. & THORNTON, S. E. 1981. Processes and rates of filling in an active transform margin: Santa Monica Basin, California Continental Borderland. *Journal of Sedimentary Petrology*, **51**, 1077–1095 .

NORMARK, W. R. & PIPER, D. J. W. 1972. Sediments and growth pattern of Navy deep-sea fan, San Clemente Basin, California. *Journal of Geology*, **80**, 198–223.

O'BRIEN, N. R. 1996. Shade lamination and sedimentary processes. *This volume*.

OTTO, G. H. 1938. The sedimentation unit and its use in field sampling. *Journal of Geology*, **41**, 569–582.

PRENSKY, S. E. 1973. *Carbonate stratigraphy and related events, California Continental Borderland*. MS Thesis, University of Southern California, Los Angeles.

REID, J. L. 1965. *Intermediate Waters of the Pacific Ocean*. Johns Hopkins Oceanographic Studies, 2.

REIMERS, C. E., LANGE, C. B., TABAK, M. & BERNHARD, J. M. 1990. Seasonal spillover and varve formation in the Santa Barbara Basin, California. *Limnology and Oceanography*, **35**, 1577–1585.

REYNOLDS, S. 1987. A recent turbidity current event, Hueneme Fan, California: reconstruction of flow properties. *Sedimentology*, **34**, 129–138.

—— & GORSLINE, D. S. 1987. Nicolas and Eel submarine fans, California Continental Borderland. *AAPG Bulletin*, **71**, 452–463.

SAVRDA, C. E., BOTTJER, D. J. & GORSLINE, D. S. 1984. Development of a comprehensive oxygen-deficient marine biofacies model: evidence from Santa Monica, San Pedro and Santa Barbara Basins, California Borderland. *AAPG Bulletin*, **68**, 1179–1192.

SCHIMMELMANN, A. & LANGE, C. 1996. Tales of 1001 varves: a review of Santa Barbara Basin sediment studies. *This volume*.

——, —— & BERGER, W. H. 1990. Climatically controlled marker layers in Santa Barbara Basin sediments and fine-scale core-to-core correlation. *Limnology and Oceanography*, **35**, 165–172.

——, ——, SIMON, A., BURKE, S. K. & DUNBAR, R. B. 1992. Extreme climatic conditions recorded in Santa Barbara Basin laminated sediments: the 1835–1840 Macoma event. *Marine Geology*, **106**, 279–300.

SCHWALBACH, J. 1982. *A Sediment Budget for the Northern Region of the California Continental Borderland*. MS Thesis, University of Southern California, Los Angeles.

—— & GORSLINE, D. S. 1985. Holocene sediment budgets for the basins of the California Continental Borderland. *Journal of Sedimentary Petrology*, **55**, 829–842.

——, EDWARDS, B. D. & GORSLINE, D. S. Depositional features produced by a series of events at high sea level: contemporary channel-levee systems in active borderland basin plains. *Sedimentary Geology*, **103**, 1–20.

SHEPARD, F. P. & EMERY, K. O. 1941. Submarine topography off the California coast: canyons and tectonic interpretation. *Geological Society of America Special Paper*, **31**.

SHOLKOVITZ, E. R. & GEISKES, J. M. 1971. A physical-chemical study of the flushing of the Santa Barbara Basin. *Limnology and Oceanography*, **16**, 479–489.

SOUTAR, A. & CRILL, P. A. 1977. Sedimentation and climatic patterns in Santa Barbara Basin during the l9th and 20th Centuries. *Geological Society of America, Bulletin*, **88**, 1161–1172.

TENG, L. S.-Y. 1985. *Seismic Stratigraphic Study of the California Continental Borderland Basins: Structure, Stratigraphy and Sedimentation*. PhD Thesis, University of Southern California, Los Angeles.

THORNTON, S. E. 1981. *Holocene stratigraphy and sedimentary processes in Santa Barbara Basin: influence of tectonics, ocean circulation, climate and mass movement*. PhD Thesis, University of Southern California, Los Angeles.

VAN WAGONER, J. C., MITCHUM, R. M., CAMPION, K. M. & RAHMANIAN, V. D. 1990. Siliciclastic sequence stratigraphy in well logs, cores and outcrops. *American Association of Petroleum Geologists, Methods in Exploration Series*, **7**.

Laminated sediments of Santa Monica Basin, California continental borderland

JAMES W. HAGADORN

Department of Earth Sciences, University of Southern California, Los Angeles, CA 90089–0740, USA

Abstract: Laminated sediments from Santa Monica Basin differ from other laminated sediments of the California Continental Borderland (such as varved sediments of Santa Barbara Basin) because lamina couplets represent a 3–6 year time interval and are affected by distinct local oceanographic and climatic processes. Unfortunately, mechanisms which control the temporal pattern of laminations in this basin are poorly understood. In addition, this laminated sediment record has been underutilized in paleoceanographic and paleoclimatic studies. In an attempt to identify mechanisms which control lamination formation and in order to obtain paleoclimatic and paleoceanographic records from this region, a sediment core from this basin was analysed using a suite of geochemical, sedimentary, and micropaleontologic techniques. This box-core (core DOE 26) was collected from the centre of the basin and was vertically sub-sampled at one millimetre increments, providing a temporal resolution of 1–3 years per sample. A variety of paleoceanographic and paleoclimatic records were constructed via analysis of these samples and include proxy records of: (a) regional rainfall – as expressed by gross variations in downcore density profiles; (b) relative contribution of terrestrial and marine organic carbon to the basin – as expressed by downcore variations in the isotopic composition of total organic carbon (TOC); (c) Primary productivity – as expressed in downcore variations in the carbon isotopic composition of planktonic foraminifera; (d) changes in sea surface temperatures –reflected in downcore variations in the oxygen isotopic composition of *N. pachyderma* (dextral); (e) large-scale changes in bottom-water oxygen conditions – as expressed by downcore changes in abundance of oxygen- sensitive benthic foraminifera. To extract additional information from these proxy records, spectral analysis was also employed in this study. Although ENSO-length signals are present in several records, decadal-length signals appear to dominate records from this basin. This dominance suggests the presence of a longer-term forcing function (other than the ENSO phenomena) which strongly influences this region. Considered together, preliminary results suggest that the flux of organic carbon is a dominant factor affecting lamination variability in Santa Monica Basin. Organic carbon fluxes in this basin are mainly affected by primary productivity, and to a lesser extent, by terrestrial input. Variations in bottom oxygen conditions may not control lamination variability; rather, long-term depression of oxygen levels appears to allow preservation of this organic matter as laminated sediments.

In the past, many studies of laminated sediments from the California Continental Borderland have focused on Santa Barbara Basin. A major focus of these studies has been to extract paleoclimatic and paleoceanographic records from sediment cores and to identify forcing mechanisms which might drive processes within the basin (e.g. Soutar & Crill 1977; Schimmelmann & Tegner 1991; Kennedy & Brassell 1992). Until recently, there have been few rigorous studies of laminated sediments from other basins, such as Santa Monica Basin (but see Christensen 1991; Berelson 1992; Christensen *et al.* 1994).

Laminations from Santa Monica Basin are visible in fresh core sections and as density contrasts in X-radiographs [see Gorsline *et al.* 1996) for a review of X-radiographs from this region]. Christensen (1991) and Christensen *et al.* (1994) examined these laminations by digitizing X-radiographs of several cores from the centre of

the basin (including core DOE 26; Fig. 1) and constructing relative density profiles downcore.

[210]Pb dating of cores reveals that laminations in Santa Monica Basin extend to at least the Little Ice Age (*c.* 1600 AD; Huh *et al.* 1987, 1990; Christensen 1991; Flocks 1993). In addition, Santa Monica lamination couplets, (consisting of one light and one dark lamination), occur at intervals of approximately 3–6 years (Christensen, 1991; Christensen *et al.* 1994). The 3 to 6 year recurrence of the laminae is conspicuously similar to observed El Niño frequencies and suggests a link between laminations and external climate forcing functions (Christensen 1991; Christensen *et al.* 1994). Christensen (1991) explores this link in depth and notes a decadal-scale correlation between the historical record of annual rainfall and the time series of density variations in Santa Monica Basin laminations (Fig. 2). Extended periods of higher rainfall correlate with intervals

From Kemp, A. E. S. (ed.), 1996, *Palaeoclimatology and Palaeoceanography from Laminated Sediments*, Geological Society Special Publication No. 116, pp. 111–120.

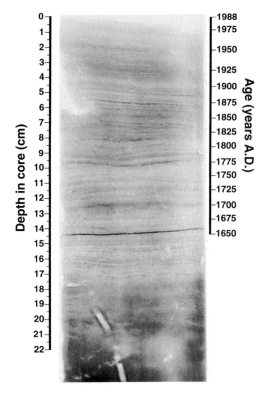

Depth in core (cm) / Age (years A.D.)

Fig. 1. Contact print of an X-radiograph of laminated sediment core DOE 26 from Santa Monica Basin. Light bands on the print reflect low density layers in the core, whereas dark bands represent higher density layers in the core.

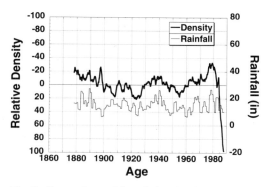

Fig. 2. Comparison of the relative density record (and hence proxy lamination record: Christensen 1991) from core DOE 26 to the historic Los Angeles annual rainfall record (from PACLIM workshop #5, 1988).

of higher density within the core and are thought to reflect increased flux of terrigenous siliciclastics and lower relative flux of marine biogenic material to the basin floor (Christensen 1991).

Whereas mechanisms of lamination variability have not yet been resolved in Santa Monica Basin, these observations suggest that the laminations are linked to climate variability (e.g. Christensen 1991; Christensen et al. 1994). Since Christensen's work, these links have been explored in greater detail by the Paleoceanography Research Group at USC. Preliminary results of these studies were presented at the Geological Society of London laminated sediment symposium and are outlined in this chapter. These studies are described in more detail in Hagadorn et al. (1995). Detailed information about the oceanographic and geological setting of Santa Monica Basin are also presented elsewhere (i.e. Gorsline 1980, 1992; Christensen 1991; Flocks 1993; Gorsline et al. (1996)) and are thus not reviewed here.

Following up on Christensen (1991) and Christensen et al.'s (1994) hypothesis that laminations are linked to the amount of organic carbon buried, possible factors that could contribute to lamination variability were explored. These factors include:

(a) variations in the flux of terrigenous versus marine organic carbon and detritus;
(b) surface water processes which affect marine productivity and organic carbon flux to the sediments;
(c) deep water processes which affect oxygen renewal and preservation or organic matter within the basin.

To study climate variability in this region, and to determine how these factors affect lamination variability, we examined six 'tracers' that respond to variations in either rainfall, surface or deep-water organic carbon fluxes, sea surface temperatures, or benthic oxygen levels within the basin. These tracers are sediment X-ray density, $\delta^{13}C_{TOC}$, TOC concentration, $\delta^{13}C$ and $\delta^{18}O$ of the planktonic foraminifera *Neogloboquadrina pachyderma* (dextral), and abundance ratios of benthic foraminifera *Bolivina argentea* and *B. spissa*. Spectral analysis was also used in order to resolve complex temporal patterns in these records in hope that they might implicate large-scale climate-forcing functions (such as the ENSO phenomena) which control basinal sedimentologic responses.

Methods

A variety of geochemical, sedimentologic, and microfossil analyses were used in this study in order to reconstruct paleoclimate records and to

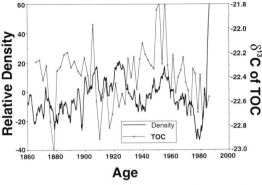

Fig. 3. Santa Monica $\delta^{13}C_{TOC}$ compared to relative density variations in core DOE 26. Note the correlation of these records from *c.* 1940 to the present. Density values are lowest in the upper portion of the core, reflecting low sediment compaction. In order to facilitate comparison to other historical and proxy records, the density records used in this study have been detrended (as in Christensen 1991 and Christensen *et al.* 1994). Density records are from Christensen (1991).

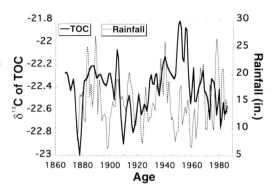

Fig. 4. Comparison of the Santa Monica $\delta^{13}C_{TOC}$ record to the historical record of Los Angeles Rainfall. Comparison of $\delta^{13}C_{TOC}$ ratios to rainfall records for this area exhibits a general trend of decreased $\delta^{13}C_{TOC}$ values in periods following years of high rainfall. However, the reverse is not true; not all troughs in the $\delta^{13}C_{TOC}$ record correspond to years of high rainfall. Rainfall data is from the Los Angeles Weather Bureau and is expressed in inches.

examine potential mechanisms of lamination variability. These methods are described in detail in Hagadorn *et al.* (1993, 1995) and are thus not repeated in this chapter.

Results and discussion

Geochemical, microfaunal, and sediment analyses

Christensen (1991) hypothesized that rainfall and/or runoff may cause variations in downcore density profiles from Santa Monica Basin. If this is correct, then the relative contribution of marine and terrestrial organic carbon may also vary in the basin. Because terrestrial and marine organic carbon each have distinct carbon isotopic signatures, a systematic relationship should exist between the $\delta^{13}C_{TOC}$, the laminations, and rainfall records in Santa Monica Basin.

To explore this possibility, the $\delta^{13}C_{TOC}$ record is compared to the lamination (i.e. density) record (Fig. 3). Variations in the isotopic signature of TOC appear to covary with density patterns from 1988 to 1940, but the magnitude of density and $\delta^{13}C_{TOC}$ changes are not correlated. Comparison of the $\delta^{13}C_{TOC}$ record to the rainfall record exhibits a similar pattern, with phase relationships trailing off prior to 1940 (Fig. 4).

Comparison of the relative concentration of TOC to the Santa Monica density profile also

reveals a systematic relationship since 1940 (Fig. 5). This pattern indicates a relationship between the amount of carbon preserved and the density record and suggests that the relative organic matter fluxes control the pattern of lamination.

With the foregoing in mind, influences of primary productivity (and hence phytoplanktic organic carbon) were examined by comparing the Santa Monica $\delta^{13}C_{TOC}$ record to the $\delta^{13}C$ record extracted from *N. pachyderma* (Fig. 6). Higher values of $\delta^{13}C$ in *N. pachyderma* tend to correlate with high $\delta^{13}C$ values of organic carbon, especially from *c.* 1900 to the present. This correlation suggests a dominance of marine productivity as the primary control on the flux of organic carbon to the sediments, except, perhaps, under unusually strong and extended ENSO conditions (e.g. the 1885, 1890 and 1901 events). If primary productivity is increased, the $\delta^{13}C$ of HCO_3 should increase due to the preferential photosynthetic removal of ^{12}C from surface waters. Time series studies of $\delta^{13}C$ of particulate organic carbon during the North Atlantic Spring bloom experiment have shown that $\delta^{13}C$ in the organic and inorganic carbon pools varies like a closed system (Rau *et al.* 1992). Inorganic and organic carbon both become progressively enriched in ^{13}C as photosynthate is progressively exported from the surface water (Rau *et al.* 1992). $\delta^{13}C$ variability in Santa Monica Basin is thought to behave in a similar manner. In a time series study conducted in San

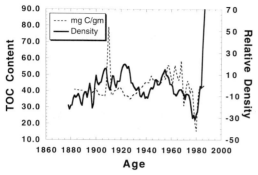

Fig. 5. Direct comparison of the downcore variations in TOC concentration (expressed as mg C/gm of sediment) to the variations in density in core DOE 26. TOC concentrations were calculated by normalizing the amount of CO_2 released from each sample.

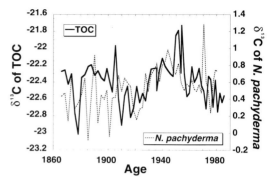

Fig. 6. The Santa Monica $\delta^{13}C_{TOC}$ record is compared to the $\delta^{13}C$ record from *N. pachyderma* (dextral). Note the close correlation of these records, especially from *c.* 1900 to the present.

Pedro Basin, the basin adjacent to Santa Monica Basin, the $\delta^{13}C$ and $\delta^{15}N$ of various plankton groups is highest during Spring months when productivity is highest. The isotopic values decrease during late Summer and Fall and are lowest during Winter months when productivity is low. The seasonal changes in $\delta^{13}C$ observed in San Pedro Basin are similar to the down-core variations in the Santa Monica Basin box-core record. The down-core variations observed in the Santa Monica Basin record thus suggest that interannual variation in photosynthate export during Spring production cycles is responsible for the 2‰ variation in organic carbon and the 1‰ variation in inorganic carbonates. As a cautionary note, however, the effects of organic carbon diagenesis have not yet been adequately addressed in this region. In particular, the influence of relatively 'fresh' surficial organic carbon versus progressively more refractory underlying organic carbon may bias patterns observed in surficial samples (Schimmelmann, pers. comm.).

In summary, these observations do not suggest that at the scale of the laminations, the flux of terrigenous organic carbon controls lamination formation. Rather, correlation of the $\delta^{13}C_{TOC}$ record with the *N. pachyderma* $\delta^{13}C$ record and the ranges values exhibited in the Santa Monica $\delta^{13}C_{TOC}$ profile suggest that phytoplankton is the dominant organic carbon contributor to the sediments. This hypothesis agrees with findings by Eppley & Holm-Hansen (1986). Because the primary components of organic carbon ratios in these Santa Monica Basin sediments are phytoplankton and terrigenous sources of organic carbon, when one of these sources is diminished or shut off, one should see a shift in the overall isotopic composition of the sediments. Weak patterns of decreased $\delta^{13}C_{TOC}$ values following

years of high rainfall suggest that when large amounts of terrigenous sediments are flushed into the basin during 'wet' years, the mass balance of the carbon flux to the sediments may be slightly decreased by the input of isotopically-light terrigenous organic carbon.

However, ENSO events may be overprinting, and thus complicating, this signal. During an ENSO event, the productivity of phytoplankton in surface waters is greatly diminished due to shut-down of upwelling (Chelton *et al.* 1982; McGowan 1985; Lange *et al.* 1987, 1990). This shut-down could affect the overall mass balance of the different carbon sources; effectively decreasing marine organic carbon fluxes without significantly changing terrestrial organic carbon fluxes. Schimmelmann & Tegner (1991) tested this in varved sediments of Santa Barbara Basin and found a correlation between the $\delta^{13}C_{TOC}$ and the historical record of storm and ENSO events.

However, no similar systematic correlation is observed between the $\delta^{13}C_{TOC}$ records from Santa Monica and Santa Barbara Basin (Fig. 7). These findings suggest that organic carbon deposits in Santa Monica Basin and Santa Barbara Basin may not be recording the same regional signal. Additionally, the ~1‰ disparity between the Santa Monica and Santa Barbara records suggests that Santa Monica Basin is more heavily influenced by terrestrial organic matter.

The $\delta^{13}C_{TOC}$ record from Santa Monica Basin is compared to the same Quinn *et al.* (1987) ENSO record used by Schimmelmann & Tegner (1991; Fig. 8). Of the 19 large decreases (or troughs) in the $\delta^{13}C_{TOC}$ record, ten of these correlate with ENSO events – of these, six are very strong or strong events. Assuming a 1–2

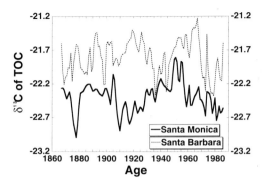

Fig. 7. The $\delta^{13}C_{TOC}$ record from Santa Monica Basin is compared to the $\delta^{13}C_{TOC}$ record from Santa Barbara Basin. Whereas no similar patterns of variation are observed, it is important to note that the Santa Monica $\delta^{13}C_{TOC}$ values are more negative, suggesting that Santa Monica Basin is more heavily influenced by terrestrial organic matter. The $\delta^{13}C_{TOC}$ measurements from Santa Barbara Basin were obtained from Arndt Schimmelmann at Scripps Institution of Oceanography (Schimmelmann & Tegner 1991).

year lag, 16 of these decreases correlate with ENSO events. The correlation of the Quinn et al. (1987) ENSO record with the Santa Monica Basin $\delta^{13}C_{TOC}$ record compares well with levels of correlation observed by Schimmelmann & Tegner (1991) and Kennedy & Brassel (1992). This mixed level of correlation suggests that not all ENSO events affect Santa Monica Basin, nor do they affect the basinal TOC inventory in a similar manner. In addition, these results suggest that: (a) the magnitude or duration of strong or very strong ENSO events does not correlate with the magnitude (and hence relative effects) of negative shifts in the

$\delta^{13}C_{TOC}$ record within Santa Monica Basin, and (b) ENSO effects may not be the dominant control on the temporal pattern of variations in TOC. However, it is possible that the lack of correlation may also result from weak or 'normal' ENSO events being overprinted or reinforced by other intra- or inter-basinal processes.

Laminated sediments from Santa Monica may also preserve a record of additional climate changes which may be forcing hydrographic or biological changes within the basin. Some of these changes can be assessed by examining records of surface and deep-water processes. In particular, reconstruction of proxy records of sea surface temperatures and bottom-water oxygen concentrations illustrate hydrographic changes which may be related to lamination formation.

To assess changes in surface water temperatures, the oxygen isotopic composition of N. pachyderma was examined. This species of planktonic foraminifera inhabits surficial waters of Santa Monica Basin and 'blooms' in the Spring (Sautter & Thunell 1991). Because the oxygen isotopic composition of their tests is thought to be controlled by a temperature dependent fractionation during test secretion, tests can be utilized to reconstruct palaeo-sea surface temperatures [palaeoSST; see Hagadorn et al. (1995) for further explanation of methodology and assumptions relevant to constructing this record]. The palaeo-SST record constructed in this study compares well with the short-term historical Spring SST record (Fig. 9A) and exhibits the same general 20th century warming trend observed the longer-term global temperature record (Fig. 9B).

To assess the impact of oxygen variations on lamination patterns, a semi-quantitative deep

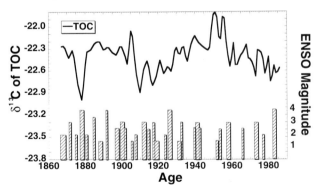

Fig. 8. The $\delta^{13}C_{TOC}$ record from Santa Monica Basin is compared to the same Quinn et al. (1987) ENSO record used by Schimmelmann & Tegner (1991). ENSO events were converted to a numerical format based on intensity of event: 0, no event; 1, weak event; 2, moderate event; 3, strong event; 4, very strong event. Whereas direct comparisons between excursions are visible, the relative magnitude of $\delta^{13}C_{TOC}$ excursions does not correlate with the relative intensity of historical ENSO events. Hence it does not appear that the $\delta^{13}C_{TOC}$ record can be used as an independent proxy for historical ENSO events.

water oxygen record was constructed based on previous work by Douglas (1979, 1981) and Douglas *et al.* (1976, 1979, 1980). Douglas (1979, 1981) illustrates that there is a critical difference in abundances of two species of benthic foraminifera, *Bolivina argentea* and *Bolivina spissa*, which occurs at oxygen levels below ~1 ml l⁻¹. This relationship is utilized to reconstruct a semi-quantitative record of oxygen variations within the deepest part of Santa Monica Basin.

Several large excursions in deep water oxygen concentrations are observed in this preliminary paleoxygen record (Fig. 10). These may result from basin flushing, bottom water overturn, or other oxygen renewal events, but do not appear to correlate with the Quinn *et al.* (1987) El Niño record. In general, an extended period of higher O_2 levels is observed from 1929 to 1939 and there appear to be fewer flushing events or lower

average oxygen levels since *c.*1940. This record indicates at least dysoxic to suboxic (Tyson & Pearson 1991) basin floor conditions dating back to 1910. The limited number of large oxygen excursions in this record may represent the true record of oxygen conditions or might suggest that *Bolivina* sp. are not sen-sitive palaeo-oxygen proxies at extremely low (≤0.1 ml l⁻¹) oxygen levels. The overall pattern of depressed oxygen levels (or fewer flushing events) toward the recent parallels post-World War II population growth in Los Angeles, however. This population increase was accompanied by increased sewage levels. Although it is unlikely that most of the LA-derived sewage is transported to the centre of the basin, it may have placed an increased demand on oxygen concentrations in the basin and thus affected sediments in this region (Flocks 1993). At the very least, these results suggest that the duration of renewal

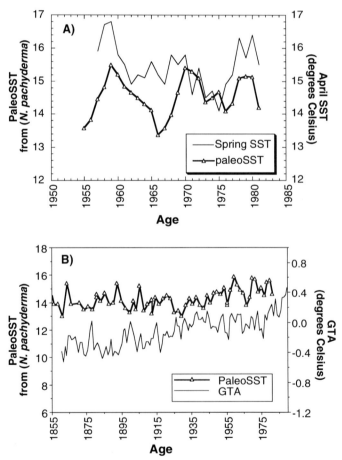

Fig. 9. (**A**) Comparison of the short-term historical record of Spring SST to the palaeoSST record generated from the δ^{18}O of *N. pachyderma*. (**B**) Comparison of a segment of the long-term palaeoSST record to the Global Temperature Anomaly (GTA).

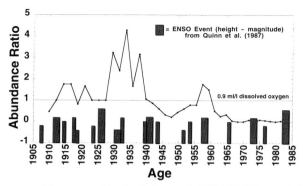

Fig. 10. The palaeoxygen record is compared to the Quinn *et al.* (1987) ENSO record. There does not, however, appear to be a strong correlation between these two records. Large excursions in deep water oxygen concentrations are observed from 1957 to 1959, from 1914 to 1916, and in 1921. An extended period of higher O_2 levels is observed from 1929 to 1939. There is a general trend of fewer flushing events or lower average oxygen levels in recent years. Because oxygen values do not change linearly toward the x-axis of the graph, we are hesitant to identify specific oxygen values below 0.9 ml/l. However, oxygen levels from *c*. 1965 to the present have likely remained ≤0.2 ml/l for extended periods of time.

events may not have been sufficient to raise oxygen levels above the threshold necessary to allow recolonization by bioturbating organisms.

Climate forcing functions

Whereas the sampling resolution available for this study approximates the thickness of individual laminae, it is difficult to correlate individual laminae with individual peaks in the proxy records because of high noise levels in the data sets. Additional uncertainties are introduced when comparing linear and non-linear time series. Thus, it is not always clear from direct observation which forcing function causes variations within these records. Several forcing functions may be present, albeit overprinted, within the Santa Monica Basin proxy records. Spectral analysis allowed us to break up the signals in these records into smaller, more readily identifiable components.

The first step necessary to identify forcing functions in this region is to find out what patterns exist in historical records. Historical records of Los Angeles rainfall, runoff, and the Quinn *et al.* (1987) ENSO record were analysed for time periods comparable to proxy data sets from Santa Monica Basin (Fig. 11). The historical rainfall and runoff records contain distinct frequency bands of high power at 10–28 years and at 4–6 years. Whereas the ENSO record does not exhibit a similar decadal length signal, it does have a strong band in the 4–6 year frequency range.

If Christensen (1991) was correct, one should see variations in the density of the Santa Monica

core which reflect short 3–6 year ENSO phenomena (Fig. 12). No ENSO-length signal appears to be dominant. The dominant frequency bands visible in the density data, at 20 and 13 years, are similar to the decadal band seen in the rainfall and runoff records. Lack of correlation of these results with Christensen (1991) and Christensen *et al.* (1994) may occur as a result of two factors. First, they compile stacked frequency distributions of laminations 'picked' from the density data by a computer algorithm, whereas our results are based on multi-taper spectral analysis of the

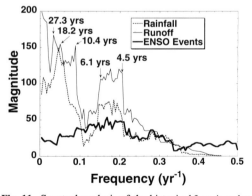

Fig. 11. Spectral analysis of the historical Los Angeles rainfall, runoff, and Quinn *et al.*'s (1987) ENSO record. The historical rainfall and runoff records have two distinct frequency bands of high power at around 10–27 years and one at around 4–6 years. Two components may exist within the longer term band, one at around 10 years and one at around 19 years. Whereas the ENSO record does not exhibit a similar decadal length signal, it does have a strong band in the 4–6 year frequency range.

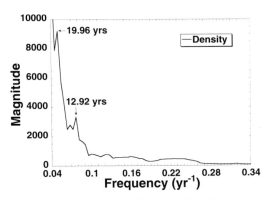

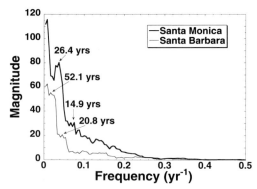

Fig. 12. Spectral analysis of the Santa Monica density record. Two distinct spectral peaks are visible at around 20 and around 13 years.

Fig. 13. Spectral analysis of the Santa Monica and Santa Barbara $\delta^{13}C_{TOC}$ records. Dominant frequency bands are visible at around 26 and 15 years in the Santa Monica record. Longer and weaker bands at around 50 and 20 years are visible in the Santa Barbara record (data from A. Schimmelmann).

actual X-ray density records. Second, we utilized a different type of spectral analysis technique (i.e. multi-taper vs. standard FFT) which hopefully is better able to resolve small-scale patterns within the studied records. Although Christensen (1991) demonstrated that there was minimal deleterious effect of using computer-picked lamination profiles, the accrued effect of small errors in computer 'picks' and lowered resolution of the standard FFTs (vs. the multi-taper approach) may combine to decrease the precision of their time series analysis.

If the variations in the $\delta^{13}C_{TOC}$ record in Santa Monica and Santa Barbara Basin are related to ENSO and strong storm events, then they should exhibit similar patterns. Analysis of these records indicates longer-term frequency bands of $c.\,26$ and 15 years dominate variations in Santa Monica Basin, whereas weaker and longer bands of $c.\,20$ and $c.\,50$ years dominate Santa Barbara Basin (Fig. 13). ENSO-length signals are not dominant in either basin. These findings suggest that these basins may be less sensitive to these ENSO-length forcing functions or that these forcing functions are obscured by a noisy signal. The absence of the ENSO-length signal may be an artifact of sampling, however, because it becomes increasingly difficult to resolve periodic signals (like the 3–6 year ENSO signal) as the signal length approaches the length of time represented by the sampling interval (i.e. 1–3 years).

Whereas comparative analysis of the variations in TOC did not illustrate any direct correlations with interannual variability in the Santa Monica density record, this record does appear to vary on interannual timescales. Spectral analysis of the record of TOC concentration reveals decadal-length signals which are similar in length to those observed in the climate and the density records (Fig. 14).

Decadal-length signals are also observed in historical and proxy SST records. Signals are centred at around 15.5 years in both the historical Spring SST record and in the palaeoSST record (Fig. 15) generated from analysis of *N. pachyderma*. The high level of correlation between these signals adds weight to direct correlations between these records (outlined above) and suggests potential for further investigation of this palaeoSST indicator.

Analysis of the palaeo-oxygen and *N. pachyderma* $\delta^{13}C$ records reveals frequency bands centred near 46 years (not figured here). However, these peaks do not fall in the centre of anomalous frequency bands, and are therefore probably not significant.

To summarize, a dominant decadal-length frequency band is observed in all of the Santa Monica and Santa Barbara proxy data sets

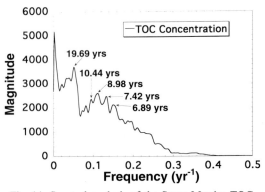

Fig. 14. Spectral analysis of the Santa Monica TOC concentration record. A decadal-length band exists between around 20 and 9 years.

(including density, Santa Monica and Santa Barbara $\delta^{13}C_{TOC}$, TOC concentration, $\delta^{13}C$ of *N. pachyderma*, palaeoSST, and palaeoxygen records) and seems to correlate with similar decadal length frequency bands in historical rainfall, runoff and SST records. However, none of these proxy records exhibit dominant variability that is similar in length to the ENSO phenomena. However, this does not mean that an ENSO-length signal does not exist, but rather suggests that this signal is difficult to resolve because the sampling interval may too closely approach the length of the ENSO signal. The source of the decadal-length signal is difficult to identify at this point, but should be the focus of future investigation. Similar frequency bands are found in tree ring and other data sets and are comparable in length to solar or Bruckner length cycles (Guiot 1987), suggesting that they represent an important forcing function in local climatic and oceanographic regimes.

Conclusions

Recent geochemical, sedimentological, micropaleontologic, and time series studies of Santa

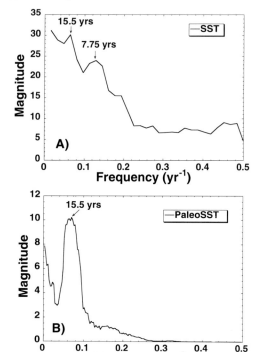

Fig. 15. (A) Spectral analysis of the short-term historical Spring SST record. **(B)** Spectral analysis of the long-term palaeoSST record. Note the similar frequency bands at around 15 yrs in both records.

Monica Basin sediments produced a number of palaeoceanographic and palaeoclimatic records for this region. These include proxy records of rainfall, organic carbon burial, primary productivity, sea surface temperatures and bottom-water oxygen concentrations. Although results are preliminary (especially since only one core has yet been analysed), many of these records correlate well with historical climatic and oceanographic records. Analysis of these records also provides clues about mechanisms of lamination formation, and suggests an influence of decadal-length climatic forcing functions on this region. At present, the interannual variability in the laminations is thought to be strongly influenced by the relative contribution of organic carbon, particularly from marine sources, such as phytoplankton.

JWH is grateful for the support and encouragement of A. E. S. Kemp; without which travel to England and involvement in the Laminated Sediment Conference and Volume would not have been possible. The work outlined in this manuscript stemmed from a project completed at USC under the supervision of L. D. Stott. A. Schimmelmann provided data from Santa Barbara Basin. T. D. Dickey, R. G. Douglas, A. G. Fischer, D. S. Gorsline, D. G. Hammond, T.-L. Ku and S. P. Lund provided helpful suggestions and comments on this project. This manuscript benefitted from suggestions by A. Schimmelmann and an anonymous reviewer.

References

BERELSON, W. M. 1991. The flushing of two deep sea basins, southern California Borderland. *Limnology and Oceanography*, **36**, 1150–1166.

CHELTON, D. B., BERNAL, P. A. & McGOWAN, J. A. 1982. Large-scale interannual physical and biological interactions in the California Current. *Journal of Marine Research*, **40**, 1095–1125.

CHRISTENSEN, C. J. 1991. *An analysis of sedimentation rates and cyclicity in the laminated sediments of Santa Monica Basin, California Continental Borderland.* MS Thesis, University of Southern California, Los Angeles, California.

——, GORSLINE, D. S., HAMMOND, D. E. & LUND, S. P. 1994. Non-annual laminations and expansion of anoxic basin-floor conditions in Santa Monica Basin, California Borderland, over the past four centuries. *Marine Geology*, **116**, 399–418.

DOUGLAS, R. H. 1979. Benthic foraminiferal ecology and paleoecology: A review of concepts and methods. *In*: LIPPS, J. H., BERGER, W. H., BUZAS, M. A., DOUGLAS, R. G. & ROSS, C. A. (eds) *Foraminiferal Ecology and Paleoecology.* SEPM Short Course, No. 6 Society of Economic Paleontologists and Mineralogists, 21–55.

DOUGLAS, R. G. 1981. Paleoecology of continental margin basins: A modern case history from the Borderland of southern California. *In: Depositional Systems of Active Continental Margin Basins.* SEPM Pacific Section Short Course Notes, San Francisco, CA, 121–156.

——, WALCH, C. & BLAKE, G. 1976. *Benthic Foraminifera from the Borderland of Southern California.* BLM Technical Report TR 76-2, Washington DC.

——, COTTON, M. L. & WALL, L. 1979. *Distributional Variability Analysis of Benthic Foraminifera in the Southern California Bight.* Southern California Baseline Study: Bureau of Land Management Report, 2/21.

——, LIESTMAN, J., WALCH, C., BLAKE, G. & COTTON, M. L. 1980. The transition from live to sediment assemblage in benthic foraminifera from the southern California Borderland. *In:* FIELD, M., BOUMA, A., COLBURN, I., DOUGLAS, R. G. & INGLE, J. (eds) *Quaternary Depositional Environment of the Pacific Coast.* Pacific Coast Paleogeography Symposium, **4**, 257–280.

EPPLEY, R. W. & HOLM-HANSEN, O. 1986. Primary production in the southern California bight. *In:* EPPLEY, R. W. (ed.) *Plankton Dynamics of the Southern California Bight.* Lecture Notes on Coastal Estuarine Studies. Springer, New York, 176–215.

FLOCKS, J. G. 1993. *Transport Mechanisms and Element Distribution in a Semi-enclosed, Anoxic Marine Basin, and the Influences of Natural and Anthropogenic Input, California Continental Borderland.* MSc Thesis, University of Southern California, Los Angeles, California.

GORSLINE, D. S. 1980. Depositional patterns of hemipelagic Holocene sediment in borderland basins on an active margin. *In:* FIELD, M. E., BOUMA, A. H., COLBURN, I. P., DOUGLAS, R. G. & INGLE, J. C. (eds) *Quaternary Environments of the Pacific Coast.* Society of Economic Paleontologists and Mineralogists Pacific Section Symposium, **4**, 185–200.

——1992. The geologic setting of Santa Monica and San Pedro Basins, California Continental Borderland. *Progress in Oceanography*, **30**, 1–36.

——, NAVA-SANCHEZ, E. & MURILLO DE NAVA, J. 1996. A study of occurrences of Holocene laminated sediments in California borderland basins: products of a variety of depositional processes. *This volume.*

GUIOT, J. 1987. Reconstruction of seasonal temperatures in central Canada since A.D. 1700 and detection of the 18.6 and 22 year signals. *Climate change*, **10**, 249–268.

HAGADORN, J. W., FLOCKS, J. G., STOTT, L. D. & GORSLINE, D. S. 1993. A simple method to obtain high-resolution (millimeter-thick) samples from laminated sediments. *Journal of Sedimentary Petrology*, **63**, 755–758.

——, STOTT, L. D., SINHA, A. & RINCON, M. 1995. Geochemical and sedimentologic variations in inter-annually laminated sediments from Santa Monica Basin. *Marine Geology*, **125**, 111–131.

HUH, C. A., AAHNLE, D. L., SMALL, L. F. & NOSHKIN, V. E. 1987. Budgets and behaviors of uranium and thorium series isotopes in Santa Monica Basin sediments. *Geochimica et Cosmochimica Acta*, **51**, 1743–1754.

——, SMALL, L. F., NIEMNIL, S., FINNEY, B. P., HICKEY, B. M., KACHEL, N. B., GORSLINE, D. S. & WILLIAMS, P. M. 1990. Sedimentation dynamics in the Santa Monica-San Pedro Basin off Los Angeles: Radiochemical, sediment trap and transmissometer studies. *Continental Shelf Research*, **10**, 137–164.

KENNEDY, J. A. & BRASSELL, S. 1992. Molecular records of twentieth-century El Niño events in laminated sediments from the Santa Barbara basin. *Nature*, **357**, 62–64.

LANGE, C. B., BERGER, W. H., BURKE, S. K., CASEY, R. E., SCHIMMELMANN, A., SOUTAR, A. & WEINHEIMER, A. L. 1987. El Niño in Santa Barbara Basin: Diatom, Radiolarian, and Foraminiferan responses to the '1983 El Niño' event. *Marine Geology*, **78**, 153–160.

——, BURKE, S. K. & BERGER, W. H. 1990. Biological production off southern California is linked to climatic change. *Climatic Change*, **16**, 319–329.

MCGOWAN, J. A. 1985. El Niño 1983 in the southern California bight. *In:* WOOSTER, W. S. & FLUHARTY, D. L. (eds) *El Niño North.* Washington Sea Grant Program, University of Washington, Seattle, Washington, 166–184.

PACLIM Workshop 1988. *Climate Variability of the Eastern North Pacific and Western North America*, 5.

QUINN, W. H., NEAL, V. T. & ANTUNEZ DE MAYOLO, S. E. 1987. El Niño occurrences over the past four and a half centuries. *Journal of Geophysical Research*, **92**, 14449–14461.

RAU, G. H., TAKAHASHI, T., DES MARAIS, D. J., REPETA, D. J. & MARTIN, J. H. 1992. The relationship between δ^{13}C of organic matter and [CO_2(aq)] in ocean surface water: Data from a JGOFS site in the northeast Atlantic Ocean and a model. *Geochimica et Cosmochimica Acta*, **56**, 1413–1419.

SAUTTER, L. R. & THUNELL, R. C. 1991, Seasonal variability in the δ^{13}O and δ^{13}C of planktonic foraminifera from an upwelling environment. Sediment trap results from the San Pedro Basin, Southern California Bight. *Paleoceanography*, **6**, 307–334.

SCHIMMELMANN, W. & TEGNER, M. J. 1991. Historical oceanographic events reflected in ^{13}C/^{12}C ratios of total organic carbon in Santa Barbara Basin sedi-ment. *Global Biogeochemical Cycles*, **5**, 173–188.

SOUTAR, A. & CRILL, P. A. 1977. Sedimentation and climatic patterns in the Santa Barbara Basin during the 19th and 20th centuries. *Geological Society of America Bulletin*, **88**, 1161–1172.

TYSON, R. V. & PEARSON, T. H. (eds) 1991. *Modern and Ancient Continental Shelf Anoxia.* Geological Society, London, Special Publication, **58**.

Tales of 1001 varves: a review of Santa Barbara Basin sediment studies

ARNDT SCHIMMELMANN[1] & CARINA B. LANGE[2]

[1] *Department of Geological Sciences, Geology 129, Indiana University,*
1005 East 10th Street, Bloomington, IN 47405-1403, USA
[2] *University of California, San Diego, Scripps Institution of Oceanography,*
Geosciences Research Division, Mailstop 0215, La Jolla, CA 92093-0215, USA

Abstract: The varved marine anoxic sediment in the deep centre of the Santa Barbara Basin off California has been studied extensively by sedimentologists, micropalaeontologists and geochemists during the past four decades. The Santa Barbara Basin is the only basin in the California Borderland Province that exhibits laminations for most of the Holocene. These sediments are essential to further our understanding of climate in that they provide a rare oceanic record of climate change with high resolution. In this overview of the complex and multidisciplinary Santa Barbara Basin sediment research, we briefly introduce the reader to the various analytical approaches and present a comprehensive collection of literature references.

The central Santa Barbara Basin (SBB) holds the most extensively studied laminated marine sediment in the world. However, the scientific literature about SBB sediments is poorly cross-referenced and often hard-to-find without expert guidance. We attempt here a brief but exhaustive overview on relevant literature up to 1994. Our title alludes to the fact that prior to 1994 most of the work was focused on the topmost sediment interval representing the last millennium of depositional history. In the Appendix we offer a collection of references indicating the various topics of SBB research. The final section 'Long-term fluctuations' gives a brief overview on current research and most recent publications that result from two long piston cores of the Ocean Drilling Program (ODP).

Laminations in SBB sediments, later proven to be annual varves (Fleischer 1972), had been recognized since the 1950s as potentially containing an excellent climatic record at high resolution (e.g. Emery 1960; Emery & Hülsemann 1962; Hallam 1967). Laminations, and sometimes varves, have also been reported from other near-by locations, such as adjacent offshore California basins, from offshore northern California, and from the Gulf of California, with ages ranging from modern to Miocene. For reviews regarding these areas, the reader is referred to Fisher (1990), Eganhouse & Venkatesan (1993), and Pike & Kemp (1966).

Andrew Soutar at Scripps Institution of Oceanography (SIO) pioneered the documentation of varves in the SBB (Soutar 1975; Sholkovitz & Soutar 1975; Soutar & Crill 1977; Soutar et al. 1981). Further work, again mostly performed at SIO, expanded the available records downcore and later emphasized an increase in resolution, up to a seasonal level (for example, Pisias 1978a, b; Dunbar 1981; Lange et al. 1987, 1990; Schimmelmann et al. 1990a, 1992; Schimmelmann & Tegner 1991; Lange & Schimmelmann 1994). The SBB has recently become a focus of international palaeo-ceanographic efforts, after the Ocean Drilling Program (ODP) made available in 1993 sediments from the 196.5 m deep hole 893A (Shipboard Party 1994). Studies of Site 893 are pioneering because no pre-Holocene sediment sequence has previously been obtained from the area.

Mechanisms of varve-formation

Varved sediments in the SBB are primarily composed of transported sedimentary particles of detrital and biologic origin. Varves are formed by settling of suspended particles transported by rivers and wind and the sinking of dead marine plankton and fecal pellets (e.g. Emery & Hülsemann 1962; Berger & Soutar 1967; Drake 1971; Crisp et al. 1979; Dunbar 1981; Dymond et al. 1981; Gorsline et al. 1984; Schwalbach & Gorsline 1985; Gorsline & Tseng 1989; articles in Angel et al. 1992). Sediment can also be supplied by turbidity currents which resuspend and transport sediment from further upslope and deposit it downslope (Fleischer 1972; Thornton 1981, 1984).

The SBB is a semi-enclosed basin with a topographic sill at 475 m depth separating it from adjacent basins to the west and preventing ventilation from below the oxygen minimum zone

From Kemp, A. E. S. (ed.), 1996, *Palaeoclimatology and Palaeoceanography from Laminated Sediments,* Geological Society Special Publication No. 116, pp. 121–141.

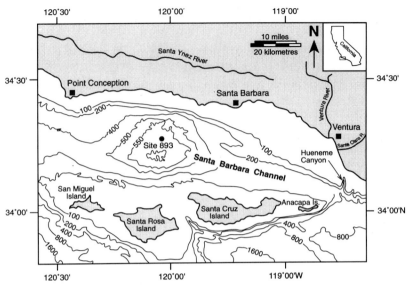

Fig. 1. Location map showing the Santa Barbara Basin and the position of the recent Ocean Drilling Program Site 893.

which at present reaches well below 500 m depth (Fig. 1). The resulting depletion of oxygen in sub-sill bottom waters in the centre of the SBB is a precondition for the preservation of laminations. The sediment surface in the central SBB is frequently covered by a mat of sulfur (H_2S)-oxidizing *Beggiatoa* spp. filamentous bacteria, especially during summer and early fall (Reimers *et al.* 1990; Grant 1991). The mucilaginous sheaths of bacterial filaments effectively bind inorganic and organic matter into a cohesive, highly porous network which traps sedimenting particles, and protects the topmost varves from erosion by weak bottom currents. Bacterial growth rate varies seasonally. Reimers *et al.* (1990) suggested that the increase in bacterial biomass during late summer/early fall was a response to changing redox conditions in the basin's surficial sediments. Oxygen supply varies seasonally in response to bottom water changes induced by spillover into the basin. Bacterial growth and associated bacterial mat thickening at the sediment–water interface during the late summer to early fall, when riverine input is low, result in the formation of the detritus-poor, high porosity layers that are typical of annual pairs of varved sediment. High numbers of diatoms are ascribed to these light-coloured laminae (Hülse-mann & Emery 1961; Emery & Hülsemann 1962; Reimers *et al.* 1990). In contrast, 'dark,' dense laminae form in winter, during times of low biomass and high terrigenous input. Thus, the formation of the laminae is due to the seasonal

rate and nature of deposition and may be enhanced by the growth of a *Beggiatoa* spp. mat on the surface sediments (Soutar & Crill 1977; but see Bull & Kemp 1995).

Individual SBB varves may contain multiple independent palaeoclimatic proxies related to changes in oceanic circulation and water masses, sea surface temperature, local precipitation on land, biological productivity, among others. Seasonally varying precipitation is responsible for pulses of terrigenous sediment input, and regional changes in wind patterns affect upwel-ling and productivity. The multifaceted nature of these records and their potential for recovery of very long time series make varves particularly useful. By correlating varves of the last few decades with known oceanographic and climatic time-series, the proxies can be calibrated, and older varve records can then be used to extend reconstruction of palaeoclimate. Palaeoceano-graphic and palaeoclimatic information derived from varves is not merely a duplication of coastal terrestrial datasets. The records off California provide an important complementary account of the Pacific marine influence on the continental climate of North America.

Sediment sampling and X-radiography

Laminated sediment is retrieved with coring devices which do not disturb the fine lamina-tions. Several common devices used for this purpose are box corers and Kasten corers (see also Chant & Cornett 1991; Leonard 1990). Box cores are best suited to sample the sediment/water interface and water-rich unconsolidated SBB sediment to a depth of about 70 cm. Sediment collected with these coring devices are sliced longitudinally into 1–2 cm thick slabs and these are then X-radiographed for accurate visual identification and counting of laminae (Schimmelmann et al. 1990a; Pike & Kemp 1996). A dark band on an X-radiograph contact print (positive image) indicates greater absorption of X-rays, mostly by mineral components of that sediment layer. Water-rich layers contain less solid material per unit volume and appear lighter coloured. The X-radiograph's dark/light pattern can therefore be associated with sediment porosity which is directly correlated to pore water content since all pore space is usually filled with water. Thin-sections of epoxy-embedded varved sediment (Grimm 1992; Lange & Schimmelmann 1994, and technical references therein; Pike & Kemp 1996) permits the evaluation of microlaminations, such as seasonal successions, bloom events, etc., within individual laminae.

Separately from the X-ray slabs, individual box cores and Kasten cores can be sampled with cylindrical subcores of various diametres (typically 5 to 10 cm), for collection of individual varve samples for microfossil, geochemical, and other studies (Schimmelmann et al. 1990a). Layers as thin as 1 mm can be separated manually. The relatively high sedimentation rate in the SBB (approx. 10 cm/100 years) permits sampling of the sediment at near annual resolution. Each sectioned sediment sample is weighed wet and then re-weighed after freeze drying in order to calculate pore water content. Comparison of X-ray images with independent water content determinations from duplicate samples permits precise association of sectioned sample intervals into the corresponding X-radiograph framework for dating (Schimmelmann et al. 1990a).

Dating of varves

To document the range and frequency of palaeoclimatic variation it is necessary to associate analytical data on varves with at least relative age and preferably absolute age. The annual varves allow us to do this at a resolution adequate for studying climatically interesting events. If the varves were perfectly deposited and preserved, an age assignment could be made by counting the number of varves from the present-day topmost sediment downward and a simple palaeoclimatic variable could be varve thickness. Several pro-

blems such as turbidites, flood deposits, bioturbation, and erosion make absolute age assignment in the varve-counting procedure uncertain. The validity of the palaeoclimatic record depends upon the care exercised to develop a proper time series of the proxy values.

Errors in assignment of age and thickness in varve data may occur due to thin (mm-scale) 'non-regular-varve' deposits that are contained within single regular varves, or that mimic regular varves, and therefore remain unidentified. Counting of such layers as regular varves in X-radiographs may lead to faulty varve-counts by overestimating the age of underlying varves. When strong bottom currents associated with the initial phase of a major submarine mudflow or turbidite erode the topmost and least compacted varves (Yemelyanova et al. 1992; see Dade et al. 1992, for forces involved in erosion processes), underestimating age is also a possibility. The 'non-regular-varves' are analogous to the 'false ring' detection problem in dendrochronology, while the underestimation of age problem is similar to the 'missing ring' problem in trees.

Turbidite layers, excluding flood and nepheloid deposits, can be considered instantaneous deposits whose frequency and local thickness are chiefly-controlled by internal mechanisms, source size, and proximity. The presence of turbidites interferes with reading the climatic record. Fortunately, only very few of the turbidites of the past 2000 years of sedimentation in the central SBB appear to have had the potential for erosion of underlying varves based on sedimentological criteria.

About 95% of the sediment accumulated in the SBB over the last 1000 years represents non-turbidite olive laminated material (Munsell Soil Colour 5Y 4/3). Within the top two metres of sediment there are several gray (5Y 5/1) and olive homogeneous layers (5Y 4/2), with thicknesses varying between 0.1 to several tens of centimetres (Schimmelmann et al. 1990b). It is interesting to note that gray layers thicker than 3 mm occur below the 1743 AD level only (Fig. 2). The reason for this shift in sedimentation pattern is unknown to us. Change in climate and in wave regime might have led to changes in supply of sediment to the coastal region and to changes in the rate and pattern of nearshore transport of terrigenous sediments.

The gray layers are characterized by light colour, low porosity, low concentrations of total organic carbon and total nitrogen, depletion of ^{13}C in organic carbon (Schimmelmann & Kastner 1993), and an order of magnitude fewer marine diatoms and silicoflagellates than in

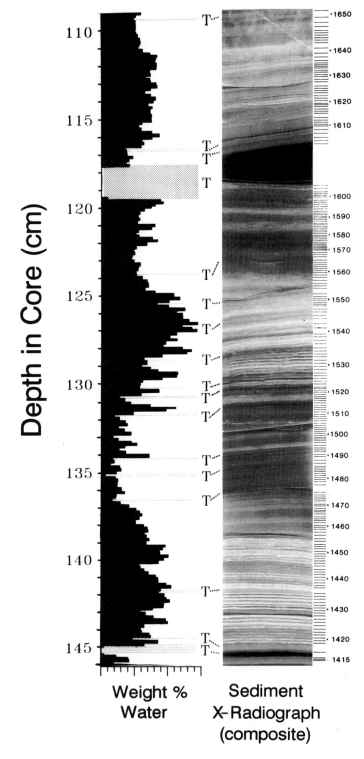

Fig. 2. Santa Barbara Basin varve record of 1415 to 1791 AD. Composite of X-radiograph contact prints of Santa Barbara Basin sediments, with ages of deposition serially assigned (on right-hand side). The 1988 to 1765 AD record is summarized in Schimmelmann *et al.* (1990*a*, 1992). Left: Pore-water profile (in salt-free wet wt%).

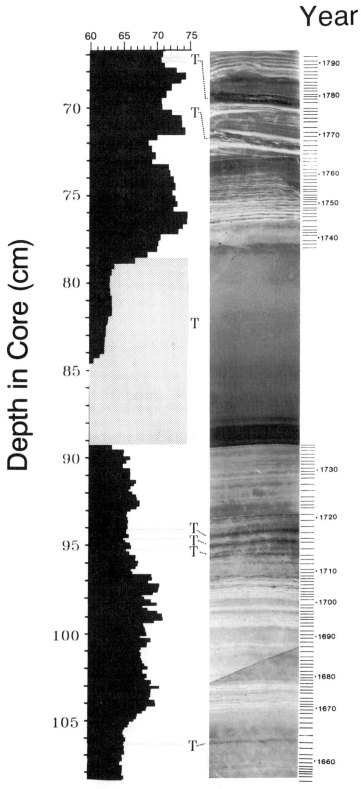

Fig. 2. (*continued*)

adjacent laminated horizons. This evidence, along with mineralogical data of Fleischer (1972), agrees with Thornton's (1984, 1986) conclusion that the gray layers are turbidites or flood suspensate deposits which originated at the upper near-shore basin slope. The extractable palaeoceanographic and palaeoclimatic information in these layers appear to be limited although their presence throughout the basin provides valuable correlation markers.

The average composition of the olive homogeneous layers, including their microfossil content, is similar to that of the laminated sediment at comparable depths (Schimmelmann & Kastner 1993). They seem to be redeposited sediments from the basin slope where average terrigenous and marine components are similar to those in varved sediments of the central basin (Thornton 1984).

Bioturbation is rare in the central SBB over the last 2000 years. A significant disturbance of the varve record is seen in the 1830s, possibly related to the aftermath of a severe volcanic event (Schimmelmann et al. 1992). The composition of bioturbated layers is largely the same as for varves. The presence or absence of bioturbation in itself, for example at some distance from the deep basin, along the basin slope of the SBB, is a measure of fluctuating benthic oxygenation conditions and faunal activity (Savrda & Bottjer 1986).

Time lag

To interpret palaeoclimatic data collected from varves, it is important to consider the time lag associated with the occurrence of a climatic/oceanographic event and the imprint of this signal in an accumulating varve. The lag is influenced by many factors, most of which are inherent to the mode of deposition of the signal-carrying material. The best studied lag mechanism in the SBB is the one between rainfall on land and the deposition of river-transported detritus in the basin (Soutar & Crill 1977). Although most of the terrigenous material carried into the SBB by river runoff arrives at the sediment–water interface in the centre of the basin within one year, smaller increments continue to accrue over the following years (Drake et al. 1972). The significance of the lag was quantified statistically by Soutar & Crill (1977), using multiple regression analysis and setting the baseline for optimal cross-dating of precipitation and varve records. They found that it may take as long as seven years for 90% of the effect of a single year's rainfall to pass through the rainfall–sediment system.

The deposition of biogenic particles is tied to biological processes and has little time lag. Due to their short life span and rapid overturn in conjunction with rapid sinking as aggregates, the time lag between a bloom of a given biota and the incorporation of its skeletal debris into the topmost accumulating varve is short (less than 10 days in the centre of the SBB, assuming a sinking velocity of 50–175 m/day; Silver & Alldrege 1981; Takahashi 1986).

Schimmelmann & Kastner (1993) observed that elemental sulphur concentrations correlated with total organic carbon, with a lag of about 1.4 years. They proposed a mechanism that involves bacterial motility which produces this apparent lag. *Beggiatoa's* ability to protectively bury itself and its intracellular sulphur into the anoxic sediment in the event of unfavourable bottom water conditions may cause selective vertical displacement of elemental sulphur, whereas significant amounts of non-*Beggiatoa*-based organic carbon would not be mobile.

Dating of varves with radioisotopes

Several isotope methods permit the assignment of absolute ages to sediment intervals. In deeper sediment radioisotope dates serve as valuable anchor points for varve counts. Sediments a few hundred to several thousand years old can be dated using radiocarbon techniques. The accuracy of ^{14}C dates in sediments from the Southern California Borderland and from the Gulf of California is limited by the deposition of ^{14}C-depleted carbon. Both the non-zero ^{14}C age of surface and deep waters, and the incorporation of old organic matter into young sediments cause a positive age bias. Biological utilization of carbon dioxide from upwelling waters includes carbon which has aged in isolation from the atmosphere during ocean transport (Emery 1960; Emery & Bray 1962; Robinson & Thompson 1981; DeMaster & Turekian 1987). Fossil-carbon cycling from nearshore natural hydrocarbon seeps (Bauer et al. 1990), and pollution of some marine environments by anthropogenic discharge of terrigenous carbon (Summers et al. 1988; Schaefer 1989; Schimmelmann & Tegner 1991) both contribute carbon not currently in ^{14}C-equilibrium with the atmosphere. To avoid contamination, it is best to analyse individual identifiable particles, such as fish skeletal debris or shells. A few milligrams of carbon are sufficient for ^{14}C-accelerator-mass spectrometry (AMS) dating, permitting dating of individual, larger collagenous fish bones, calcareous otoliths or other biogenic debris in a given varve. Ingram & Kennett (1995) present a radiocarbon chronology for piston cores from the SBB.

Geochronology based on ^{210}Pb is limited to varves close to the sediment/water interface, with ages up to 100 years. ^{228}Th/^{232}Th geochronologies may be applicable for times up to a decade (Koide et al. 1972, 1973, 1976; Krishnaswami et al. 1973; Bruland 1974). These rapidly decaying isotopes were instrumental for the development and quality control of 20th century varve counts in the SBB, by proving that the results of all methods are in close agreement (Soutar & Crill 1977; Bruland et al. 1979; see also articles in Angel et al. 1992, for other basins of the California Borderland).

Cross-dating with tree-ring records

Soutar & Crill (1977) established statistical correlations between local tree-ring indices and varve thicknesses with the historical Santa Barbara precipitation record (see 'Time lag' above). Both tree-ring growth and varve thickness were shown to be significantly controlled by precipitation. Wet years favour tree growth and supply greater terrigenous detritus to the basin via rivers. For the purpose of cross-dating, one should rely on standardized tree ring records which are sensitive to precipitation, comparable in time interval to the varve records, and which are located as close as possible to the SBB. Recent applications for the SBB are shown in Schimmelmann et al. (1992).

Recognition of historic events in distinct varves

For the past 200 years, there are distinctive 'marker layers' within the SBB sediment record that are helpful in verifying age assignments. These layers relate to historical climatic/meteorological events. For example, the heavy rainfall and record floods during January and February 1969 caused a massive influx of terrigenous sediment into the SBB, via the Ventura and Santa Clara Rivers (Drake et al. 1979). Major oil-spills during January to December 1969 (Straughan 1971) added ^{13}C-depleted carbon to the basin sediments resulting in a geochemically distinct varve with regard to its carbon isotopic composition (Schimmelmann & Tegner 1991). Historic El Niño events between 1940 and 1987 are recognized in the sediment by diagnostic warm water diatom assemblages (Lange et al. 1987, 1990) and by the biogeochemical alkenone palaeotemperature index U_{37}^{K} (Kennedy & Brassell 1992a, b). Major El Niño events accompanied by severe storms between 1844 and 1987 destroyed kelp beds along the coast (Schimmelmann & Tegner 1992), and caused deposition of ^{13}C-enriched organic material in the basin. The stable carbon isotope stratigraphy can be used to reconstruct the occurrence of such events (Schimmelmann & Tegner 1991). An 1811 gray layer may represent a turbidite that resulted from the severe earthquake centering around Santa Barbara in 1811 (Schimmelmann et al. 1992). However, other details of the 200-year historical record of seismicity in the Santa Barbara region (Hamilton et al. 1969) do not match well with the sedimentary record of turbidites (Schimmelmann et al. 1990a, 1992).

Precision of the SBB varve chronology

The resolution of climatic/oceanographic signals in the varved sedimentary record depends mostly on the quality of varve preservation. In the SBB, the best conditions are found in the deep centre where turbidites are less frequent and thinner than along and near the slope, bottom waters have the least amount of dissolved oxygen, and the combination of a relatively flat stable bottom topography together with the presence of a bacterial mat preserves laterally uniform varve thicknesses.

Soutar & Crill (1977) constructed the original varve chronology for the SBB, spanning approx. from 1820 to 1970 AD. The original chronology was revised and extended (Schimmelmann et al. 1990a), and several undetected thin gray layer problems were identified. Finally, additional X-radiographs of higher quality from several new sediment cores were obtained and permitted a more reliable and extended varve-count, especially for the 19th century (Schimmelmann et al. 1992). Our detection and discounting of thin gray layers yielded the revised varve-chronology down to 1415 AD (see Fig. 2). Based on varve count comparisons among several sediment cores, we estimate the precision of the latest time scale to be ±1 year for 1900–1987 and ±2 years down to 1840 AD. Further downcore the precision deteriorates to an estimated 10 years in the 15th century level. Within any single continuously varved sequence regardless of absolute age, the relative age relationships between any two layers is accurate to about ±2% (for example ±2 years for two samples that differ in age by 100 years).

Palaeoclimatic proxies in SBB varves

The Appendix gives an overview of the numerous studies (sedimentological stratigraphies, geochemical depth- and time-series, microfossils, etc.) that are available from the SBB. It is beyond the scope of this contribution to more than briefly introduce the main categories of data.

Varve thickness

Varve records can be correlated across the central SBB (e.g. Soutar & Crill 1977; Schimmelmann et al. 1990a); distinct marker layers that are laterally continuous can be traced from core to core. For quantitative comparison between deep and shallow sediment layers, the varve thicknesses must be corrected for the substantial effects of compaction with increasing depth of burial. At a typical (non-gray layer) sedimentation rate of 1 cm/year at the sediment/water interface a 1 cm thick, highly porous surface sediment would be compacted to about 3 to 4 mm after five years and to 1 to 2 mm by the time it is buried 2 m deep, after 1000 years of ongoing sedimentation (Doose 1980; Schimmelmann & Kastner 1993). The extremely rapid compaction during the first few years of diagenesis is probably related to the fact that benthic living biomass is concentrated, and largely limited to, the uppermost few centimetres of the sediment. The non-linearity over depth in the degree of compaction is discussed in detail by Doose (1980).

Geochemical parameters

A vast array of geochemical parameters has been studied in SBB sediments (Appendix). Geochemical parameters with palaeoclimatic significance for the SBB have been found in the concentrations of sterols and fatty acids (Lajat et al. 1990), in the isotopic composition of organic carbon (Schimmelmann & Tegner 1991), and in concentrations of organic carbon and reduced species of sulfur (Schimmelmann & Kastner 1993). The degree of unsaturation of C_{37} alkenones in the bulk sediment is a measure of sea surface temperature. A time-series 1910–1987 of C_{37} alkenones from the SBB confirms the usefulness of the parameter for reconstruction of El Niño events (Kennedy & Brassell, 1992a, b). In addition, the oxygen stable isotope ratios in calcium carbonate from foraminiferal tests have been used to infer trends in palaeo-temperature (Dunbar 1983). Recent abstracts feature the distribution and speciation of sulphur (Fischer 1993; Brüchert & Pratt 1994).

Time-series of toxin concentrations in SBB sediment are valuable tracers of anthropogenic pollution of the coastal environment through the 20th century (Chow et al. 1973; Young et al. 1973; Hom et al. 1974; Schmidt & Reimers 1991, and references therein). Geochemical results from an 18th century marine mammalian coprolite recovered from the centre of the SBB are reported in Schimmelmann et al. (1994).

Marine biogenic components

Santa Barbara Basin sediments are mostly hemipelagic in nature, being almost entirely composed of silt and clay-sized terrigenous and biogenic grains. The biogenic component includes varying quantities of diatoms, nannofossils, sponge spicules, silicoflagellates, radiolarians, foraminifers, ostracods, pteropods, and fish debris.

Microfossils found within varves contain a wealth of palaeoclimatic data. The dominant preserved shell materials are the two mineral polymorphs of calcium carbonate ($CaCO_3$), calcite and aragonite, and silica (SiO_2) or opal. In addition, sporopollen-like material (a unique class of biopolymers; summarized by Dale 1983) as the one found in resistant pollen grains and in the wall of some organic dinoflagellate cysts is also present in sediments.

Diatoms have long been accepted as good indicators of productivity patterns, because of their position at the base of the food chain (Berger et al. 1989). Diatom spores (e.g. Chaetoceros spp.) account for a major fraction of the opaline sediment component within coastal upwelling regions, and have been interpreted as an indication of high primary production during spring in the SBB (Lange et al. 1990; Reimers et al. 1990; Bull & Kemp 1995).

Among microfossils, foraminifera are probably the group most widely used for palaeoclimate reconstruction in the California Borderland. Their relative abundances are useful in determining palaeotemperature (plankton), palaeoproductivity (both plankton and benthos; Berger et al. 1989), and changing oxygen conditions in the basin (Reimers et al. 1990; Burke et al. 1995; Shipboard Party 1994). For an exhaustive overview on radiolarian record in SBB sediment, see Weinheimer (1994).

Fish scales and other skeletal fragments are useful in tracing water-mass movements and aspects of palaeo-productivity (Soutar & Isaacs

1969). In the SBB, Soutar (1967) and Soutar & Isaacs (1969) carried out the first studies of the economically and ecologically important pelagic fish species of the California Current System over the past 150 years. Their scales are regularly found in sediments of the SBB. The proportional abundance of sedimented fish scales to fishery-derived population estimates permitted the generation of an estimate of pelagic fish biomass in pre-fishery times (Soutar & Isaacs 1974). Recently, Baumgartner et al. (1992) presented a composite time series of Pacific sardine and northern anchovy fish scale deposition rates over the past two millennia. They showed nine major recoveries and subsequent collapses of the sardine population.

Land-derived components

Pollen grains and spores from the continent have been used to interpret drainage changes, wind-direction and strength, aridity, and recent human alteration of stream flow. According to Heusser (1978), the SBB pollen record is primarily a record of changes in the dominance of the various plant communities in the Santa Barbara region. Climatic change during the last 12 000 years caused the decrease in coniferous upland vegetation and the increase in the lowland coastal sage scrub and chaparral. On the other hand, Byrne et al.'s (unpublished report to the Forest Service) observations suggest that variations in pollen influx are more a function of sediment transport mechanisms rather than of vegetation change *per se*. However, a comparison of the prehistoric and modern pollen data indicate that oaks and pines are more common now than 500 years ago. Mensing (1993) evaluated the impact of European settlement near Santa Barbara through pollen and charcoal evidence in a high-resolution sediment sequence. Preliminary results from ODP Site 893A show that interglacial periods were dominated by oak, while conifers dominated glacial episodes (Shipboard Party 1994; Heusser 1995).

Charcoal in laminated sediment in the SBB was interpreted as evidence for large forest fires in the drainage area of near-by rivers (Byrne et al., unpublished report to the Forest Service). Byrne and collaborators used the abundance of river- and wind-transported charcoal in SBB sediment to reconstruct prehistoric fire frequencies in the Los Padres National Forest. Two major peaks in charcoal influx were recorded and attributed to the 1955 Refugio Fire and the 1964 Coyote Fire, respectively. During the period 730 to 1505 AD

major fires in the coastal area occurred on average once every 65 years, and once every 30 to 35 years further inland.

Reconstructing the biological seasonal signal

Palaeoceanographic/palaeoclimatic proxies need to be calibrated against an instrumental record, or at least a historical, qualitatively known record of climatic events during modern time. For the California Current System, the past five decades are especially suited as a calibration period, because the California Cooperative Oceanic Fisheries Investigations (CalCOFI) program has compiled a large amount of relevant oceanographic data (see articles of Beers 1986; Eppley 1986a, b, 1992; Eppley & Holm-Hansen 1986; for an overview of collected data, see Hewitt 1988).

Major seasonal differences in the aspect of intercepted sediment were observed by Soutar et al. (1977). Particulate matter observed accumulating in sediment traps deployed during the rainy season (winter) is made up of clay flocculates with abundant silt particles. In contrast, material accumulating in the traps during the spring and summer seasons may be made up almost entirely of faecal pellets (Soutar et al. 1977, 1990; Dunbar & Berger 1981).

A detailed study of the benthic environment in the central SBB showed a clear difference in diatom composition, abundance and diversity in the surficial bacterial mat throughout the 1988 yearly cycle. Greatest diatom abundances (mainly *Chaetoceros* spores) were observed in the spring surface bacterial mat, suggesting recent sedimentation after the chlorophyll maximum (Reimers et al. 1990). Intermediate abundances were observed in the fall when the assemblage was composed of large forms. Benthic diatoms that were presumably carried into the central basin (e.g. by resuspension of upslope material) from the shelf were more abundant in winter. The central basin floor has a large standing crop of benthic foraminifera ($>100 \, \text{m}^{-2}$, Phleger & Soutar 1973). This may be due to the abundance of food derived from the high organic carbon flux and benthic bacterial mat (Hülsemann & Emery 1961), and the lack of predators. During the 1988 seasonal cycle, live benthic foraminifera were abundant in October and were found down to 4 cm in core (Bernhard & Reimers 1991; Bernhard 1992).

The 1988 data indicate that seasonality is strong in the SBB with episodic events of the order of days to weeks (Reimers et al. 1990). These events are preserved in the sedimentary record as microlaminations within seasonal laminae, and can be analysed in detail by means of backscattered electron imagery (Lange & Schimmelmann 1994).

Stable isotope ratios $^{18}O/^{16}O$ and $^{13}C/^{12}C$ in calcium carbonate from planktonic foraminifera collected in sediment traps in the San Pedro Basin indicate that at the onset of upwelling, *Globigerina bulloides* migrates to the surface from depths below the thermocline and remains in surface waters throughout upwelling, where conditions are optimal for increased production.

This migration results in specimens with less-calcified, more open tests that may be preferentially removed by dissolution, and thus an actual record of upwelling itself is not preserved in the sediments. Instead, conditions before and just after are more likely recorded by *Neogloboquadrina pachyderma* and *N. dutertrei*, respectively, both of which are species highly resistant to dissolution (Sautter & Thunell 1991a, b).

Reconstruction of past California Current fluctuations

The dominant hydrographic features of the SBB are the cold California Current, and the nearshore warmer countercurrent. These currents control the water masses and circulation in the region and supply the nutrients necessary to maintain high biological productivity.

An estimate of the course of primary production based on diatom flux fluctuations observed in modern (1954–1986 AD) sediments from the SBB is given by Lange et al. (1987, 1990). Minor fluctuations between 1954 and 1972, and a large, statistically significant decrease around 1972/1973 were observed. They proposed that this decrease in plankton fluxes in this coastal basin was not just a local phenomenon, but could reflect large-scale climatic changes in the eastern North Pacific related to an intensification of the Aleutian Low in the North Pacific over the past 14 years, providing for a weakening of the California Current, and an overall reduction of mixing and upwelling.

Dunbar (1983) produced a foraminiferal oxygen isotope record of 230 years from a box core collected in the SBB. The oxygen stable isotope record for the planktonic foraminifer *G. bulloides* exhibits good correlation with the historical record of sea surface temperature in the area since 1870. This indicates relatively cool conditions between 1850 and 1880, 1895 and 1925, and for a period of ±30 years prior to 1780 (Dunbar 1983; Anderson et al. 1992). A study by Anderson (1990) showed that changes in $^{18}O/^{16}O$ after 1750

consistently and closely parallel the frequency of occurrence of El Niño events.

Santa Barbara Basin sediment has been used in several studies to reconstruct the history of El Niño (EN) events. The average period of EN events in historic data is 3 to 4 years. EN events may occur as close as one or two years apart, or longer than six years apart; they may last from a few months to a full year (Deser & Wallace 1987; Quinn & Neal 1992). In the eastern Pacific, EN conditions are represented by anomalously warm sea surface temperature (SST), negative swings of the Southern Oscillation, high sea level along the coast and reduced upwelling (McGowan 1985). Decreases in total diatom flux and increases in the relative abundance of certain warm-water diatoms are observed in the SBB sedimentary record (Lange et al. 1987, 1990). Some warm water planktonic foraminifera (*Globigerinoides ruber*, *Globoturborotalita rubescens*) also increase their abundance during EN years (Lange et al. 1990). Thus, the characteristic microfossil assemblages of strong EN events in the varves of, for example, 1942, 1959, and 1984 may also serve as reference for age assignment (Schimmelmann et al. 1990a).

In contrast, radiolarian fluxes determined from Santa Monica and SBB varves show considerable increases in all environmental groups during the California El Niño that occurred between 1950 and 1985 (reviewed by Casey et al. 1989). Strong El Niños, such as the 1983 event, appear to cause anoxic basins to become oxic, which may be interpreted from the phaeodarian radiolarian record. Weinheimer (1994) presents the most comprehensive study of radiolaria in the SBB.

Schimmelmann & Tegner (1991) correlated the isotopic composition of sedimentary organic carbon with historic El Niño and other severe weather events from 1844 to 1987. The destruction of kelp forests in the SBB area by severe wind and wave events (during EN) produces strongly ^{13}C-enriched kelp-derived particles which contribute significantly to sedimentary organic matter. Typically, an EN event coincides with or immediately precedes ^{13}C-enriched varves.

A model for palaeoceanographic reconstruction of the California Current was presented by Pisias (1978b, 1979). He used radiolarian assemblages as a tracer of SST changes, and calibrated this tracer using the transfer function method of Imbrie & Kipp (1971). He produced a record of radiolarian-based SST for the past 8000 years, and showed that the relative strengths of the California Current and the Countercurrent varied temporally during that period. Prior to 5400 years BP, and during intervals from 800 to 1800, and 3600 to 3800 years BP temperatures were higher than present.

Diagenetic studies

Biological and diagenetic geochemical change in very young sediments near the sediment/water interface (for example, Craven & Jahnke 1992, and references therein) makes quantitative comparison with more mature, deeper buried sediment difficult. Dated varve sequences offer ideal conditions for the study of chemical kinetic transformations, such as on the diagenesis of sterol and fatty acid biomarkers (Lajat *et al.* 1990; Grant 1991) and the diagenesis of sulphur species (Schimmelmann & Kastner 1993).

The significance of the *Macoma* Event of the 1830s

The *'Macoma* Event' is the brief flourishing during 1835–1840 AD of a macrofaunal community in an environment, from which large organisms are normally excluded by the lack of oxygen. Shells of the pelecypod *Macoma leptonoidea* Dall and associated fossils, which congregate in a thin layer in the central SBB dated 1840 AD, provide evidence for a sudden local extinction of macrobenthos by suffocation. No other shell layer is known in the same sedimentary record, covering the known record of the past 2000 years. The evidence suggests that *M. Ieptonoidea* colonized the basin floor for a few years prior to 1840 as a result of temporarily decreased productivity and increased oxygen content in the water spilling over the sill into the basin (Schimmelmann *et al.* 1992; Burke *et al.* 1995). The taxonomic and quantitative composition of the diatom assemblage, as well as the δ^{18}O of *G. bulloides* (Dunbar 1983), essentially support the general increase in upwelling postulated at the end of the *Macoma* Event. Historical eyewitness reports indicate unusually severe winter storms from southeasterly directions during the 1830s in the SBB (Schimmelmann & Tegner 1991) which may have contributed to increased oxygenation of bottom waters. Alternatively, the winter storms may have been symptomatic of a different oceanographic regime with decreased coastal productivity and a shallower oxygen minimum zone, permitting oxygenated water to spill over the sill into the SBB. The *Macoma* Event is apparently tied into large-scale climatic anomalies felt throughout North America, possibly linked to the major volcanic eruption of Cosiguina, Nicaragua, in 1835 AD. Quinn's (1992) latest historical compilation of El Niño records indicates several events clustering in the 1830s, possibly in causal relationship to the temporary survival of *M. Ieptonoidea* at that time. The *Macoma* Event may have a

counterpart in the laminated sediments of the Black Sea, where a distinct black marker horizon was dated to 1838 ± 8 AD (Crusius & Anderson 1992).

The 'Little Ice Age'

The 'Little Ice Age' (LIA, roughly from 1450 to 1850) covers the period of a few centuries between the Middle Ages and the warm period of the first half of the 20th century, during which glaciers in many parts of the world expanded and fluctuated about more advanced positions (Grove 1988). It is characterized by repeated rapid climatic change and is of special palaeoclimatic interest, because it immediately predates the industrial revolution and the growing anthropogenic impact on global climate.

The LIA is reflected in the composition of planktonic foraminifer assemblages. Kipp & Towner (1975) used the foraminiferal assemblages in the SBB and elsewhere, in combination with other palaeoclimatic, dendrochronological information, to reconstruct climatic regimes during the last 625 years. The data suggest a decrease in SST from 1350 to 1625 AD (evidence based on increasing abundance of *G. quinqueloba)*, and a cool period from 1625 to 1875 (*G. quinqueloba* dominant). From 1875 to 1925, which marks the end of the LIA, warming water permitted dominance of *Globigerina bulloides*; this warming trend continued since 1925 with *G. bulloides* decreasing, and *Neogloboquadrina dutertrei* becoming more abundant (Fig. 2 in Berger 1971).

As mentioned above, the oxygen isotope record (^{18}O/^{16}O ratio) in the planktonic foraminifer *G. bulloides* in the SBB reflect a prolonged cool period from 200 to 230 years ago equated with the LIA. Post-1930 deposits indicate warmer surface water, greater stratification of the water column and less intense upwelling (Dunbar 1983).

Long-term fluctuations

Few piston cores have been recovered and analysed from the SBB. Up to 12 m long cores from various basins of the California Borderland with bottom ages up to 48 000 years were used by Kalil (1976) to reconstruct time-series of concentrations of carbon, sulphur and uranium (among others). The uranium content correlated with the occurrence of glacial events. A breakthrough occurred in 1992 when at Site 893 the Ocean Drilling Program (ODP) employed advanced piston coring techniques to recover two long cores, termed Holes A and B. Subsequently, the almost 200 m long core from Hole A revealed a near-complete record spanning the last 160 000 years with a near-linear sedimentation rate of about 15 cm/100 years (Shore-based Scientific Party 1994). This triggered a vast array of multidisciplinary analytical and palaeoclimatic work which is newly published (Kennett *et al.* 1995). The reader is referred to the reference list for a selected listing of relevant contributions: Behl (sediment facies, palaeo-oxygenation), Bull & Kemp (nature of laminations), Herbert *et al.* (palaeotemperature), Heusser (pollen), Ingram & Kennett (radiocarbon dating), Kennett (carbon and oxygen isotopes), Kennett & Venz (foraminifera assemblage), Lange & Schimmelmann (X-radiology), Marsaglia *et al.* (coarse grained sediment components). Studies are currently continuing and plans are developed for additional sites to recover laminated sediments in the California Borderland and further north along the Continental margin of North America during the forthcoming ODP Leg 167.

This work was supported by National Science Foundation grant OCE-9301438 to W. H. Berger and by NATO Collaborative Research Grant No. 930391 to A. Schimmelmann. We thank R. J. Behl and D. S. Gorsline for valuable suggestions.

Appendix

Geochronology

[14]C ages of organic and inorganic carbon from various basins in the California Borderland; time-series, up to 4.25 m. Emery & Bray (1962).

[210]Pb time-series 1850–1962; depth 37 cm; 17 two-year intervals. Koide *et al.* (1972).

[210]Pb, [226]Ra, [232]Th, [55]Fe. Time-series 2–30 years. Also other basins in California Borderland; Gulf of California; other locations. Krishnaswami *et al.* (1973).

[228]Th/[232]Th, [210]Pb. Santa Barbara Basin and other localities. Santa Barbara Basin: Time-series as long as 1925–1971, up to 19.1 cm deep; resolution 2–5 years. Koide *et al.* (1973).

[210]Pb (excess). In Santa Barbara Basin: time-series of 112 years; resolution 2–30 years. Bruland (1974).

[241]Pu and [241]Am in sediment, Santa Barbara and Soledad Basins. Santa Barbara Basin: Time-series 1954–1973, depth to 7.6 cm; resolution approximately 1 cm. Koide *et al.* (1980).

Other geochemical depth- and time-series

Grain size distribution, porosity, calcium carbonate, nitrogen, pH, in various basins in the California Borderland; depth-series of a few metres; resolution approximately 10 m. Emery & Rittenberg (1952).

Porosity, pH, Eh, concentrations of nitrogen and phosphorous, pore water composition, in Santa Barbara, Santa Monica, and Santa Catalina Basins; depth-series of 2 m; resolution 10–20 m. Rittenberg *et al.* (1955).

Grain size distribution, porosity, calcium carobonate, nitrogen, and chlorophyll derivatives, in various basins in the California Borderland; depth-series of up to 5 m; resolution approximately 30–50 m. Orr *et al.* (1958).

Amino acids in sediment cores, Santa Barbara Basin; depth-series, 4.3 m; resolution 10–15 cm. Degens *et al.* (1961).

Carbohydrates in sediment cores, Santa Barbara Basin, depth-series, 4.3 m; resolution approximately 0.5 m. Prashnowsky *et al.* (1961).

Porosity, carbonate, N, organic C, pH, Eh, sulphate, pyrite in total sulphur, sulphur stable isotope ratios, and other geochemical parameters in Santa Barbara Basin and other basins, in water column and in sediment; depth-series of isolated 10 cm intervals, reaching 4–5 m depth. Kaplan *et al.* (1963).

Amino acids and sugars in sediment cores, Santa Barbara Basin and several shelf locations; depth-series up to 170 m; resolution approximately 25 m. Degens *et al.* (1964).

Mineralogy, grain size distribution, Santa Barbara Basin 6.6 m depth-series, resolution approximately 0.4 m; indication of thick turbidites. Fleischer (1972).

Pb concentration in sediment, in several basins in the Southern California Borderland; Santa Barbara Basin: Time-series 1926–1970; resolution 2–12 years. Chow *et al.* (1973).

Interstitial water geochemistry, total carbon, organic carbon, calcium carbonate, pyrite in sediment cores from Santa Barbara Basin; depth-series to 55 cm; resolution 5–15 cm. Sholkovitz (1973).

Hg in sediment, Santa Barbara Basin, time-series 1820–1971; decadal resolution; two samples 1400 and 1500 years. Young *et al.* (1973).

DDE and polychlorinated vinyls, Santa Barbara Basin, time-series 1890–1967; resolution 3–15 years. Hom *et al.* (1974).

Nitrogen and argon concentrations, depth-series, up to 480 cm, resolution 20 cm to 2 cm. Barnes *et al.* (1975).

Total carbon, sulphur, uranium, total organic carbon, calcium carbonate, and pore water analyses in some intervals; Santa Barbara, Catalina, and Tanner Basins; Gulf of California, and other locations. In Santa Barbara Basin: depth-series (up to 5.3 m) with approximate dating; resolution between 3 and 10 cm; ^{14}C age of 3920 + 160 years at 5 m depth. Kalil (1976).

Organic carbon in sediment, Santa Barbara Basin, depth-series 8.1 m; approximate dating, with deepest sample 10 200 a BP; resolution approximately 25 years. Heath *et al.* (1977).

Stable isotope ratios of total organic carbon and of total nitrogen, from Santa Barbara, Santa Catalina, and Tanner Basins, and elsewhere. Santa Barbara and Santa Cruz Basins, San Diego Trough: depth-series, up to 5.3 m; resolution 0.5–2.0 m. Peters *et al.* (1978).

Pore water sulphate and ammonia, total nitrogen in sediment from Santa Barbara Basin; depth-series, up to 5.5 m; resolution approximately 20 m. Sweeney *et al.* (1978).

Methane production and oxidation in sediment, also concentrations of interstitial sulphate and hydrogen sulphide, Santa Barbara Basin; depth-series of 0–340 m; resolution approximately 35 cm. Kosiur & Warford (1979).

Porosity, concentrations of methane, sulphate, total and organic carbon, total nitrogen, ammonia, and calcium carbonate; nitrogen stable isotope ratios, Santa Barbara Basin; depth-series (up to 5.5 m) with approximate dating; resolution between 1 and 10 cm. Doose (1980).

Concentrations of ammonia, calcium carbonate, and total nitrogen, nitrogen stable isotopic composition, Santa Barbara Basin sediment; depth-series of several cores, up to 5.5 m; resolution 10–50 cm. Sweeney & Kaplan (1980).

Calcium carbonate, stable isotope ratios of foraminiferal carbonate, Santa Barbara and San Clemente Basins; depth-series; Santa Barbara Basin: up to 62 cm; resolution approximately 1 cm. Dunbar (1981).

Various organic geochemical and isotopic parameters from DSDP Sites 467–469, California Borderland; depth-series, 7.5–1035 m; resolution 10–200 m. Simoneit & Mazurek (1981).

Inorganic elemental composition (Al, Cl, P S, K Si, Ca, Fe, Ti, Mg, Mn) of Santa Barbara Basin sediment; depth-series 0–46.5 cm; resolution approximately 35 cm. Shiller (1982).

Textural and mineralogical characteristics of turbidite layers; also shows X-radiograph, Santa Barbara Basin; 3.56 m long core; resolution 2–10 m. Thornton (1984).

Fluorescence in pore water from sediment, Santa Barbara Basin; depth-series of 0–195 cm, resolution approximately 5–10 m. Chen & Bada (1989).

Concentrations of sterols and fatty acids, Santa Barbara Basin; time-series 1835–1987, resolution approximately 10 years. Lajat et al. (1990).

Fe and sulphide in pore water, extractable ATP, carbon, nitrogen, phosphorus, porosity in surface-near sediment from Santa Barbara Basin; 1988, seasonal resolution. Reimers et al. (1990).

Fatty acid concentrations, Santa Barbara Basin, Depth-series, 0–27 cm; resolution 2.5 cm. Grant (1991).

Stable isotope ratios of total organic carbon, Santa Barbara Basin; time-series 1844–1987; approximate annual resolution. Schimmelmann & Tegner (1991).

Accumulation rates of Cd, Cu, Ni, Pb, and Zn, Santa Barbara Basin; time-series 1931–1986; approximately annual resolution. Schmidt & Reimers (1991).

U_{37}^K alkenone unsaturation index, palaeotemperature, Santa Barbara Basin, depth-series, with selected marker ages between 1918 and 1984. Kennedy & Brassell (1992a).

U_{37}^K alkenone unsaturation index, palaeotemperature, Santa Barbara Basin, also concentrations of dinosterol, cholesterol, brassicasterol and alkenones. Depth-series, with selected marker ages between 1918 and 1984. Kennedy and Brassell (1992b).

Total organic carbon and its stable isotope ratio, porosity, total nitrogen, and organic sulphur, in Santa Barbara Basin sediment; time-series c. 1800–1855, with about annual resolution. Schimmelmann et al. (1992).

Concentrations of reduced sulphur species (elemental, organic, mineral sulphide), total organic carbon, total nitrogen; porosity; Santa Barbara Basin, intervals 1801–1847 and 1932–1987, most parameters with approximate annual resolution; additional shorter intervals and isolated samples as old as 1020 AD. Schimmelmann & Kastner (1993).

Sterol concentrations, C and N stable isotope ratios, inorganic compositional data, for coprolite and host sediment, 18th century Santa Barbara Basin sediment. Schimmelmann et al. (1994).

Biological/microfossil depth- and time-series

Benthic Recent foraminifers from California Borderland. Cushman (1927).

Benthic foraminifers from California Borderland; shallow-water foraminifers, Channel Islands; no age assignment. Cushman & Valentine (1930).

Benthic foraminifers from California Borderland; comparison of Recent with Pliocene foraminifers. Crouch (1952).

Benthic foraminifers from California Borderland; Recent samples; faunal gradations in 3 profiles off San Francisco, Point Conception, San Diego. Bandy (1953).

Benthic foraminifers from California Borderland; Pleistocene and Recent; benthic depth assemblages. Uchio (1960).

Foraminifers and diatoms, Santa Barbara Basin; summary, no age assignment. Hülsemann & Emery (1961).

Foraminifers, Santa Barbara Basin; summary, no age assignment. Emery & Hülsemann (1962).

Foraminifers, from grab samples and piston cores, Santa Barbara Basin. Recent samples compared with Pliocene and Miocene data. Harman (1964).

Fish scales, Santa Barbara Basin box cores; 1780–1969, with 5-year resolution. Soutar & Isaacs (1969, 1974).

Planktonic foraminifers, radiolarians, diatoms, and pteropods, Santa Barbara Basin, from box cores and gravity cores. Berger & Soutar (1970).

DDT in myctophid fish from California Borderland; CalCOFI 1949–1972; annual resolution. Macgregor (1974).

Pollen, Santa Barbara Basin; 12 000 year record at 20 m interval. Heusser (1978).

Radiolarians, Santa Barbara Basin; surface sediment and 8000 year record. Pisias (1978a, b; 1979).

Benthic foraminifers from California Borderland; 1977–78, seasonal sampling of topmost sediment. Cotton (1979).

Foraminifers from California Borderland. Summary; large-scale pattern distribution since 1975. Douglas (1979, 1981).

Benthic foraminifers from California Borderland box cores; samples 0–1 cm interval; approx. 300 years BP. Douglas & Heitman (1979).

Diatoms, silicoflagellates, California Borderland; sampling interval 1–3 cm; no timescale. Arends & Damassa (1980).

Benthic foraminifers from California Borderland; transition from live to sediment assemblage; selective preservation. Douglas *et al.* (1980).

Oxygen isotopes in planktonic foraminifers from Santa Barbara Basin, 1750–1970; sampling every 4–10 years. Dunbar (1983).

Stable isotopes in foraminifers from California Borderland. Live benthics; carbon and oxygen stable isotope ratios. Grossman (1984).

Stable isotopes in foraminifers from California Borderland. 20 000 year record, no indication of sampling interval. Gao *et al.* (1985).

Living coccolithophores from California Current; March and June 1982. Winter (1985).

Diatoms and foraminifers from Naples Bluff, Neogene. Arends & Blake (1986).

Diatoms, DSDP sites 173, 469, 470; Miocene. Barron (1986).

Radiolarians in plankton, California Current; July, August and November 1983. Carson (1986).

Radiolarians, 10 Ma BP; reconstruction of California Current. Domack (1986).

Radiolarians, Santa Barbara Basin; 1950–1968, biennial resolution. Weinheimer (1986).

Radiolarians, Santa Barbara Basin, from box cores; 1964 and 1983, plankton; also Neogene sediments. Weinheimer *et al.* (1986).

Radiolarians in plankton, California Current; 1972, 1–3 months resolution. Boltovskoy & Riedel (1987).

Diatoms, radiolarians, and foraminifers, Santa Barbara Basin; selected intervals (1977–1985), about annual resolution. Lange *et al.* (1987).

Radiolarians and varve thickness, Santa Barbara Basin; 1935–1986, biennial resolution. Baumgartner *et al.* (1989).

Radiolarians, Santa Barbara Basin; 1874–1985, biennial resolution. Casey *et al.* (1989).

Benthic foraminifers from California Borderland box cores; downcore distribution of live and dead species. Mackensen & Douglas (1989).

Bacteria, nematodes, harpacticoida, and other meiofauna, Santa Barbara Basin; 1985–1986, seasonal resolution. Montagna *et al.* (1989).

Diatoms and planktonic foraminifers; Santa Barbara Basin 1950–1986 yearly resolution. Lange *et al.* (1990).

Diatoms, benthic foraminifers, and porewater analyses, Santa Barbara Basin; 1988, seasonal resolution. Reimers *et al.* (1990).

Fish scales, Santa Barbara Basin; time-series of scale-deposition from 270 to 1970 AD. Baumgartner *et al.* (1992).

Pollen from from Newport Bay. 7000-year chronology. Davis (1992).

Foraminifers and diatoms, Santa Barbara Basin; time-series 1800 to 1850 AD, annual resolution. Schimmelmann *et al.* (1992).

Pollen and sedimentary chemical parameters from Santa Rosa Island; 5200 year record. Cole & Liu (1994).

Foraminifers, Santa Barbara Basin; annual resolution, approx. 1815 to 1850. Burke *et al.* (1995).

Charcoal, Santa Barbara Basin; approx. 1.5 ka record with about decadal resolution; X-radiographic varve record; Byrne *et al.* (unpublished).

References

ANDERSON, R. Y. 1990. Solar-cycle modulations of ENSO: a possible source of climatic change. *In*: BETANCOURT J. L. & MACKAY A. M. (eds) *Proceedings of the Sixth Annual Pacific Climate (PACLIM) Workshop,* March 5–8, 1989. Interagency Ecological Studies Program for the Sacramento-San Joaquin Estuary Technical Report, **23**, 77–81.

——, SOUTAR, A. & JOHNSON, T. C. 1992. Long-term changes in El Niño/Southern Oscillation: evidence from marine and lacustrine sediments. *In*: DIAZ, H. F. & MARKGRAF, V. (eds) *El Niño. Historical and Paleoclimatic Aspects of the Southern Oscillation,* Chapter 20. Cambridge University Press, 419–433.

ANGEL, M. V., SMITH, R. L. & SMALL, L. F. 1992. California basin studies: physical, geological, chemical and biological attributes. *Progress in Oceanography,* **30**, 1–398.

ARENDS, R. G. & BLAKE, G. H. 1986. Biostratigraphy and paleoecology of the Naples Bluff coastal section based on diatoms and benthic foraminifera. *In*: CASEY, R. E. & BARRON, J. A. (eds) *Siliceous Microfossil and Microplankton of the Monterey Formation and Modern Analogs.* Pacific Section, Society of Economic Paleontologists and Mineralogists, Los Angeles, 39–54.

—— & DAMASSA, S. P. 1980. Diatoms, silicoflagellates and palynomorphs from Holocene sediments of basins in the southern California Continental Borderland. *In*: FIELD, M. E. BOUMA, A. H., COLBURN, I. P., DOUGLAS, R. G. & INGLE, J. C. (eds) *Quaternary Depositional Environments of the Pacific Coast.* Pacific Coast Paleogeography Symposium 4. Pacific Section Society of Economic Paleontologists and Mineralogists, Los Angeles, 313–324.

BANDY, O. L. 1953. Ecology and paleoecology of some California foraminifera. Part I. The frequency distribution of recent foraminifera off California. *Journal of Paleontology,* **27**, 161–183.

BARNES, R. O., BERTINE, K. K. & GOLDBERG, E. D. 1975. N_2 : Ar, nitrification and denitrification in southern California Borderland basin sediments. *Limnology and Oceanography,* **20**, 962–970.

BARRON, J. A. 1986. Updated diatom biostratigraphy for the Monterey Formation of California. *In*: CASEY, R. E. & BARRON, J. A. (eds) *Siliceous Microfossil and Microplankton of the Monterey Formation and Modern Analogs.* Pacific Section Society of Economic Paleontologists and Mineralogists, Los Angeles, 39–54.

BAUER, J. E., SPIES, R. B., VOGEL, J. S., NELSON, D. E. & SOUTHON, J. R. 1990. Radiocarbon evidence of fossil-carbon cycling in sediments of a nearshore hydrocarbon seep. *Nature,* **348**, 230–232.

BAUMGARTNER, T. R., MICHAELSON, J., THOMPSON, L. G., SHEN, G. T., SOUTAR, A. & CASEY, R. E. 1989. The recording of interannual climatic change by high-resolution natural systems: tree-rings, coral bands, glacial ice layers, and marine varves. *In*: PETERSON, D. H. (ed.) *Aspects of*

Climate variability in the Pacific and the Western Americas. American Geophysical Union, Geophysical Monograph, **55**, 1–14.

——, SOUTAR, A. & FERREIRA-BARTRINA, V. 1992. Reconstruction of the history of the Pacific sardine and northern anchovy populations over the past two millennia from sediments of the Santa Barbara basin, California. *California Cooperative Oceanic Fisheries Investigations Report*, **33**, 24–40.

BEERS, J. R. 1986. Organisms and the food web. *In*: EPPLEY, R. W. (ed.) *Plankton Dynamics of the Southern California Bight*. Lecture Notes on Coastal and Estuarine Studies, **15**, 84–175.

BEHL, R. J. 1995. Sedimentary facies and sedimentology of the late Quaternary Santa Barbara Basin, ODP Site 893. *In*: KENNETT, J. P., BALDAUF, J. G. & LYLE, M. (eds) *Proceedings of the Ocean Drilling Program (Scientific Results) 146 (Part 2)*, College Station, TX (Ocean Drilling Program), 295–308.

——1995. Detailed lithostratigraphy of Hole 893A, Santa Barbara Basin, southern California continental borderland (USA). *In*: KENNETT, J. P., BALDAUF, J. G. & LYLE, M. (eds) *Proceedings of the Ocean Drilling Program (Scientific Results) 146 (Part 2)*, College Station, TX (Ocean Drilling Program), 347–351.

—— & KENNETT, J. P. 1996. Brief interstadial events in the Santa Barbara basin, NE Pacific, during the past 60 kyr. *Nature*, **379**, 243–246.

BERGER, W. H. 1971. Sedimentation of planktonic foraminifera. *Marine Geology*, **11**, 325–358.

—— & SOUTAR, A. 1967. Planktonic foraminifera: field experiment on production rate. *Science*, **156**, 1495–1497.

—— & ——1970. Preservation of planktonic shells in an anaerobic basin off California. *Bulletin Geological Society of America*, **81**, 275–282.

——, SMETACEK, V. S. & WEFER, G. (eds) 1989. *Productivity of the Oceans: Present and Past*. Report of the Dahlem Workshop, Berlin 24–29 April 1988. Wiley, Chichester.

BERNHARD, J. M. 1992. Benthic foraminiferal distribution and biomass related to porewater oxygen content: central California continental slope and rise. *Deep-Sea Research*, **39**, 585–605.

—— & REIMERS, C. E. 1991. Benthic foraminiferal population fluctuations related to anoxia: Santa Barbara basin. *Biogeochemistry*, **15**, 127–149.

BOLTOVSKOY, D. & RIEDEL, W. R. 1987. Polycystine radiolaria of the California Current region: seasonal and geographic patterns. *Marine Micropaleontology*, **12**, 65–104.

BRÜCHERT, V. & PRATT, L. M. 1994. Stratigraphic variation in abundance and isotopic composition of sulfur species from ODP Site 893, Santa Barbara basin (abstract). *EOS, Transactions, American Geophysical Union, 1994 Spring Meeting*, **75**(16), 201.

BRULAND, K. W. 1974. *Pb-210 Geochronology in the Coastal Marine Environment*. PhD Dissertation, University of California, San Diego.

——, FRANKS, R. P., BELLOWS, J. & SOUTAR, A. 1979. Age Dating of Box Cores. *Southern California Baseline Study, Year 2 vol. 2 Final Report*, Report No. 21.0, SAI-77-917-LJ, Bureau of Land Management, U. S. Dept. Interior.

BULL, D. & KEMP, A. E. S. 1995. Composition and origins of laminae in late Quaternary and Holocene sediments from the Santa Barbara Basin. *In*: KENNETT, J. P., BALDAUF, J. G. & LYLE, M. (eds) *Proceedings of the Ocean Drilling Program, (Scientific Results) 146 (Part 2)*. College Station, TX (Ocean Drilling Program), 77–87. (reprinted in this volume)

BURKE, S. K., DUNBAR, R. B. & BERGER, W. H. 1995. Benthic and pelagic foraminifera of the *Macoma* layer, Santa Barbara Basin. *Journal of Foraminiferal Research*, **25**, 117–133.

BYRNE, R., MICHAELSON, J. & SOUTAR, A. (unpublished). Prehistoric Wildfire Frequencies in the Los Padres National Forest: Fossil Charcoal Evidence from the Santa Barbara Channel. Unpublished report to the U.S. Dept. of Agriculture, Forest Service, University of California at Berkeley.

CALIFORNIA COOPERATIVE OCEANIC FISHERIES INVESTIGATIONS (CalCOFI) Reports.

CARSON, T. L. 1986. Radiolarian response to the 1983 California El Niño. *In*: CASEY, R. E. & BARRON, J. A. (eds) *Siliceous Microfossil and Microplankton of the Monterey Formation and Modern Analogs*. Pacific Section Society of Economic Paleontologists and Mineralogists, Los Angeles, 9–19.

CASEY, R. E., WEINHEIMER, A. L. & NELSON, C. O. 1989. California El Niños and related changes of the California Current system from recent and fossil radiolarian records. *In*: PETERSON, D. H. (ed.) *Aspects of Climate Variability in the Pacific and the Western Americas*. Geophysical Monograph 55, American Geophysical Union, 85–92.

CHANT, L. A. & CORNETT, R. J. 1991. Smearing of gravity core profiles in soft sediments. *Limnology and Oceanography*, **36**, 1492–1498.

CHEN, R. F. & BADA, J. L. 1989. Seawater and porewater fluorescence in the Santa Barbara Basin. *Geophysical Research Letters*, **16**, 687–690.

CHOW, T. J., BRULAND, K. W., BERTINE, K., SOUTAR, A., KOIDE, M. & GOLBERG, E. D. 1973. Lead pollution: records in southern California coastal sediments. *Science*, **181**, 551–552.

COLE, K. L. & LIU, G. 1994. Holocene paleoecology of an estuary on Santa Rosa Island, California. *Quaternary Research*, **41**, 326–335.

COTTON, M. L. 1979. *Seasonality in Benthic Foraminiferal Populations of the Southern California Borderland*. MSc thesis, University of Southern California.

CRAVEN, D. B. & JAHNKE, R. A. 1992. Microbial utilization and turnover of organic carbon in Santa Monica Basin sediments. *In*: ANGEL, M. V., SMITH, R. L. & SMALL, L. F. (eds) *California Basin Studies: Physical, Geological, Chemical, and Biological Attributes. Progress in Oceanography*, **30**, 313–333.

CRISP, P. T., BRENNER, S., VENKATESAN, M. I., RUTH, E. & KAPLAN, I. R. 1979. Organic chemical characterization of sediment-trap particulates from San Nicolas, Santa Barbara, Santa Monica and San Pedro Basins, California. *Geochimica et Cosmochimica Acta*, **43**, 1791–1801.

CROUCH, R. W. 1952. Significance of temperature on foraminifera from deep basins off southern California coast. *AAPG Bulletin*, **36**, 807–843.

CRUSIUS, J. & ANDERSON, R. F. 1992. Inconsistencies in accumulation rates of Black Sea sediments inferred from records of laminae and [210]Pb. *Paleoceanography*, **7**, 215–227.

CUSHMAN, J. A. 1927. Recent foraminifera off the west coast of America. *Bulletin Scripps Institution of Oceanography, Technical Series*, **1**, 119–188.

—— & VALENTINE, W. W. 1930. Shallow-water foraminifera from the Channel Islands of southern California. *Contributions from the Department of Geology of Stanford University*, **1**, 1–51.

DADE, W. B., NOWELL, A. R. M. & JUMARS, P. A. 1992. Predicting erosion resistance of muds. *Marine Geology*, **105**, 285–297.

DALE, B. 1983. Dinoflagellate resting cysts: "benthic plankton". *In*: FRYXELL, G. A. (ed.) *Survival Strategies of the Algae*. Cambridge University Press, 49–68.

DAVIS, O. K. 1992. Rapid climatic change in coastal southern California inferred from pollen analysis of San Joaquin marsh. *Quaternary Research*, **37**, 89–100.

DEGENS, E. T., PRASHNOWSKY, A., EMERY, K. O. & PIMENTA, J. 1961. Organic materials in recent and ancient sediments. Part II: Amino acids in marine sediments of Santa Barbara Basin, California. *Neues Jahrbuch für Geologie und Paläontologie, Monatshefte*, 413–426.

——, REUTER, J. H. & SHAW, K. N. F. 1964. Biochemical compounds in offshore California sediments and sea waters. *Geochimica et Cosmochimica Acta*, **28**, 45–66.

DEMASTER, D. J. & TUREKIAN, K. K. 1987. The radiocarbon record in varved sediments of Carmen basin, Gulf of California: a measure of upwelling intensity variation during the past several hundred years. *Paleoceanography*, **2**, 249–254.

DESER, C. & WALLACE, J. M. 1987. El Niño events and their relation to the Southern Oscillation: 1925–1986. *Journal of Geophysical Research*, **92**, 14 189–14 196.

DOMACK, C. R. 1986. Reconstruction of the California Current at 5, 8, and 10 million years b.p. using radiolarian indicators. *In*: CASEY R. E. & BARRON, J. A. (eds) *Siliceous Microfossil and Microplankton of the Monterey Formation and Modern Analogs*. Pacific Section Society of Economic Paleontologists and Mineralogists, Los Angeles, 39–54.

DOOSE, P. R. 1980. *The Bacterial Production of Methane in Marine Sediments*. PhD dissertation, University of California, Los Angeles.

DOUGLAS, R. G. 1979. Benthic foraminiferal ecology and paleoecology: a review of concepts and methods. *In*: *Foraminiferal Ecology and Paleoecology*. Society of Economic Paleontologists and Mineralogists Short Course No. 6 Houston, Texas, 21–53.

——1981. Paleoecology of continental margin basins: a modern case history from the borderland of southern California. *In*: *Depositional Systems of Active Continental Margin Basins*. Society of Economic Paleontologists and Mineralogists, Pacific Section, Short Course Notes (San Francisco, May 1981), 121–156.

—— & HEITMAN, H. L. 1979. Slope and basin benthic foraminifera of the California Borderland. *Society of Economic Paleontologists and Mineralogists Special Publication*, **27**, 231–246.

——, LIESTMAN, J., WALCH, C., BLAKE, G. & COTTON, M. L. 1980. The transition from live to sediment assemblage in benthic foraminifera from the southern California Borderland. *In*: FIELD, M. E., BOUMA, A. H., COLBURN, I. P., DOUGLAS, R. G. & INGLE, J. C. (eds) *Quaternary Depositional Environments of the Pacific Coast*. Pacific Coast Paleogeography Symposium 4. Pacific Section Society of Economic Paleontologists and Mineralogists, Los Angeles, 257–280.

DRAKE, D. E. 1971. Suspended sediment and thermal stratification in Santa Barbara Channel, California. *Deep-Sea Research*, **18**, 763–769.

——, FLEISCHER, P. & KOLPACK, R. L. 1971. Transport and deposition of flood sediment, Santa Barbara Channel, California. *In*: KOLPACK, R. L. (ed.) *Survey of the Santa Barbara Channel Oil Spill 1969–1970*, Vol. 2. Alan Hancock Foundation, University of Southern California, Los Angeles, 181–217.

——, KOLPACK, R. L. & FISCHER, P. J. 1972. Sediment transport on the Santa Barbara-Oxnard shelf, Santa Barbara channel, California. *In*: SWIFT, D. J. P., DUANE, D. B. & PILKEY, O. H. (eds) *Shelf Sediment Transport: Process and Pattern*. Dowden, Hutchinson & Ross, Stroudsburg, Pennsylvania, 307–331.

DUNBAR, R. B. 1981. *Sedimentation and the History of Upwelling and Climate in High Fertility Areas of the Northeastern Pacific Ocean*. PhD dissertation, University of California, San Diego.

——1983. Stable isotope record of upwelling and climate from Santa Barbara basin, California. *In*: THIEDE, J. & SUESS, E. (eds) *Coastal Upwelling, its Sedimentary Record. Part B: Sedimentary Records of Ancient Coastal Upwelling*. Plenum, New York, 217–246.

—— & BERGER, W. H. 1981. Fecal pellet flux to modern bottom sediment of Santa Barbara Basin (California) based on sediment trapping. *Bulletin Geological Society of America*, **92**, 212–218.

DYMOND, J., FISHER, K., CLAUSON, M., COBLER, R., GARDNER, W., RICHARDSON, M. J., BERGER, W., SOUTAR, A. & DUNBAR, R. 1981. A sediment trap intercomparison study in the Santa Barbara basin. *Earth and Planetary Science Letters*, **53**, 409–418.

EGANHOUSE, R. P. & VENKATESAN, M. I. 1993. Chemical oceanography and geochemistry. *In*: DAILEY, M. D., REISH, D. J. & ANDERSON, J. W. (eds) *Ecology of the Southern California Bight: A Synthesis and Interpretation*. University of California Press, Berkeley, 71–189.

EMERY, K. O. 1960. *The Sea off Southern California: A Modern Habitat of Petroleum*. Wiley, New York.

—— & BRAY, E. E. 1962. Radiocarbon dating of California basin sediments. *AAPG Bulletin*, **46**, 1839–1856.

—— & HÜLSEMANN, J. 1962. The relationships of sediments, life and water in a marine basin. *Deep-Sea Research*, **8**, 165–180.

—— & RITTENBERG, S. C. 1952. Early diagenesis of California basin sediments in relation to origin of oil. *AAPG Bulletin*, **36**, 735–806.

EPPLEY, R. W. (ed.) 1986a. *Plankton dynamics of the Southern California Bight*. Lecture Notes on Coastal Estuarine Studies, **15**, Springer, Berlin.

——1986b. People and the plankton. *In*: EPPLEY, R. W. (ed.) *Plankton Dynamics of the Southern California Bight*. Lecture Notes on Coastal and Estuarine Studies, **15**, 289–303.

——1992. Chlorophyll, photosynthesis and new production in the Southern California Bight. *In*: ANGEL, M. V., SMITH, R. L. & SMALL, L. F. (eds) *California Basin Studies: Physical, Geological, Chemical and Biological Attributes, Progress in Oceanography*, **30**, 117–150.

—— & HOLM-HANSEN, O. 1986. Primary productivity in the Southern California Bight. *In*: EPPLEY, R. W. (ed.) *Plankton Dynamics of the Southern California Bight. Lecture Notes on Coastal and Estuarine Studies*, **15**, 176–215.

FISCHER, K. M. 1993. The microfabric of Santa Barbara basin surface sediment. *American Association of Petroleum Geologists Hedberg Conference Abstract Volume*, April 22nd–23rd 1993, 21.

FISHER, C. G. 1990. *Bibliography and Inventory of Holocene Varved and Laminated Marine Sediments*. NOAA Paleoclimate Publications Series Report No. 1 Boulder, Colorado.

FLEISCHER, P. 1972. Mineralogy and sedimentation history, Santa Barbara basin, California. *Journal of Sedimentary Petrology*, **42**, 49–58.

GAO, L., EMERY, K. O. & KEIGWIN, L. D. 1985. Late Quaternary stable isotope paleoceanography off southern California. *Deep-Sea Research*, **32**, 1469–1484.

GORSLINE, D. S. & TSENG, L. S. Y. 1989. The California Continental Borderland. *In*: WINTERER, E. L., HUSSONG, D. M. & DECKER, R. W. (eds) *The Geology of North America. The Eastern Pacific Ocean and Hawaii*, Vol. N, Chapter 24, Geological Society of America, Boulder, Colorado, 471–487.

——, KOLPACK, R. L., KARL, H. A., DRAKE, D. E., FLEISCHER, P., THORNTON, S. E., SCHWALBACH, J. R. & SAVRDA, C. E. 1984. Studies of fine-grained sediment transport processes and products in the California Continental Borderland. *In*: STOW, D. A. V. & PIPER, D. J. W. (eds) *Fine-Grained Sediments: Deep-Water Processes and Facies*. Blackwell, Oxford, 395–415.

GRANT, C. W. 1991. *Distribution of Bacterial Mats (Beggiatoa spp.) in Santa Barbara Basin, California: A Modern Analog for Organic-rich Facies of the Monterey Formation*. MS Thesis, California State University, Long Beach.

GRIMM, K. A. 1992. Preparation of weakly consolidated, laminated hemipelagic sediment for high-resolution visual microanalysis: an analytical method. *In*: PISCIOTTO, K. A., INGLE, J. C., JR., VON BREYMANN, M. T. & BARRON, J. (eds) *Proceedings of the Ocean Drilling Program, (Scientific Results)*, **127/128**, Part I College Station (TX) 57–62.

GROSSMAN, E. L. 1984. Stable isotope fractionation in live benthic foraminifera from the southern California Borderland. *Palaeogeography, Palaeoclimatology, Palaeoecology*, **47**, 301–327.

GROVE, J. M. 1988. *The Little Ice Age*. Routledge, London.

HALLAM, A. 1967. The depth significance of shales with bituminous laminae. *Marine Geology*, **5**, 481–493.

HAMILTON, R. M., YERKES, R. F., BROWN, R. D., JR., BURFORD, R. O. & DeNOYER, J. M. 1969. Seismicity and associated effects, Santa Barbara region. *U.S. Geological Survey Professional Paper* no. 679-D, 47–71.

HARMAN, R. A. 1964. Distribution of foraminifera in the Santa Barbara basin, California. *Micropaleontology*, **10**, 81–96.

HEATH, G. R., MOORE, T. C., JR. & DAUPHIN, J. P. 1977. Organic carbon in deep-sea sediments. *In*: ANDERSON, N. R. & MALAHOFF, A. (eds) *The Fate of Fossil Fuel CO_2 in the Oceans*. Plenum, New York, 605–625.

HERBERT, T. D., YASUDA, M. & BURNETT, C. 1995. Glacial-interglacial sea–surface temperature record inferred from alkenone unsaturation indices; Site 893, Santa Barbara Basin. *In*: KENNETT, J. P., BALDAUF, J. G. & LYLE, M. (eds) *Proceedings of the Ocean Drilling Program, (Scientific Results)* **146** (Part 2), College Station, TX (Ocean Drilling Program), 257–264.

HEUSSER, L. E. 1995. Pollen stratigraphy and paleoecologic interpretation of 160-k.y. record from Santa Barbara Basin, Hole 893A. *In*: KENNETT, J. P., BALDAUF, J. G. & LYLE, M. (eds) *Proceedings of the Ocean Drilling Program, (Scientific Results)* **146** (Part 2), College Station, TX (Ocean Drilling Program), 265–280.

——1978. Pollen in Santa Barbara basin, California: A 12,000-yr record. *Geological Society of America Bulletin*, **89**, 673–678.

HEWITT, R. P. 1988. Oceanographic Approach to Fishery Research. *California Cooperative Oceanic Fisheries Investigations Report*, **29**, 27–41.

HOM, W., RISEBROUGH, R. W., SOUTAR, A. & YOUNG, D. R. 1974. Deposition of DDE and polychlorinated biphenyls in dated sediments of the Santa Barbara basin. *Science*, **184**, 1197–1199.

HÜLSEMANN, J. & EMERY, K. O. 1961. Stratification in recent sediments of Santa Barbara Basin as controlled by organisms and water character. *Journal of Geology*, **69**, 279–290.

IMBRIE, J. & KIPP, N. G. 1971. A new micropaleonto-logical method for quantitative paleoclimatology. Application to a late Pleistocene Caribbean core. *In*: TUREKIAN, K. (ed.) *The Late Cenozoic Glacial Ages*, Chapter 5, Yale University Press, New Haven, Connecticut, 71–181.

INGRAM, B. L. & KENNETT, J. P. 1995. Radiocarbon chronology and planktonic benthic foraminiferal ^{14}C age differences in Santa Barbara Basin sediments, Hole 893A. *In*: KENNETT, J. P., BALDAUF, J. G. & LYLE, M. (eds) *Proceedings of the Ocean Drilling Program, (Scientific Results)* **146** (Part 2), College Station, TX (Ocean Drilling Program) 19–27.

KALIL, E. K. 1976. *The Distribution and Geochemistry of Uranium in Recent and Pleistocene Muds*. PhD dissertation, University of California, Los Angeles.

KAPLAN, I. R., EMERY, K. O. & RITTENBERG, S. C. 1963. The distribution and isotopic abundance of sulphur in recent marine sediments off southern California. *Geochimica et Cosmochimica Acta*, **27**, 297–331.

KENNEDY, J. A. & BRASSELL, S. C. 1992a. Molecular records of twentieth century El Niño events in laminated sediments from the Santa Barbara basin. *Nature*, **357**, 62–64.

—— & ——1992b. Molecular stratigraphy of the Santa Barbara basin: comparison with historical records of annual climate change. *Organic Geochemistry* **19**, 235–244.

KENNETT, J. P. 1995. Latest Quaternary benthic oxygen and carbon isotope stratigraphy Hole 893A, Santa Barbara Basin, California. *In*: KENNETT, J. P., BALDAUF, J. G. & LYLE, M. (eds) *Proceedings of the Ocean Drilling Program, (Scientific Results)* **146** (Part 2), College Station, TX (Ocean Drilling Program), 3–18.

—— & BEHL, R. J. 1993. The Channel Islands Marine Sanctuary through the last ice ages. *Alolkoy*, **6**(2), 14–15.

—— & INGRAM, B. L. 1995. Paleoclimatic evolution of Santa Barbara Basin during the last 20 k.y.: marine evidence from Hole 893A. *In*: KENNETT, J. P., BALDAUF, J. G. & LYLE, M. (eds) 1995. *Proceedings of the Ocean Drilling Program, (Scientific Results)* **146** (Part 2), College Station, TX (Ocean Drilling Program), 309–325.

—— & VENZ, K. 1995. Late Quaternary climatically related planktonic foraminiferal assemblage changes: Hole 893A, Santa Barbara Basin, California. *In*: KENNETT, J. P., BALDAUF, J. G. & LYLE, M. (eds) 1995. *Proceedings of the Ocean Drilling Program, (Scientific Results)* **146** (Part 2), College Station, TX (Ocean Drilling Program), 281–294.

——, BALDAUF, J. G. & LYLE, M. (eds) 1995. *Proceedings of the Ocean Drilling Program, (Scientific Results)* **146** (Part 2), College Station, TX (Ocean Drilling Program).

KIPP, N. G. & TOWNER, D. P. 1975. The last millennium of climate: foraminiferal records from coastal basin studies. *In*: *Proceedings of the WMO/IAMAP Symposium on Long-term Fluctuations, Norwich, 18–23 August 1975*. World Meteorological Organization Contribution No. 421, Geneva, Switzerland, 119–126.

KOIDE, M., SOUTAR, A. & GOLDBERG, E. D. 1972. Marine geochronology with ^{210}Pb. *Earth and Planetary Science Letters*, **14**, 422–446.

——, BRULAND, K. W. & GOLDBERG, E. D. 1973. Th-228/Th-232 and Pb-210 geochronologies in marine and lake sediments. *Geochimica et Cosmochimica Acta*, **37**, 1171–1187.

——, —— & ——1976. ^{226}Ra chronology of a coastal marine sediment. *Earth and Planetary Science Letters*, **31**, 31–36.

——, GOLDBERG, E. D. & HODGE, V. F. 1980. ^{241}Pu and ^{241}Am in sediments from coastal basins off California and Mexico. *Earth and Planetary Science Letters*, **48**, 250–256.

KOSIUR, D. R. & WARFORD, A. L. 1979. Methane production and oxidation in Santa Barbara Basin sediments. *Estuarine and Coastal Marine Science*, **8**, 379–385.

KRISHNASWAMI, S., LAL, D., AMIN, B. S. & SOUTAR, A. 1973. Geochronological studies in Santa Barbara Basin: ^{55}Fe as a unique tracer for particulate setting. *Limnology and Oceanography*, **18**, 763–770.

LAJAT, M., SALIOT, A. & SCHIMMELMANN, A. 1990. Free and bound lipids in recent (1835–1987) sediments from Santa Barbara Basin. *Organic Geochemistry*, **16**, 793–803.

LANGE, C. B. & SCHIMMELMANN, A. 1994. Seasonal resolution of laminated sediments in Santa Barbara Basin: Its significance in paleoclimatic studies. *In*: REDMOND, K. T. & THARP, V. L. (eds) *Proceedings of the Tenth Annual Pacific Climate (PACLIM) Workshop*, April 4–7, 1993, California Department of Water Resources, Interagency Ecological Studies Program Technical Report **36**, 83–92.

——, BERGER, W. H., BURKE, S. K., CASEY, R. E., SCHIMMELMANN, A., SOUTAR, A. & WEINHEIMER, A. L. 1987. El Niño in Santa Barbara Basin: diatom, radiolarian and foraminiferan responses to the "1983 El Niño" event. *Marine Geology*, **78**, 153–160.

——, BURKE, S. K. & BERGER, W. H. 1990. Biological production off southern California is linked to climatic change. *Climatic Change*, **16**, 319–329.

—— & SCHIMMELMANN, A. 1995. X-radiography of selected, predominantly varved intervals at Hole 893A. *In*: KENNETT, J. P., BALDAUF, J. G. & LYLE, M. (eds) *Proceedings of the Ocean Drilling Program, (Scientific Results)* **146** (Part 2), College Station, TX (Ocean Drilling Program), 333–346.

LEONARD, E. 1990. An assessment of sediment loss and distortion at the top of short gravity cores. *Sedimentary Geology*, **66**, 57–63.

MACGREGOR, J. S. 1974. Changes in the amount and proportions of DDT and its metabolites, DDE and DDD, in the marine environment off southern California 1949–72. *Fishery Bulletin*, **72**, 275–293.

MACKENSEN, A. & DOUGLAS, R. G. 1989. Down-core distribution of live and dead deep-water benthic foraminifera in box cores from the Weddell Sea and the California Continental Borderland. *Deep-Sea Research*, **36**, 879–900.

MARSAGLIA, K. M., RIMKUS, K. C. & BEHL, R. J. 1995. Provenance of sand deposited in the Santa Barbara Basin. *In*: KENNETT, J. P., BALDAUF, J. G. & LYLE, M. (eds) *Proceedings of the Ocean Drilling Program, (Scientific Results)* **146** (Part 2), College Station, TX (Ocean Drilling Program), 61–76.

MCGOWAN, J. A. 1985. El Niño 1983 in the Southern California Bight. *In*: WOOSTER, W. & FLUHARTY, D. L. (eds) *El Niño North. Niño Effects in the Eastern Subarctic Pacific Ocean.* Washington Sea Grant Program, University of Washington, Seattle, 166–184.

MENSING, S. 1993. *The Impact of European Settlement on Oak Woodlands and Fire: Pollen and Charcoal Evidence from the Transverse Ranges, California.* PhD Thesis. University of California–Berkeley, Berkeley, CA.

MONTANGNA, P. A., BAUER, J. E., HARDIN, D. & SPIES, R. B. 1989. Vertical distribution of microbial and meiofaunal populations in sediments of a natural coastal hydrocarbon seep. *Journal of Marine Research*, **47**, 657–680.

ORR, W. L., EMERY, K. O. & GRADY, J. R. 1958. Preservation of chlorophyll derivatives in sediments off southern California. *AAPG Bulletin*, **42**, 925–962.

PETERS, K. E., SWEENEY, R. E. & KAPLAN, I. R. 1978. Correlation of carbon and nitrogen stable isotope ratios in sedimentary organic matter. *Limnology and Oceanography*, **23**, 598–604.

PHLEGER, F. B. & SOUTAR, A. 1973. Production of benthic foraminifera in three east Pacific oxygen minima. *Micropaleontology*, **19**, 110–115.

PIKE, J. & KEMP, A. E. S. 1996. Preparation and analysis techniques for studies of laminated sediments. *This volume.*

PISIAS, N. G. 1978a. *Paleoceanography of the Santa Barbara Basin and the California Current During the Last 8,000 Years.* PhD dissertation, University of Rhode Island.

——1978b. Paleoceanography of the Santa Barbara basin during the last 8000 years. *Quaternary Research*, **10**, 366–384.

——1979. Model for paleoceanographic reconstructions of the California Current during the last 8000 years. *Quaternary Research*, **11**, 373–386.

PRASHNOWSKY, A., DEGENS, E. T., EMERY, K. O. & PIMENTA, J. 1961. Organic materials in recent and ancient sediments. Part I: sugars in marine sediment of Santa Barbara Basin, California. *Neues Jahrbuch für Geologie und Paläontologie, Monatshefte*, 400–413.

QUINN, W. H. 1992. A study of southern oscillation-related climatic activity for A.D. 622–1900 incorporating Nile River flood data. *In*: DIAZ, H. F. & MARKGRAF, V. (eds) *El Niño: Historical and Paleoclimatic Aspects of the Southern Oscillation.* Cambridge University Press, New York, 119–149.

—— & NEAL, V. T. 1992. The historical record of El Niño events. *In*: BRADLEY, R. S. & JONES, P. D. (eds) *Climate Since A.D. 1500.* Routledge, London, 623–648.

REIMERS, C. E., LANGE, C. B., TABAK, M. & BERNHARD, J. M. 1990. Seasonal spillover and varve formation in the Santa Barbara Basin, California. *Limnology and Oceanography*, **35**, 1577–1585.

RITTENBERG, S. C., EMERY, K. O. & ORR, W. L. 1955. Regeneration of nutrients in sediments of marine basins. *Deep-Sea Research*, **3**, 23–45.

ROBINSON, S. W. & THOMPSON, G. 1981. Radiocarbon corrections for marine shell dates with application to southern Pacific Northwest Coast prehistory. *Syesis*, **14**, 45–57.

SAUTER, L. R. & THUNELL, R. C. 1991a. Seasonal variability in the $\delta^{18}O$ and $\delta^{13}C$ of planktonic foraminifera from an upwelling environment: sediment trap results from San Pedro Basin, Southern California Bight. *Paleoceanography*, **6**, 307–334.

—— & —— 1991b. Planktonic foraminiferal response to upwelling and seasonal hydrographic conditions: sediment trap results from San Pedro Basin, Southern California Bight. *Journal of Foraminiferal Research*, **21**, 347–363.

SAVRDA, C. E. & BOTTJER, D. J. 1986. Trace-fossil model for reconstruction of paleo-oxygenation in bottom waters. *Geology*, **14**, 3–6.

SCHAEFER, H. 1989. Improving southern California's coastal waters. *Journal of the Water Pollution Control Federation*, **61**, 1395–1401.

SCHIMMELMANN, A. & KASTNER, M. 1993. Evolutionary changes over the last 1000 years of reduced sulfur phases and organic carbon in varved sediments of the Santa Barbara Basin, California. *Geochimica et Cosmochimica Acta*, **57**, 67–78.

—— & TEGNER, M. J. 1991. Historical oceanographic events reflected in $C/^{12}C$ ratio of total organic carbon in Santa Barbara Basin sediment. *Global Biogeochemical Cycles*, **5**, 173–188.

—— & ——1992. Historical evidence of abrupt coastal climatic change in Southern California, 1790–1880. *In*: K. T. REDMOND (ed.) *Proceedings of the Eighth Annual Pacific Climate (PACLIM) Workshop, March 10–13, 1991*, California Department of Water Resources, Interagency Ecological Studies Program Technical Report, **31**, 47–56.

——, LANGE, C. B. & BERGER, W. H. 1990a. Climatically controlled marker layers in Santa Barbara Basin sediments, and fine-scale core-to-core correlation. *Limnology and Oceanography*, **35**, 165–173.

——, ——, MICHAELSEN, J. & BERGER, W. H. 1990b. Climatic changes reflected in laminated Santa Barbara sediments. *In*: BETANCOURT, J. L. & MACKAY, A. M. (eds) *Proceedings of the Sixth Annual Pacific Climate (PACLIM) Workshop, March 5–8, 1989*, Interagency Ecological Studies Program for the Sacramento-San Joaquin Estuary Technical Report, **23**, 97–99.

——, ——, BERGER, W. H., SIMON, A., BURKE, S. K. & DUNBAR, R. B. 1992. Extreme climatic conditions recorded in Santa Barbara Basin laminated sediments: the 1835–1840 *Macoma* event. *Marine Geology*, **106**, 279–299.

——, SCHUFFERT, J. D., VENKATESAN, M. I., LEATHER, J., LANGE, C. B., BATURIN, G. N. & SIMON, A. 1994. Biogeochemistry and origin of a phosphoritized coprolite from anoxic sediment of the Santa Barbara Basin. *Journal of Sedimentary Research*, **A64**, 771–777.

SCHMIDT, H. & REIMERS, C. E. 1991. The Recent history of trace metal accumulation in the Santa Barbara Basin, southern California Borderland. *Estuarine, Coastal and Shelf Science*, **33**, 485–500.

SCHWALBACH, J. R. & GORSLINE, D. S. 1985. Holocene sediment budgets for the basins of the California Continental Borderland. *Journal of Sedimentary Petrology*, **55**, 829–842.

SHILLER, A. M. 1982. *The Geochemistry of Particulate Major Elements in the Santa Barbara Basin and Observations on the Calcium Carbonate Carbon Dioxide System in the Ocean*. PhD dissertation, University of California, San Diego.

Shore-based Scientific Party 1994. Site 893. *In*: *Proceedings of the Ocean Drilling Program, Initial Reports*, **146** (Part 2), College Station, TX (Ocean Drilling Program), 15–50.

SHOLKOVITZ, E. R. 1973. Interstitial water chemistry of the Santa Barbara Basin sediments. *Geochimica et Cosmochimica Acta*, **37**, 2043–2073.

—— & SOUTAR, A. 1975. Changes in the composition of the bottom water of the Santa Barbara Basin: effect of turbidity currents. *Deep-Sea Research*, **22**, 13–21.

SILVER, M. W. & ALLDREDGE, A. L. 1981. Bathypelagic marine snow: deepsea algal and detrital community. *Journal of Marine Research*, **39**, 501–530.

SIMONEIT, B. R. T. & MAZUREK, M. A. 1981. Organic geochemistry of sediments from the southern California Borderland. *In*: YEATS, R. S., HAQ, B. U. *et al.* (cds) *Initial Reports of the Deep Sea Drilling Project*, **63**, United States Government Printing Office, Washington, 837–852.

SOUTAR, A. 1967. The accumulation of fish debris in certain California coastal sediments. *California Cooperative Oceanic Fisheries Investigations Report*, **11**, 136–139.

——1975. Historical fluctuations of climate and bioclimatic factors as recorded in varved sediment deposits in a coastal sequence. *Proceedings of the WMO/IAMAP Symposium on Long-term Climatic Fluctuations*, World Meteorological Organization, Publication, **421**, 147–158.

—— & CRILL, P. A. 1977. Sedimentation and climatic patterns in the Santa Barbara Basin during the 19th and 20th Centuries. *Geological Society of America Bulletin*, **88**, 1161–1172.

—— & ISAACS, J. D. 1969. History of fish populations inferred from fish scales in anaerobic sediments off California. *California Cooperative Oceanic Fisheries Investigations Report*, **13**, 63–70.

—— & ——1974. Abundance of pelagic fish during the 19'th and 20'th centuries as recorded in anaerobic sediment off the Californias. *Fishery Bulletin*, **72**, 257–273.

——, KLING, S. A., CRILL, P. A., DUFFRIN, E. & BRULAND, K. W. 1977. Monitoring the marine environment through sedimentation. *Nature*, **266**, 136–139.

——, JOHNSON, S. & BAUMGARTNER, T. 1981. In search of modern depositional analogs to the Monterey Formation. *In*: GARRISON, R. E. & PISCIOTTO, K. (eds) *The Monterey Formation and Related Siliceous Rocks of California*. Society of Economic Paleontologists and Mineralogists Special Publication, Pacific Section, 123–147.

——, BAUMGARTNER, T. R., CASEY, R. E., SINGLETON, J., FERREIRA-BARTRINA & NELSON, C. O. 1990. Sediment flux at selected coastal sites: proposed time series measurement by particle traps. *In*: BETANCOURT, J. L. & MacKAY, A. M. (eds) *Proceedings of the Sixth Annual Pacific Climate (PACLIM) Workshop, March 5–8, 1989*, Interagency Ecological Studies Program for the Sacramento-San Joaquin Estuary Technical Report, **23**, 91–95.

STRAUGHAN, D. (ed.) 1971. *Survey of the Santa Barbara Channel Oil Spill 1969–1970*, Volume 1, Allan Hancock Foundation, University of Southern California, Los Angeles.

SUMMERS, J. K., STROUP, C. F., DICKENS, V. A. & GAUGHAN, M. 1988. *A Retrospective Pollution History of Southern California*. Report for NOAA, Natl. Ocean Survey, Ocean Assessments Div., Seattle, Wash., and National Marine Fisheries Service, Southwest Fisheries Center, La Jolla, Calif., VERSAR Inc., Columbia, Maryland.

SWEENEY, R. E. & KAPLAN, I. R. 1980. Natural abundance of ^{15}N as a source indicator for near-shore marine sedimentary and dissolved nitrogen. *Marine Chemistry*, **9**, 81–94.

——, LIU, K. K. & KAPLAN, I. R. 1978. Oceanic nitrogen isotopes and their uses in determining the source of sedimentary nitrogen. *In*: ROBINSON, B. W. (ed.) *Stable Isotopes in the Earth Sciences*, New Zealand Department of Scientific and Industrial Research, Wellington, New Zealand, DSIR Bulletin, **220**, 9–26.

TAKAHASHI, K. 1986. Seasonal fluxes of pelagic diatoms in the subarctic Pacific, 1982–1983. *Deep-Sea Research*, **33**, 1225–1251.

THORNTON, S. E. 1981. Suspended sediment transport in surface waters of the California Current: 1977–78 floods. *Geo-Marine Letters*, **1**, 23–28.

——1984. Basin model for hemipelagic sedimentation in a tectonically active continental margin: Santa Barbara basin, California Continental Borderland. *In*: STOW, D. A. V. & PIPER, D. J. W. (eds) *Fine-Grained Sediments: Deep-Water Processes and Facies*. Blackwell, Oxford, 377–394.

——1986. Origin of mass flow sedimentary structures in hemipelagic basin deposits: Santa Barbara basin, California Borderland. *Geo-Marine Letters*, **6**, 15–19.

UCHIO, T. 1960. Ecology of living benthonic foraminifera from the San Diego, California, area. *Special Publication Cushman Foundation for Foraminiferal Research*, **5**, 1–72.

WEINHEIMER, A. L. 1994. *Radiolarian and Diatom Fluxes in Two California Borderland Basins as Indices of Climate Variability*. PhD Thesis, University of California-Santa Barbara.

——1986. Radiolarian indicators of El Niño and anti-El Niño events in the recent sediment of the Santa Barbara basin. *In*: CASEY, R. E. & BARRON, J. A. (eds) *Siliceous Microfossil and Microplankton Studies of the Monterey Formation and Modern Analogs*. Pacific Section Society of Economic Paleontologists and Mineralogists, Los Angeles, 31–37.

——, CARSON, T. L., WIGLEY, C. R. & CASEY, R. E. 1986. Radiolarian responses to recent and Neogene California El Niño and anti-El Niño events.

Palaeogeography, Palaeoclimatology, Palaeoecology, **53**, 3–25.

WINTER, A. 1985. Distribution of living coccolithophores in the California Current system, southern California Borderland. *Marine Micropaleontology*, **9**, 385–393.

YEMELYANOVA, M. YU., KOVACHEV, S. A. & KONTAR, YE. A. 1992. Generation of bottom storms by turbidite (mud) flows. *Transactions (Doklady) of the Russian Academy of Sciences*, **223**(2), 207–211.

YOUNG, D. R., JOHNSON, J. N., SOUTAR, A. & ISAACS, J. D. 1973. Mercury concentrations in dated varved sediments off southern California. *Nature*, **244**, 273–275.

Composition and origins of laminae in late Quaternary and Holocene sediments from the Santa Barbara Basin

DAVID BULL & ALAN E. S. KEMP

Department of Oceanography, University of Southampton,
Southampton Oceanography Centre, European Way, Southampton SO14 3ZH, UK

Abstract: Late Quaternary and Holocene sediments from the Santa Barbara Basin have been analyzed using backscattered electron imaging (BSEI) techniques to identify the origins of laminae and document intra- and inter-annual ocean/ climate variability over the last glacial cycle. Laminae observed in X-ray radiography and BSEI are entirely defined by variations in the relative abundance and grain size of biogenic and terrigenous components. There is no evidence to suggest that bacterial mat mediation may have played a role in lamina formation. The simplest lamination type comprises silt-rich, silt-poor couplets which originate from seasonal variation in the supply of silt to the basin. The second lamination type contains an additional lamina of diatom ooze to form terrigenous sediment-diatomaceous triplets. The thicker diatom ooze laminae commonly exhibit a vertical succession of diatom floras which represents intra-annual variation in productivity. The variation in thickness and assemblage of diatom ooze laminae is probably a function of the amount of diatom flux, which is controlled by upwelling-driven primary productivity in the basin. Laminated sediments deposited during cooler periods generally contain thicker diatom ooze laminae, consistent with more vigorous regional circulation and increased upwelling.

Laminated marine sediments provide the highest resolution information available for reconstruction of ancient ocean and climate variability. Recent development and application of scanning electron microscope (SEM)-based analytical techniques has led to a fuller realization of the potential of laminated sediments as palaeoceanographic indicators (Kemp 1990; Kemp & Baldauf 1993; Brodie & Kemp 1994) and include the potential to record information on intra-annual variability comparable to sediment time trap series (Pike *et al.* 1993; Pike & Kemp 1996).

Holocene laminated sediments from the Santa Barbara Basin have been the subject of palaeoceanographic study for some time (see Lange & Schimmelmann 1994, and references therein). The new sedimentary record from Site 893 presents the first opportunity to study the late Quaternary sedimentary record over the last Glacial cycle. Although there have been several studies of high-resolution variability in the Holocene laminated sequences (e.g. Soutar & Crill 1977; Pisias 1978; Baumgartner *et al.* 1992), few have carried out studies at the scale of individual lamina (Lange *et al.* 1987). A pre-

requisite for correct interpretation, including timing of lamina formation, is an adequate understanding of lamina fine structure and origins, including timing of lamina formation. The laminae observed in Holocene sediments of the Santa Barbara Basin are widely interpreted as annual in nature (Krishawani *et al.* 1973). Various mechanisms of lamina formation have been proposed, including seasonally varying sources of input (Emery & Hülsemann 1962), bacterial mat response to seasonal changes in deposition (Soutar & Crill 1977) and annual cycles of bottom water oxygen replenishment and depletion (Reimers *et al.* 1990).

Although some conventional secondary electron imagery has been undertaken (Reynolds & Gorsline 1992), which is incapable by itself of resolving laminations, little high-resolution backscattered electron imagery (BSEI) of these sediments has been undertaken (Lange & Schimmelmann 1994). The purpose of this paper is to report the results of preliminary analysis of laminated sediments from the Santa Barbara Basin at ODP Site 893 using BSEI techniques.

General sedimentology

The sediments from ODP Site 893 consist of two three-stage sedimentary cycles. Each cycle is composed of a basal intermittently laminated sequence, a massive non-laminated sequence, and

Previously published as: BULL, D. & KEMP, A. E. S., 1995. Composition and origins of Late Quaternary and Holocene sediments from the Santa Barbara Basin, California. *In*: KENNETT, J. P., BALDAUF, J. G. & LYLE, M. (eds) 1995. *Proceedings of the Ocean Drilling Programme, Scientific results. Leg146 (Pt. 2)*: College Station, TX(Ocean Drilling Program), 77–87.

From Kemp, A. E. S. (ed.), 1996, *Palaeoclimatology and Palaeoceanography from Laminated Sediments*, Geological Society Special Publication No. 116, pp. 143–156.

a relatively thinner laminated sequence (Kennett *et al.* 1994). The laminated to massive clays, silty clays, and diatomaceous muds are punctuated by discrete sand beds and by distinctive grey, some-times graded, clayey silts, that occur in a variety of thicknesses from lamina scale (see 'Event Beds' below) to thicker beds. The general sedimentol-ogy and depositional history of Site 893 is sum-marized by Behl (1995).

Source of material

The sediments examined in this study were selected from laminated Subunits IA and ID, and laminated sequences within intermittently laminated Subunits IC and IF. A total of 68 polished thin sections (PTS), containing in excess of 2000 laminae, were prepared from material obtained from 12 core intervals. In addition, more than 100 raw sediment samples were also pre-pared for topographic work.

Methods

BSEI work was carried out on both PTS and raw sediment samples. Raw sediment samples were split in half and one half processed to produce a PTS, whereas the counterpart was retained to provide material for subsequent topographic work.

Batches of sediment for PTS processing were vacuum-dried prior to being placed in a Logi-tech vacuum impregnator. The chamber was pumped down to 10^{-6} mb prior to the introduc-tion of low-viscosity epoxy resin. Once impreg-nated, the blocks were sectioned, mounted on glass microscope slides, and polished to a sub-micron finish. Slides were coated with 5Å carbon film prior to analysis in a JEOL JSM-6400 SEM fitted with a solid state BSE detector.

Diatom identification was aided by topo-graphic work carried out in BSEI mode in a JEOL 5300 low-vacuum SEM. The advantage of using a low-vacuum SEM was that 'wet', un-coated sediment samples could be analyzed, elim-inating the time-consuming preparation required for studying topographic specimens in high-vac-uum conditions. The use of low-vacuum techni-ques incurs slight penalties in terms of reduced resolution and increased susceptibility to speci-men charging. These were not found to be serious handicaps. Optimal results were obtained when individual laminae of interest were split horizon-tally, prior to mounting on standard pin stubs. This made the maximal surface area of each lamina available for study.

Lamina components

Laminae may be composed of terrigenous couplets or mixed terrigenous–biogenic triplets. Couplets consist of alternate silt-rich, silt-poor laminae whereas triplets contain an additional lamina of diatom ooze. In triplet sequences the thickness of the diatom ooze lamina is the main variable, and ranges from thin ($<20\,\mu$m thick) discontinuous stringers and ribbons to discrete laminae up to 2 mm in thickness. Variability in thickness is less pronounced in couplet seq-uences in which variation is largely confined to variations in the thickness and grain size of the silt-rich component.

Terrigenous laminae

In terrigenous laminae clay is the domi-nant component. Silt-rich laminae are distin-guished by a 20% to 40% silt component and are generally poorly sorted, with the silt grains evenly distributed within the clay. Grading of the silt component is only exceptionally observed. Mineralogical work carried out by Fleischer (1972) suggests that the Santa Clara river is the major source of the silt component. River-borne silt is deposited on the shelf prior to remobilization, and subse-quent deposition in the Basin (Drake *et al.* 1972). Typical couplets are about 0.5 mm thick. In the silt-rich laminae the coarsest material occurs in laminated sediments deposited closest to glacial maxima, where material of very fine- to fine-sand grade is common. In the majority of silt-rich laminae however, material is of the medium- to coarse-silt grade. A degree of internal variation in the thickness of silt-rich, silt-poor laminae is observed within terrigenous sequences (Fig. 1). In well devel-oped terrigenous sequences groups of coup-lets form several light–dark series. These are controlled by the abundance of material of the fine and medium-silt grade. Groups of laminae richer in material of the fine-silt grade appear lighter than those richer in coarse-silt grade material. In the sediments examined, well developed terrigenous couplet sequences are the less common laminae type, accounting for 40% of observed laminae. Terrigenous couplet sequences however, are prevalent in interglacial stages. In general appearance, the Santa Bar-bara Basin silt-rich, silt-poor couplets closely resemble similar couplets described in late Quaternary sediments from the Peru margin (Brodie & Kemp 1994).

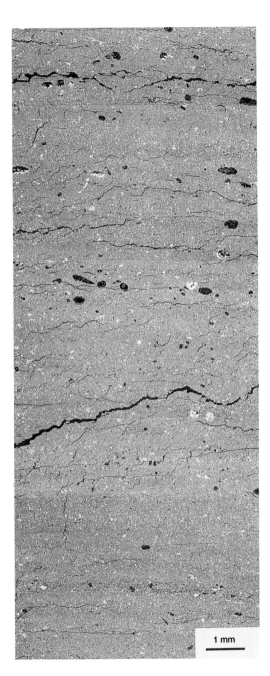

Fig. 1. Silt-rich and silt-poor in laminated sediments from the Santa Barbara Basin. Variability recorded in the silt content of a terrigenous couplet laminated sequence deposited near the end of isotope stage 6. Sample 146–893A-18H-5, 36–41 cm. The lighter packets contain a greater abundance of fine-grade material.

'Event beds'

Some thin sections examined contain thin laminae that were clearly not one of the normal lamina components described above. Their sharp contacts with the subjacent laminae suggest that these beds are generated by rapid depositional events. Two types are apparent. The first type are composed of terrigenous sediment ranging in grade from the coarser clay fractions to coarse silt; in contrast, the second type contains abundant diatom frustules. The fact that terrigenous event beds most commonly occur in terrigenous couplet laminated sequences, and that diatom containing event beds are restricted to triplet laminated sequences leads us to the conclusion that these beds are mainly derived from the reworking of pre-existing basin deposits, and were not derived from exotic sources. In diatom-containing beds there is a link between the amount of diatomaceous material in the event bed and that in the underlying laminae.

Diatom ooze laminae

The diatom component of the laminated sequences is highly variable: it may be reduced to individual valves and scattered fragmented debris or discontinuous lenses within one of the terrigenous laminae (Fig. 2); alternatively, thick laminae of diatom ooze may incorporate the bulk of the surrounding terrigenous couplet. In exceptional cases diatom ooze laminae may be up to 2 mm thick, and are clearly visible to the naked eye in the raw sediment samples. The more usual thickness range is from 0.25 to 0.75 mm.

Some of the characteristic diatom species observed are displayed in Figs 3 & 4. The diatom component consists primarily of *Chaetoceros* spp. in the form of setae and/or resting stages. The weakly silicified vegetative stages of *Chaetoceros* diatoms are very rare (cf. Pike & Kemp 1996, fig. 6). The resting-stage assemblage is invariably dominated by *Chaetoceros radicans*. *Chaetoceros vanheurckii*, *C. diadema*, and *C. didymus* are less abundant constituents of ooze laminae. Preservation of resting spores is good. Within many *Chaetoceros* laminae, other centric and pennate diatoms form a lesser part of the assemblage. The most commonly observed species belong to the *Thalassiosiracae* and *Coscinodiscacae* groups. The quality of preservation of these centric diatoms is largely size dependent. The valves of larger species generally occur as broken valves and detached girdles, whereas those of smaller species such as

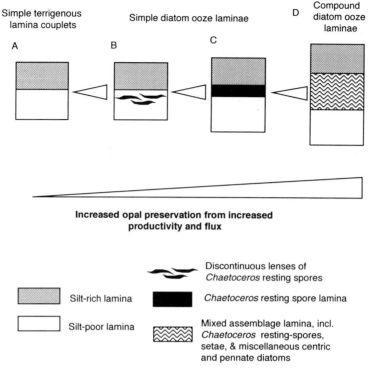

Fig. 2. Schematic diagram illustrating the various stages of diatom ooze laminae development observed in Santa Barbara Basin sediments. A, simple terrigenous silt-rich, silt-poor couplets containing fragmented diatoms. B, scattered lensoidal accumulations of *Chaetoceros* resting-spores within the silt-poor lamina. C, continuous diatom ooze lamina formed from resting-spores. D, examples of complex ooze laminae exhibiting vertical grading of the floral assemblage and formed from various centric and pennate species.

Thalassiosira oestrupii are usually preserved intact. Less commonly observed are species belonging to the *Stephanopyxis*, *Actinoptychus* and *Cocconeis* groups. In exceptional cases broken chains of *Skeletonema costatum* are also present. Pennate diatoms are less common than centrics within *Chaetoceros* laminae. The most commonly observed pennate species is *Thalassionema nitzschioides*. In exceptional cases *T. nitzschioides* may be sufficiently abundant to form micro-laminae in excess of 100 μm in thickness.

Ooze laminae composed of species other than *Chaetoceros* occur as isolated laminae within sequences of thick *Chaetoceros* laminae and, with greater frequency, in sequences containing thinner *Chaetoceros* laminae. Typically, these laminae are composed of *Thalassiosiracae* and *Coscinodiscacae* group diatoms.

Most diatom ooze laminae also contain varying amounts of terrigenous material. Coarse material, ranging from medium-silt to very fine-sand, occurs as individual grains. This is commonly most abundant in the uppermost section of the

diatom ooze laminae. Finer, clay-grade material occurs as lensoidal inclusions and, less frequently, may be sufficiently abundant to form a terrigenous-rich sub-lamina within the ooze lamina. In some cases this material is compacted in aggregates with varying amounts of fragmented biogenic material.

Mode of deposition of diatom ooze laminae

Before any palaeoceanographic interpretation can be attempted it is essential to know what exactly the preserved assemblage represents, and how the observed assemblage relates to the original live surface assemblage. Work conducted by Sancetta (1992) indicates that in terms of diversity, the diatom assemblage reaching the ocean floor is impoverished with respect to the surface community; and Reimers *et al.* (1990) showed that further dissolution of the diatom assemblage occurs at and immediately below the

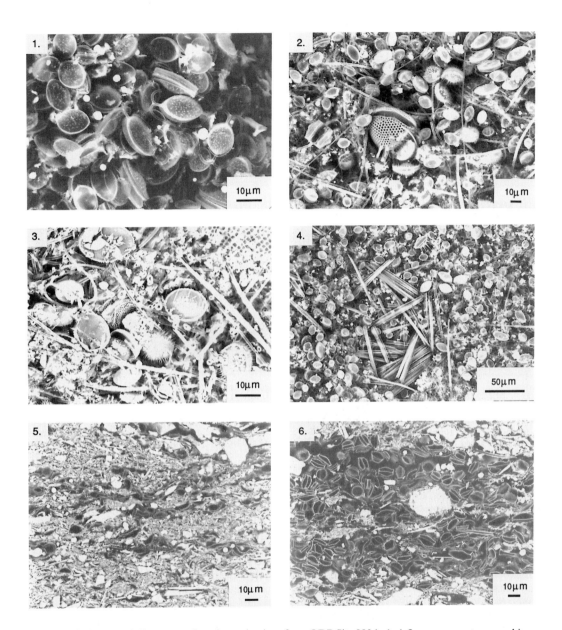

Fig. 3. SEM images of *Chaetoceros*-based ooze laminae from ODP Site 893A. 1–4, Low-vacuum topographic backscatter images. 1, resting-spores of *Chaetoceros radicans* showing the characteristic high-porosity of near monospecific resting spore laminae. Sample 146–893A-9H-2, 116–121 cm. 2, mixed resting-stage and setae laminae. Resting-spores are primarily *C. radicans* and *C. vanheurckii*. A single *Stephanopyxis turris* centric diatom can be seen in the centre of the image. Sample 146–893A-11H-4, 63–67 cm. 3, detail of an ooze lamina of varied assemblage. The resting-stage component includes *C. radicans*, and *C. vanheurckii*. Small centric diatoms are represented by an example of *Thalassiosira oestrupii*. Debris from a larger centric diatom, and abundant *coccoliths* are also apparent. Sample 146–893A-21H-5, 112–125 cm. 4, 'bundle' of *Thalassionema nitzschioides* in a mixed setae and resting-stage lamina. Sample 146–893A-15H-2, 96–101 cm. 5–6, high-vacuum backscatter images of polished thin sections. 5, individual *Chaetoceros* resting spores within the silt-poor lamina. 6, discrete diatom ooze laminae formed from *Chaetoceros* resting spores. Both from sample 146–893A-9H-2, 116–121 cm.

Fig. 4. SEM images of non-Chaetoceros ooze laminae. 1–3, topographic backscatter images. 1, ooze lamina composed of *Thalassionema nitzschioides*. 2, detail of 1. Both sample 146–893A-15H-6, 111–114 cm. 3, terrigenous-rich ooze lamina composed of small centric species, including *Thalassiosira oestrupii*. Sample 146–893A-15H-6, 42–50 cm. 4, high-vacuum backscatter image of a polished thin section showing a thick lamina composed of *Coscinodiscus* spp. diatoms. Sample 146–893A-1H-2, 34–46 cm.

sediment/water interface. Dissolution is minimized by rapid descent from surface waters to the subjacent seafloor. Rapid sedimentation of diatoms may be effected by faecal pellets (Schrader 1971), mass flocculation by way of marine snow (Alldredge & Gottschalk 1989) or transparent exopolymer particles (Alldredge *et al.* 1993), and mass sinking of diatom mats (Sancetta 1991; Kemp & Baldauf 1993).

Although the role of faecal pellets in the sedimentation of Santa Barbara Basin sediments has been documented in the sediment trap studies of Dunbar & Berger (1981), the intact nature of much of the diatomaceous material within the diatom ooze lamina suggests that this was not the principal mechanism of deposition. In a combined field and laboratory study of the faecal pellets of copepods (the most abundant metazoan oceanic zooplankton) Honjo & Roman (1978) showed that only skeletal material less than 10 μm in size was preserved intact within faecal pellets, and that larger material was fragmented. Mixed

terrigenous-diatomaceous pellets inferred to be faecal pellets do occur in diatom ooze laminae, but only as a minor component. The fabric of this pelleted material is markedly different from that of the surrounding diatom ooze lamina material. The non-pelleted material, which comprises the bulk of the diatom ooze laminae, is characterized by a higher porosity, reduced terrigenous content, and in some cases, reduced abundance of the coccolith component relative to the pelleted material. In view of the lack of extensive fragmentation of diatoms it is considered that the characteristics of diatom ooze laminae are incompatible with deposition by faecal pelleting as the major mechanism of deposition.

The bulk of biogenic material in diatom ooze laminae is derived from resting-spore forming *Chaetoceros* species which characterize the second bloom stage of Guillard & Kilham (1977). Weakly silicified stage 1 and residual stage 3 assemblages are rarely preserved. Mass flocculation of *Chaetoceros* species diatoms in the modern Santa Barbara Basin has been described by Alldredge & Gottschalk (1989). The commonest species identified in modern flocs, *Chaetoceros radicans,* is also the commonest species identified in the upper Pleistocene diatom ooze laminae. In certain diatom ooze laminae, high-porosity, tangled masses of setae are observed. Typically, these are in the order of 100 μm in length, although occasionally these masses may exceed a millimetre in length. These masses frequently contain attached resting spores and a small amount of entrained terrigenous detritus. The high porosity of these structures and the intact nature of the diatomaceous material precludes faecal pelleting as an origin.

The occurrence of *Thalassionema* within *Chaetoceros* dominated laminae provides further evidence for a mass-flocculation of the *Chaetoceros* component. The pennate diatom *Thalassionema nitzschioides* is a colonial species in which small numbers of individuals are joined by mucilage pads (Round *et al.* 1990). Where *Thalassionema* occurs as a lesser constituent within *Chaetoceros* laminae, *Thalassionema* valves occur as small bundles of a dozen or more individuals. We suggest that this bundling occurs as a result of the *Thalassionema* colonies being entrained within sinking *Chaetoceros* flocs. The greater strength of the setae and mucilage-bound *Chaetoceros* floc prevents the disruption of the *Thalassionema* colony.

From the above evidence, we believe that mass flocculation of bloom-forming diatoms produced most of the diatom ooze laminae. the rapid descent of flocculated and aggregated diatoms at rates probably exceeding 100 m/day (Alldredge & Gottschalk 1989), by-passing grazing, explains the intact nature of diatom frustules within laminae.

Vertical succession in diatom ooze lamina

In many *Chaetoceros*-rich diatom ooze laminae vertical successions were observed. In some cases, a basal setae-rich sub-lamina is overlain by a sub-lamina dominated by resting spores. Setae-rich sub-lamina may correspond to the flocculation of setae-bound *Chaetoceros* vegetative stages during the second stage of the diatom succession of Guillard & Kilham (1977). The absence of the vegetative stage frustules from this portion can be attributed to dissolution. Resting spores are produced in response to increased nutrient depletion following seasonal reduction of upwelling, and corresponds to the third stage of Guillard & Kilham (1977). Thicker diatom ooze laminae may contain four or more identifiable

sub-laminae (Fig. 5), and appear to indicate a more complex sequence of bloom events. A detailed analysis of the internal successions in diatom species within laminae is currently in progress.

Foraminifers

Benthic foraminifers are a common minor constituent of laminites. Rotaline types, dominated by *Bolivina seminuda,* are most abundant. In addition, a small number of textularian types were also observed. The latter are most abundant in the Holocene and isotope stage 5e sediments. The abundance of rotaline foraminifers is highly variable.

In laminated sequences benthic foraminifer abundance appears to be related to diatom abundance. Foraminifers are rare in triplet sequences in which well developed diatom ooze laminae occur. Conversely, in triplet sequences in which diatom laminae are poorly developed, benthic foraminifers commonly occur in great profusion. The scarcity of benthic foraminifers in well-developed ooze laminated sequences may reflect development of more fully anoxic conditions than those which currently occur in the basin related to increased organic flux.

On a lamina scale, foraminifers are largely confined to terrigenous laminae. In terrigenous couplet sequences, foraminifers tend to be concentrated in the summer clay-rich laminae. This is consistent with the work of Reimers *et al.* (1990), who found that, in the modern basin, the maximal abundance of foraminifers coincided with minima in basin oxygenation created by the input of bloom material into the basin and the cessation of Pacific Intermediate Water flow into the basin.

Bacterial mats

The role of bacterial mats in promoting laminae formation in the present-day Santa Barbara Basin has been highlighted by several authors (e.g. Soutar & Crill 1977; Williams & Reimers 1982; Reimers *et al.* 1990). However, these mats have poor long-term preservation potential. There is no evidence from the scanning electron microscope fabric studies of this research for the preservation of mats in the sediments examined. We find no need to appeal to bacterial mat mediation to explain the formation of laminae described herein. Physical characteristics of laminae such as porosity are controlled by variations in lamina constituents, i.e., fluctua-

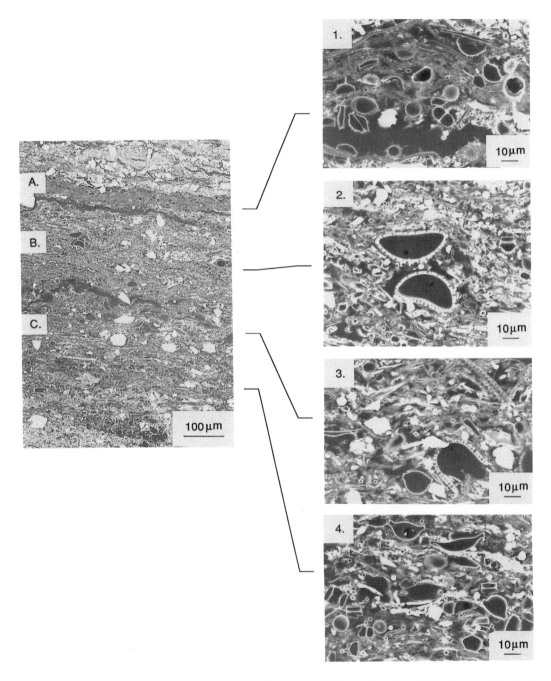

Fig. 5. BSEI images of a composite ooze laminae (left) composed of four sub-laminae A–D, illustrated in detail in the right-hand images. 1, detail of sub-laminae A composed largely of *Chaetoceros* resting-spores and setae. 2, detail of sub-lamina B composed largely of *Chaetoceros* setae and pennate diatoms, with occasional larger centric diatoms. 3, detail of sub-lamina C. composed largely of *Chaetoceros* setae and debris derived from large centric diatoms. 4 detail of sub-lamina D composed largely of resting-spores and setae derived from various *Chaetoceros* species. Sample 146–893A-15H-3, 21–29 cm.

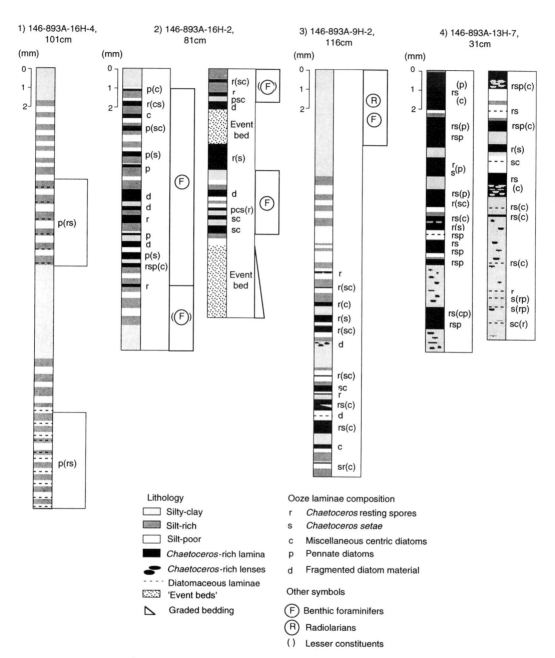

Fig. 6. Logs of characteristic examples of laminites examined by this study. 1, brief packets of discontinuous diatomaceous laminae dominated by non-upwelling characteristic species in a sequence of variable quality laminae preservation. 2, quite extensive development of thin diatom ooze laminae, containing both upwelling and non-upwelling characteristic species. 3, development of thin diatom ooze lamina composed of upwelling characteristic species. 4, sharp upwards change from thin diatomaceous laminae to thick ooze laminae associated with a general increase in assemblage diversity.

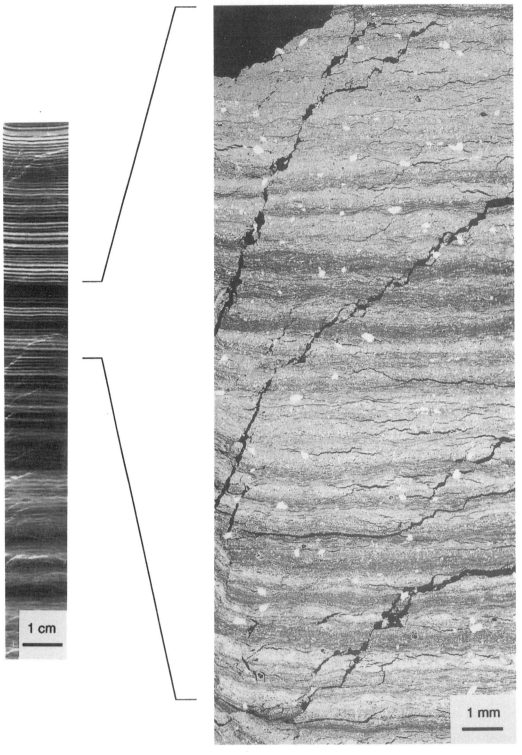

Fig. 7. Comparison of X-radiograph (left), and BSE (right) images of a laminated sequence. Sample 146–893A-21H-5, 112–127 cm. Periodic alternations between terrigenous couplet and mixed terrigenous biogenic triplet laminae styles are evident. On the X-radiograph, white laminae consist of diatom ooze, light grey laminae of silt, and darker laminae of clay. On the BSE image, dark layers correspond to the diatom ooze laminae. The deterioration in resolution towards the base of the sequence is a reflection of a downwards increase in thickness of the raw sedimentary section, rather than qualitative changes in laminae preservation. The more detailed BSE mosaic shows that the X-radiograph image primarily reflects changes in the degree of development of diatom ooze laminae.

tions in the relative proportions, grain size, and sorting of terrigenous and biogenic material.

Laminae groupings

The two end-members of laminae style are terrigenous couplets and terrigenous–ooze triplets. If an annual periodicity for silt-rich, silt-poor couplets and terrigenous–diatomaceous triplets is assumed, couplet sequences may persist for at least 50 years, and triplet sequences may persist for at least 120 years. Using data gathered from all the specimens examined a series of features can be documented that characterizes the evolution from couplet to triplet sequences. Logs of laminated intervals displaying some of these features are displayed in Fig. 6.

Regardless of diatom abundance, ooze laminae always occur in sequences. In sediments where diatom abundance is too low to form discrete ooze laminae, diatoms are concentrated in lensoidal accumulations or ribbons and stringers that, superimposed on the background clay deposition, form thin diatomaceous laminae. In these sediments terrigenous-diatomaceous triplet sequences are short-lived, spanning 3 to 7 years, and occur sporadically within terrigenous couplet sequences.

In sediments where there is a sufficient abundance of diatoms to form thin, discrete, ooze laminae the triplet sequences are closer spaced and persist over longer periods. A certain amount of internal variation is apparent in these sequences, which frequently contain marked minima in the thickness of ooze laminae.

A range of internal features is apparent in sequences of well-developed ooze laminae. These sequences can be divided into a succession of packets, each defined by a particular range of diatom ooze laminae thickness. Episodic shifts in the intensity of upwelling are indicated by changes in the thickness of ooze laminae between packets (Fig. 7). A feature common to all well-developed triplet laminated sequences is the presence of distinct minima in the thickness of ooze laminae, or complete absence of ooze laminae. The spacing of these minima shows some relation to sediment age. In isotope stage 5 sediments two groups of minima occur. The more common group is centred around a 10-year periodicity. The less common group most commonly occurs over a 5-to-7 year periodicity. In isotope stage 6 sediments the 10-year periodicity is largely absent, and the shorter periodicity is manifest over a greater range of 4 to 8 years. In certain isotope stage 6 sediments the transition

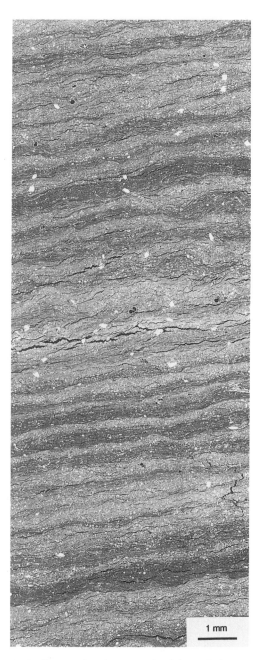

Fig. 8. Short-scale primary productivity variations recorded in laminated sediments from the Santa Barbara Basin. Pronounced cyclical variation in the opal content of diatom ooze laminae from stage 6 sediments. Sample 146–893A-21H-6, 70–75 cm. Minima in opal content occur over a period of 4 to 8 years and may indicate reduced primary productivity during ENSO-related conditions.

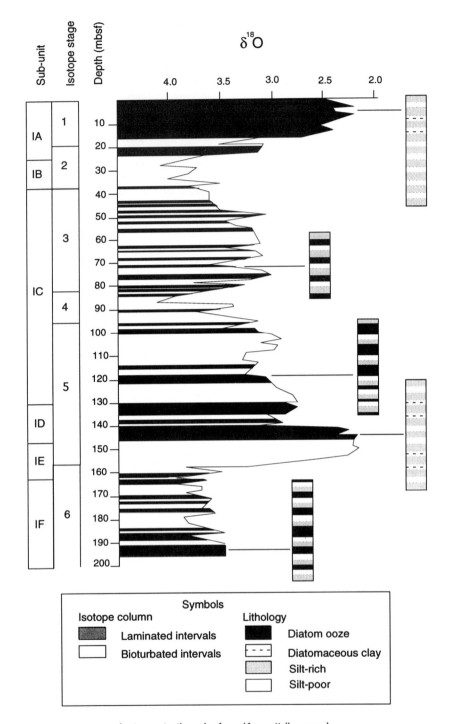

Fig. 9. Chronology of laminated intervals within ODP Hole 893A showing the general relationship between $\delta^{18}O$ values and the expression of laminated sequences. Also shown are typical examples of laminated sequences from differing stages in the glacial cycle. Diatom ooze laminae show the greatest development in isotopically cooler periods. Diatom ooze laminae from isotope stage 3 tend to be rather thin. Diatom ooze laminae from isotope stages 5 and 6 both show marked variations in the thickness of ooze laminae: in the case of isotope stage 5 sediments this is manifest as episodic shifts, whereas in the case of isotope stage 6 sediments a pronounced rhythmic variation is sometimes observed. (Lithology from Kennett *et al.* (1994); stable isotope data from Kennett (1995); age determinations by Ingram & Kennett (1995)).

from maximal to minimal ooze laminae thickness exhibits a striking rhythmic pattern (Fig 8). The significance of these minima requires further work. The shorter-term range of periodicities covers the frequencies of recurrence commonly cited for El Niño-Southern Oscillation (ENSO) events (Douglas 1981). In the modern day Santa Barbara Basin ENSO events are characterized by the replacement of the normal *Chaetoceros*-dominated upwelling assemblage by a more diverse assemblage of warmer water, more oceanic affinities (Lange *et al.* 1987) and, flushing of basin waters (Berelson 1991). In some of the sediments that date from isotope stage 3 minima in the thickness of the ooze lamina appear to be associated with a non-*Chaetoceros* assemblage. These lamina consist of the valves of large centric diatoms of *Coscinodiscacae* and *Thalassiosiracae* groups distributed within the clay-rich terrigenous laminae. In the majority of minima however, there is no assemblage change, merely a reduction in the amount of *Chaetoceros* derived material present.

Conclusions

The examination of late Quaternary and Holocene laminated sediments from the Santa Barbara Basin using BSEI has been used to elucidate the compositional variation and origins of laminae and to relate variations in lamina style to the palaeoceanographic and palaeoclimatic history.

We have found no evidence for the preservation of bacterial mats, or indeed, any evidence to suggest that bacterial mats have played a role in lamina formation. The physical characteristics of laminae are controlled by variations in lamina constituents (i.e. fluctuations in the relative proportions, grain size and sorting of terrigenous and biogenic material).

The simplest laminae are couplets of silt-rich, silt-poor sediment, around 0.5 mm thick, whose origins are best ascribed to the seasonal cycle of silt input from winter rains. Superimposed upon these terrigenous couplets is a highly variable diatom ooze component. This varies from a discontinuous series of thin lenses and stringers of diatom ooze within silt-poor laminae to discrete, millimetre thick, ooze laminae that give an overall triplet lamina bundle. The thicker diatom laminae are commonly composite, showing vertical succession in diatom species that records intra-annual variation in productivity.

Variation in the thickness of the diatom ooze laminae probably relates to the amount of diatom flux, which is a function of upwelling-driven productivity in the Basin. As a general pattern,

laminated sediments deposited in cooler periods contain thicker diatom ooze laminae (Fig. 9), which is consistent with greater wind-driven upwelling intensity and productivity (although there are no laminae preserved in glacial periods owing to increased basin oxygenation, Behl & Kennett 1996. This evidence is supported by the scarcity of benthic foraminifers in sequences of thick ooze laminae, indicative of more fully anoxic conditions related to increased organic flux. There is considerable potential for further quantitative study of opal flux and palaeoproductivity variation within these sediments.

DB acknowledges the receipt of a NERC Research Studentship. This study was supported by the award of NERC grants GST/02/820 and GR9/1347 to AESK. Attendance of AESK at the Site 893 description/ sampling meeting was funded by the NERC ODP Science Programme. We gratefully acknowledge the assistance of C. Lange in diatom identification and in the provision of sediment samples from the Holocene portion of ODP Site 893A; and S. King in the identification of foraminifers. Reviews by Hans Schrader and Don Gorsline improved the quality of this manuscript.

References

ALLDREDGE, A. L., & GOTTSCHALK, C. G. 1989. Direct observations of the mass flocculation of diatom blooms: Characteristics, settling velocities and formation of diatom aggregates. *Deep Sea Research*, **36**, 159–171.

——, PASSOW, U. & LOGAN, B. E. 1993. The abundance and significance of a large class of large, transparent organic particles in the ocean. *Deep Sea Research*, **40**, 1131–1140.

BAUMGARTNER, T. R. SOUTAR, A. & FERREIRA-BARTRINA, V. 1992. Reconstructions of the history of Pacific sardine and northern anchovy populations over the past two millennia from sediments of the Santa Barbara Basin, California. *CalCOFI Report*, **33**, 24–40.

BEHL, R. 1995. Sedimentary facies and sedimentology of the late Quaternary Santa Barbara Basin, California, Site 893. *In*: KENNETT, J. P., BALDAUF, J. *et al. Proceedings of the Ocean Drilling Programme, Scientific results. Leg146 (Pt. 2)*: College Station, TX (Ocean Drilling Program).

—— & KENNETT, J. P. 1996. Brief interstadial events in the Santa Barbara basin, N.E. Pacific, during the past 60 kyr. *Nature*, **379**, 243–246.

BERELSON, W. M. 1991. The flushing of two deep sea basins, Southern California Borderland. *Limnology and Oceanography*, **36**, 1150–1166.

BRODIE, I. & KEMP, A. E. S. 1994. Variation in biogenic and detrital fluxes and formation of lamina in late Quaternary sediments from the Peruvian coastal upwelling zone. *Marine Geology*, **116**, 385–398.

DOUGLAS, R. 1981. Paleoecology of continental basins: A modern case history from the borderland of Southern California. *In*: DOUGLAS, R., GORSLINE, D. S. & COLBURN, I. (eds) *Depositional Systems of Active Continental Margin Basins*. Society of Economic Paleontologists and Mineralogists, Pacific Section, Short course, 121–156.

DRAKE, D. E., KOLPACK, R. L., & FISCHER, P. J. 1972. Sediment transport on the D. J. P. *et al.* (eds) *Shelf Sediment Transport*, Dowden, Hutchinson and Ross, Stroudsburg, PA, 307–331.

DUNBAR, R. B. & BERGER, W. B. 1981. Faecal pellet flux to modern sediments of the Santa Barbara Basin based on sediment trapping. *Geological Society of America Bulletin*, **92**, 212–218.

EMERY, K. O. & HÜLSEMANN, J. 1962. The relationships of sediments, life and water in a marine basin. *Deep Sea Research*. Part A, 165–180.

FLEISCHER, P. 1972. Mineralogy and sedimentation history, Santa Barbara Basin, California. *Journal of Sedimentary Petrology*, **42**, 49–58

GUILLARD, R. R. L. & KILHAM, P. 1977. Ecology of marine planktonic diatoms. *In*: WERNER, D. (ed.) *The Biology of Diatoms*. Botanical monographs. Vol. 13. University of California Press, Berkeley and Los Angeles, 470–483.

HONJO, S. & ROMAN, M. R. 1978. Marine copepod faecel pellets: production, preservation and sedimentation. *Journal of Marine Research*, **36**, 45–57.

INGRAM, B. L. & KENNETT, J. P. 1995. Radiocarbon chronology and planktonic-benthic foraminiriferal ^{14}C differences in Santa Barbara Basin sediments. Hole 893A. *In*: KENNETT, J. P., BALDAUF, J. *et al. Proceedings of the Ocean Drilling Programme, Scientific results. Leg146 (Pt. 2)*: College Station, TX(Ocean Drilling Program).

KEMP, A. E. S. 1990. Sedimentary fabric and variations in lamination style in Peru continental margin upwelling sediments. *In*: SUESS, E. *et al. Proceedings of the Ocean Drilling Programme, Scientific Reports, leg 112*: College Station, TX (Ocean Drilling Program), 43–58.

—— & BALDAUF, J. 1993. Vast Neogene laminated diatom mat deposits from the eastern equatorial Pacific Ocean. *Nature*, **362**, 141–144.

KENNETT, J. P. 1995. Latest Quaternary benthic oxygen and carbon isotope stratigraphy: Hole 893A, Santa Barbara Basin, California. *In*: KENNETT, J. P., BALDAUF, J. *et al. Proceedings of the Ocean Drilling Programme, Scientific Results. Leg146 (Pt. 2)*: College Station, TX(Ocean Drilling Program).

——, BALDAUF, J. *et al.* 1994. *Proceedings of the Ocean Drilling Programme, Initial results. Leg146 (Pt. 2)*: College Station, TX (Ocean Drilling Program).

KRISHNASWANI, S., LAL, D., AMIN, B. S. & SOUTAR, A. 1973. Geochronological studies of the Santa Bar-

bara Basin: Iron-55 as a unique tracer of particle settling. *Limnology and Oceanography*, **18**, 763–770.

LANGE, C. B. & SCHIMMELMANN, A. 1994. Seasonal resolution of laminated sediments in the Santa Barbara Basin: Its significance in paleoclimatic studies. *In*: REDMOND, K. T. & THARP, V. L., (eds) *Proceedings of the Tenth Annual Pacific Climate (PACLIM) Workshop*. California department of water resources, Interagency ecological studies program, Technical report 36, 83–91.

, BERGER, W. H., BURCKE, S. K., CASEY, R. E., SCHIMMELMANN, A., SOUTAR, A. & WEINHEIMER, A. L. 1987. El Niño in the Santa Barbara Basin: diatom, radiolarian and foraminiriferan response to the "1983 El-Niño" event. *Marine Geology*, **78**, 153–160.

NYGAARD, K. & HESSEN, D. O. 1994. Diatom kills by flagellates. *Nature*, **367**, 520.

PIKE, J. & KEMP, A. E. S. 1996. Records of seasonal flux in Holocene laminated sediments, Gulf of California. *This volume*.

——, —— & SANCETTA, C. 1993. Complex signals from laminated sediments, Gulf of California. *Eos*, **74**, 363

PISIAS, N. 1978. Palaeoceanography of the Santa Barbara Basin during the last 8000 Years. *Quaternary Research*, **10**, 366–384.

REIMERS, C. E. LANGE, C. B., TABAK, M. & BERNHARD, J. M. 1990. Seasonal spillover and varve formation in the Santa Barbara Basin, California. *Limnology and Oceanography*, **35**, 1577–1585

REYNOLDS, S. & GORSLINE, D. S. 1992. Clay microfabrics of deep sea mud(stones), California Continental Borderland. *Journal of Sedimentary Petrology*, **62**, 41–53.

ROUND, F. E., CRAWFORD, R. M. & MANN, D. G. 1990. *The Diatoms: Biology and Morphology of the Genera*. Cambridge University Press.

SANCETTA, C. 1991. Mediterranean sapropels: Seasonal stratification yields high production and carbon flux. *Paleoceanography*, **9**, 195–196.

——1992. Comparison of phytoplankton in sediment time trap series and surface sediments along a productivity gradient. *Palaeoceanography*, **7**, 183–190.

SCHRADER, H. 1971. Faecal pellets: their role in the sedimentation of pelagic diatoms. *Science*, **174**, 55–57.

SOUTAR, A. & CRILL, P. A. 1977. Sedimentation and climate patterns in the Santa Barbara Basin during the 19th and 20th centuries. *Geological Society of America Bulletin*, **88**, 1161–1172.

WILLIAMS, L. A. & REIMERS, C. 1982. Recognizing organic mats in deep sea environments. *Geological Society of America Bulletin*, **14**, 697–697

Records of seasonal flux in Holocene laminated sediments, Gulf of California

JENNIFER PIKE & ALAN E. S. KEMP

Department of Oceanography, University of Southampton,
Southampton Oceanography Centre, European Way,
Southampton, SO14 3ZH, UK

Abstract: Holocene laminated sediments from Guaymas Basin (Gulf of California, Mexico) have, in the past, been described as annual couplets of sedimentation. Couplets comprise an alternation of diatom- and terrigenous-rich sediment reflecting winter and summer flux to the sediment respectively. New data from backscattered electron imagery (BSEI) show that a three-component pattern of sedimentation is preserved, comprising (1) a lithogenic lamina consisting of clays, silt and minor diatomaceous material, deposited during the summer and autumn wet season; (2) a mixed flora diatomaceous lamina deposited during the early winter collapse of the thermocline and onset of water column mixing; and (3) a near-monospecific flora lamina of *Chaetoceros* resting spores or, more rarely, *Skeletonema costatum*, representing deposition from, or at the end of, the spring coastal upwelling bloom. Recent sediment trap biogenic and terrigenous flux data also show three to four major flux events per year. Direct comparisons are possible between BSEI records and recent sediment trap flux data to facilitate high-resolution palaeoceanaographic and palaeoclimatic reconstructions throughout the Holocene.

The Gulf of California is a marginal, semi-enclosed sea in northwest Mexico, in which laminated marine sediments are preserved (Calvert 1966; Kelts & Niemitz 1982; Baumgartner 1987). Approximately 1200 km long and 150 km wide, the Gulf is divided bathymetrically into two sections by the Midriff Islands; the northern, and central and southern Gulf (Dauphin & Simoneit 1991). The central and southern Gulf comprises a series of basins silled at progressively greater depths, from northwest to southeast, which are in open communication with the Pacific Ocean (Rusnak *et al.* 1964). The largest of these basins is Guaymas Basin in the central Gulf (Fig. 1). The close proximity of the arid Sonora Desert and Baja California gives the central Gulf a continental, rather than oceanic climate (Roden 1964) and, unlike most evaporitic marine basins, there is an inflow of cooler, fresher water at depth (greater than 250 m) and a shallower outflow of warmer, more saline water (between 50 and 250 m) (Bray 1988*a*).

Upwelling

Global atmospheric circulation determines the regional wind patterns and thus, the intensity and duration of upwelling in the Gulf of California (Schrader & Baumgartner 1983). Advanced very high resolution radiometer (AVHRR) satellite sea surface temperature (SST) data show that the lowest SSTs typically occur between December and May (Thunell *et al.* 1993), coinciding with the season of northwesterly winds and greatest upwelling of cool, nutrient-rich water from depth. Upwelling brings new nutrients into the euphotic zone leading to extremely high primary productivity of the diatom flora (Calvert 1966; Alvarez-Borrego & Lara-Lara 1991). Upwelling cells are located in the lees of capes and promontories along both coastlines of the Gulf (Bray & Robles 1991) and upwelling is more intense along the mainland coast during winter when the prevailing wind is from the northwest, although SST records are essentially in phase on both sides of the central Gulf (Thunell *et al.* 1993). One of the strongest of these cells is located offshore of the Mexican town of Guaymas (Juillet-Leclerc & Schrader 1987) (Fig. 1), and a cool-water plume from this cell has been tracked using AVHRR satellite infrared imagery (Badan-Dangon *et al.* 1985), flowing along the coast in the direction of the prevailing wind until deflected out towards the centre of the Gulf by a major cape, Cabo Lobos. The plume then flows across the Gulf to Bahia Concepción on the Baja California coastline and completes a Gulf-wide cyclonic circulation pattern (Badan-Dangon *et al.* 1985). In the summer and autumn, relatively weak southerly winds produce northerly surface circulation. Badan-Dangon *et al.* (1985) report reduced intensity upwelling occurring mainly on the

From Kemp, A. E. S. (ed.), 1996, *Palaeoclimatology and Palaeoceanography from Laminated Sediments*,
Geological Society Special Publication No. 116, pp. 157–169.

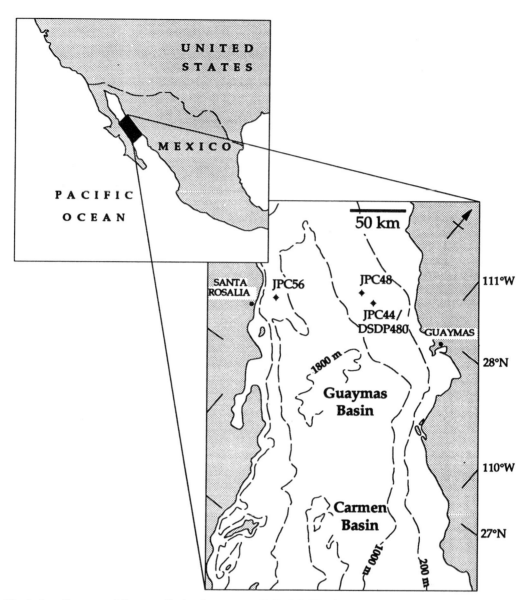

Fig. 1. Location map of Guaymas Basin piston cores, Gulf of California.

Table 1. *Piston and box core location, bathymetric and sample data*

Core	Basin	Water depth (m)	Core Length (m)	Latitude (north)	Longitude (west)	Samples (depth in core, m)
JPC44	E. Guaymas	655	16.46	27°54.09′	111°39.30′	1.51–1.55
						2.91–2.96
BC43	E. Guaymas	655	0.35	27°54.09′	111°39.30′	0–0.35
DSDP480	E. Guaymas	655	152.00	27°54′	111°39′	visual description
JPC48	E. Guaymas	530	17.99	27°56.40′	111°47.90′	2.55–2.61
						4.59–4.69
BC50	E. Guaymas	745	0.35	27°47.15′	111°42.51′	0–0.35
JPC56	W. Guaymas	818	19.12	27°28.16′	112°06.26′	2.05–2.10
						5.02–5.52

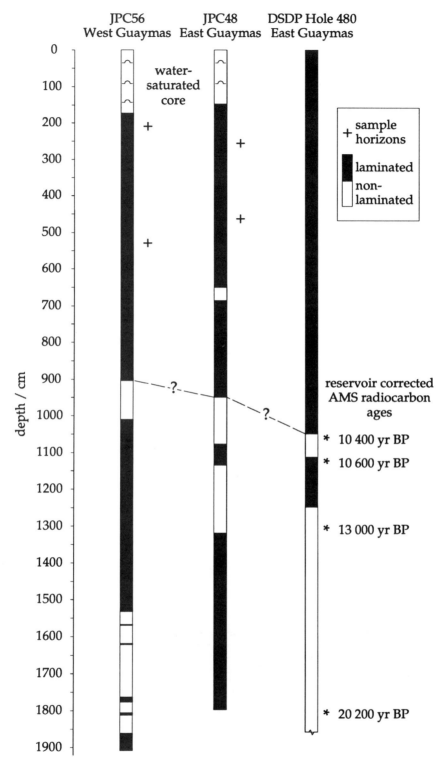

Fig. 2. Log of piston cores showing laminated and non-laminated horizons. AMS radiocarbon dates from Keigwin & Jones (1991).

Baja California side of the Gulf. However, AVHRR SST and coastal zone colour scanner surface pigment concentration data reported by Thunell *et al.* (1993) and Santamaría-del-Angel *et al.* (1994) show no indication of upwelling-induced cooling or enhanced productivity on the western side of the Gulf during summer and early autumn. This suggests that upwelling fails to entrain the cool, nutrient-rich water from depth which enhances productivity.

Oxygen minimum zone

Bacterial oxidation of organic matter is responsible for the existence of the oxygen minimum zone (OMZ) in the eastern equatorial Pacific Ocean, and the sluggish intermediate depth circulation is responsible for its position within the water column (Wyrtki 1962, 1967). In the Gulf of California, wind-induced coastal upwelling supports extremely high primary productivity (Zeitzschel 1969). However, there have been no systematic studies of the temporal and spatial variability of this productivity (Alvarez-Borrego & Lara-Lara 1991). Within the central Gulf, over Guaymas Basin, values of less than 0.2 ml/l dissolved oxygen are recorded from between 400 to 800 m water depth, and at the Gulf mouth, values of less than 0.1 ml/l are recorded from between depths of 200 to 600 m (Roden 1964).

Sedimentation

Where the OMZ impinges on the sediment surface within the Gulf of California, it restricts benthic faunal activity and facilitates preservation of a primary laminated sediment fabric on the upper slopes of the basins (Calvert 1964; van Andel 1964), dominated by diatoms (Bandy 1961). The sediment fabric comprises an alternation of pale olive and olive-brown laminae, the lighter component being diatomaceous-rich sediment and the darker being lithogenic. Sections from jumbo piston cores JPC44, 48 and 56, box cores BC43 and 50, and examination of hydraulic piston core DSDP Hole 480 (Table 1), reveal that

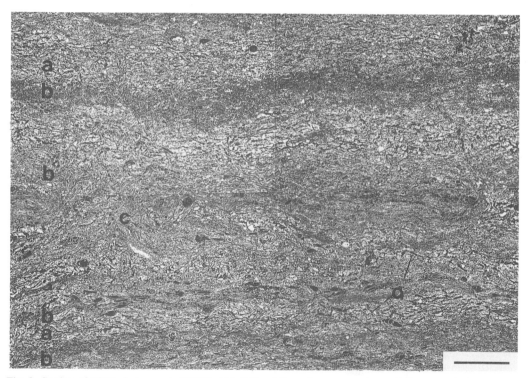

Fig. 3. Section of BSEI photomosaic (JPC56, 205–210 cm) from the western slope of Guaymas Basin showing dominant two-component alternation of lithogenic (**a**) and diatomaceous (**b**) laminae. Note slight bioturbation (**c**) disturbing the laminae, and micro-cracking in the lithogenic laminae (**d**) caused by vacuum impregnation. Scale bar = 1 mm.

a laminated fabric dominates Holocene sedimentation within the central Gulf (Fig. 2).

Methodology

Sediment preparation

Two methods of sample preparation were used to prepare polished thin sections from unconsolidated Holocene sediment. The first technique involved the vacuum drying (Murphy 1986), and subsequent resin impregnation of soft sediment using a Logitech IU-20 vacuum impregnation chamber and Epotech 301 epoxy resin, following the methods of Kemp (1990) and Patience et al. (1990). Vacuum resin impregnation is a quick, relatively straightforward method of embedding unconsolidated sediment. However, the physical drying of the sediment leads to (sometimes severe) desiccation, or micro-cracking (Fig. 3) of clay-rich layers making high-resolution fabric studies of highly water-saturated sediments very difficult. The second technique was a low viscosity resin, fluid-displacive embedding technique. This involved soaking the samples in

acetone (regularly replaced to chemically remove water from the unconsolidated sediment) followed by soaking the sample in low-viscosity Spurr epoxy resin (regularly replaced to remove the acetone) to embed the sample. After the final addition of resin, the sample was left to soak for up to one month and was then cured using multiple temperature steps (Jim 1985; Polysciences 1986; Pike & Kemp 1996). The fluid-displacive technique dramatically improves the quality of the polished thin section as desiccation and therefore fabric disturbance, does not occur within the clay-rich laminae facilitating much higher resolution analysis (Fig. 4).

Backscattered electron imagery

The many geological applications of backscattered electron imagery are discussed in detail by Krinsley et al. (1983) and Pye & Krinsley (1984), and backscattered electron image (BSEI) analysis is now a well proven technique for the study of laminated sediments (Kemp 1990; Kemp & Baldauf 1993; Brodie & Kemp 1994; Aplin et al.

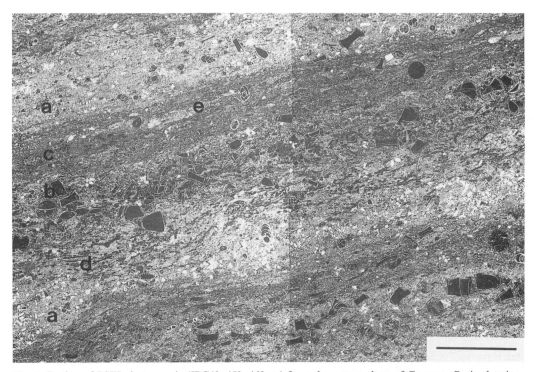

Fig. 4. Section of BSEI photomosaic (JPC48, 459–469 cm) from the eastern slope of Guaymas Basin showing dominant three-component pattern of lithogenic (**a**), mixed-flora (**b**) and near-monospecific flora (**c**) laminae. Note the presence of *Rhizosolenia* within the top of a lithogenic lamina (**d**) and the lithogenic sub-lamina within the near-monospecific flora lamina (**e**). Also note the absence of micro-cracking, prevented by using a fluid-displacive impregnation technique. Scale bar = 1 mm.

1992). High values of backscatter coefficient (related to average atomic number of the target) give bright signals or images, such as the signals produced by mineral grains of pyrite, carbonate, quartz and feldspar, whereas low values of backscatter coefficient give darker images such as those produced by carbon-based resin (i.e. black) (Fig. 4). Therefore, not only do BSEI photomosaics give compositional information but may also be regarded as porosity maps. Low porosity areas of densely packed mineral grains give very bright images and high porosity areas of resin-filled voids, such as those dominated by diatom frustules, produce dark images. The contrasting signals seen in backscattered electron images facilitate high-resolution sedimentary fabric studies and low magnification (e.g. ×20) BSEI photomosaics provide excellent base-maps for higher-magnification studies (Fig. 4).

Secondary electron imagery

A detailed description of secondary electron (SE) scanning electron microscopy may be found in Goldstein *et al.* (1981). Conventional scanning electron microscope stubs were prepared from (lamina-parallel) broken surfaces of unconsolidated sediment. These were either dried and gold-coated for analysis using a JEOL 6400SEM microscope to produce topographical SE images, or analyzed directly (still water-saturated) using a JEOL 5300LV low-vacuum microscope (producing topographical BSEI). The topographic images produced enable comparison with, and identification of sediment components noted from BSEI photomosaics, particularly diatom species.

Light microscopy

Thin sections were also examined using the light microscope, as were standard smear slides. Data obtained were used in conjunction with those from the topographical SE images for the identification of diatom species present, and matching these to the frustules cross sections seen in the BSEI photomosaics.

Results

BSEI analysis of laminated sediments from the central Gulf of California reveals a fabric of alternating bright and dark laminations. High-magnification investigations show excellent preservation of components within the sediment and little evidence of dissolution. Laminations

may be classified as (1) lithogenic, (2) mixed-flora and (3) near-monospecific flora.

Lamina classification

Lithogenic laminae. This lamina type appears very bright on the BSEI photomosaics (Fig. 4). Comprising clay and silt (predominantly quartz

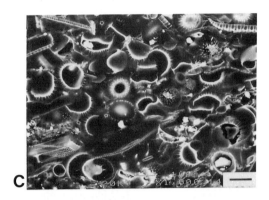

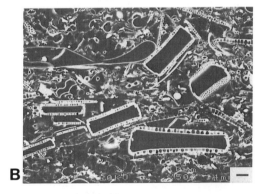

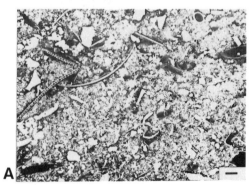

Fig. 5. High magnification backscattered electron images of the major lamina types. **A** – lithogenic lamina (scale bar = 10 μm), **B** – mixed-flora lamina (scale bar = 10 μm), **C** – near-monospecific flora lamina of *Chaetoceros* resting spores (scale bar = 10 μm).

and feldspar) there is a minor, but ubiquitous, diatomaceous component (Fig. 5a) and an average lamina thickness of 0.612 mm ($n = 268$). The lower boundary of silt laminae are gradational whereas the upper ones are sharper. The diatom flora is composed of intact *Chaetoceros* spp. resting spores (e.g. *C. van heurckii*) and other taxa such as *Actinoptychus* spp. (e.g. *A. vulgaris*) with fragments of large centric diatom (e.g. *Coscinodiscus* spp.) valves and girdle bands. Occasionally, large intact diatom taxa, such as *Rhizosolenia bergonii* and *Coscinodiscus asteromphalus*, and also *Stephanopyxis palmeriana*, occur in numbers sufficient to form sub-laminae at the top of mud laminations (Fig. 4).

Mixed-flora laminae. Dark and intermediate backscatter coefficient backscattered electron images characterize this lamina type (Fig. 4), comprising a diverse diatom flora including *Roperia tesselata*, *Coscinodiscus* spp., *Thalassiosira* spp., *Actinocyclus* spp., *Asteromphalus* spp., *Hemidiscus cuneiformis*, *Chaetoceros* spp., *Thalassionema nitzschioides*, *Fragilariopsis doliolus* (Fig. 5b). Frustules may be either dominantly intact or dominantly fragmented. There is a minor lithogenic component dominated by clays with little silt. Brighter images are either of more fragmented frustules

(therefore more closely packed) or have a higher clay content.

Near-monospecific flora laminae. This lamina type is recognised by a dark backscatter coefficient signal caused by resin-filled diatom frustules (Fig. 4). The diatomaceous component as a whole (mixed flora and near-monospecific flora laminae) has an average thickness of 1.031 mm ($n = 289$). Contacts between the two lamina types are often gradational. Taxa forming near-monospecific laminae are *Chaetoceros* spp. (Fig. 5c) comprising both *setae*-rich (from the vegetative stage) and resting spore-rich horizons (Fig. 6), *Skeletonema costatum*, *Thalassiothrix longissima*, *Coscinodiscus* spp., *Hemidiscus cuneiformis* and *Stephanopyxis palmeriana*. Laminations of *Coscinodiscus* spp. (e.g. *C. granii* and *C. asteromphalus*) and *Stephanopyxis palmeriana* tend to be found directly above lithogenic laminations and do not usually show the great number of individuals that laminae of *Chaetoceros* resting spores attain. *Chaetoceros* resting spores and *Skeletonema costatum* are usually found above a mixed-flora lamina and below a lithogenic lamina. This lamina type contains very little clay or silt, however, lithogenic material is occasionally concentrated in sub-laminae within the monospecific (and also within the mixed flora) horizons (Fig. 4).

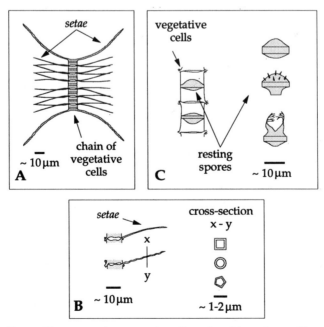

Fig. 6. Sketch of the diatom *Chaetoceros*. **A** – vegetative cells and position of *setae*, **B** – resting spores, **C** – cross-sections of *setae* as seen in high magnification images. Redrawn from Rines & Hargraves (1988).

Laminated sediment fabric

BSEI analysis shows that laminated sediment fabric from Guaymas Basin may be divided into two broad categories consisting of a two-component alternation and a three-component alternation. The two-component alternation comprises a dark, diatomaceous laminae and bright, lithogenic laminae (Fig. 3). The diatomaceous component consists of either the mixed-flora, or near-monospecific flora lamina type, and has relatively sharp boundaries. The three-component alternation consists of a lithogenic lamina, followed by a mixed-flora lamina and then a near-monospecific flora lamina, usually *Chaetoceros* spp. resting spores or *setae*, or more rarely *Skeletonema costatum* (Fig. 4). Laminae boundaries are generally sharp, but the two diatomaceous lamina types cannot be differentiated by visual examination alone.

Occasionally, the three component signal is complicated by sub-laminae of either silt, within diatomaceous components, or diatoms, within the lithogenic laminae, resulting in a four (or more) component alternation in the sediment record (Fig. 4).

Discussion

The Holocene sediments of the Gulf of California have been described as two-component marine varves, reflecting the summer/autumn wet season and winter/spring dry season oscillation of climate in this area of northwest Mexico (Byrne & Emery 1960; Calvert 1966; Donegan & Schrader 1982; Baumgartner 1987). New high-resolution data from BSEI analysis of these sediments has revealed that a more complex record of seasonal flux is present.

Lithogenic laminae

Volcanic rocks from the Baja California peninsula (Fensby & Gastil 1991), and sediment from the Sonora Desert provide the terrestrial sources for the majority of the clay, quartz and feldspar grains (Kelts & Niemitz 1982) which comprise the lithogenic flux in the central Gulf. A number of different processes could be responsible for the concentration of the lithogenic component into laminae. Calvert (1966) suggested that seasonally increased fluvial discharge from the many large rivers flowing into the central and southern Gulf (Rios Yaqui, Mayo and Feurte) caused deposition of lithogenic laminae during the summer rainy season. Construction of dams along many of these rivers began in the 1940s and such an event should have resulted in 'deposition of very high purity diatomaceous sediments' (Calvert 1966). However, a study of the Twentieth Century laminated sediment record from Guaymas Basin has shown no significant change in the appearance of the sediment fabric or the average thickness of the lithogenic laminae, at least in the central Gulf (Baumgartner *et al.* 1991). As an alternative to fluvial influx, Baumgartner *et al.* (1991) suggested aeolian transport of material from the surrounding Sonora Desert as a mechanism for the formation of lithogenic laminae, an idea supported by the fact that fine-grained sediment accumulation rates in the central Gulf remained constant after damming (Baba *et al.* 1991). Aeolian transport of dust from this desert is also suggested as the mechanism for terrigenous mineral deposition in the southwestern United States, north of the Gulf (Péwé 1981). Increased convective thunderstorm activity over the desert during summer (Idso 1976) carries dust for many kilometres, and deposits it on land to the north of the Gulf, and into the central Gulf during the season of lowest productivity, and therefore lowest flux of biogenic material to the basin slopes (Alvarez-Borrego & Lara-Lara 1991; Thunell *et al.* 1993). The greatest number of tropical storms and hurricanes enter the Gulf during the months of September and October (Roden 1964; Schrader & Baumgartner 1983). These have their origins further south along the Mexican coast in the inter-tropical convergence zone (Reyes & Mejía-Trejo 1991). Coastally trapped, long period waves (generated by these severe storms and hurricanes) are propagated northwards (Enfield & Allen 1983) and dominate shelf circulation in the eastern Gulf during this period (Merrifield & Winant 1989). Therefore, reworking of shelf deposits during storms that enter the Gulf, and also by these remotely generated waves could also be responsible for the formation of the lithogenic laminae (Baumgartner *et al.* 1991). The presence of *Actinoptychus* spp. and other coastal neritic and benthic taxa (C. Sancetta, written comm.) would support the theory that at least some material is redeposited onto the basin slopes from the shelf. Other diatoms present within the lithogenic laminae are either generally robust forms (i.e. *Chaetoceros* spp. resting spores), or occur within sub-laminae at the top of the lithogenic component (i.e. *Rhizosolenia* spp., *Coscinodiscus* spp. and *Stephanopyxis palmeriana*) (Fig. 4). *Rhizosolenia* spp. tend to occur within the lithogenic laminae and are characteristic of stratified, stable water column conditions, and low production environments (C. Sancetta, written comm.). *Coscinodiscus* spp. are open ocean taxa and sub-laminae typically occur at the top of the lithogenic laminae and could represent deposition from early blooms as the SST decreases (Thunell *et al.* 1993) and the thermocline decays in November (Robles & Marinone 1987) resulting in mixing of new nutrients into the surface layer. Surface pigment concentrations (from coastal zone colour scanner (CZCS) satellite data) in the central Gulf begin to increase in November before the strongest northwesterly winds have initiated upwelling cells, therefore supporting the presence of a phase of early winter production due to decay of thermal stratification (Thunell *et al.* 1993).

Mixed-flora laminae

By analogy with modern sediment trap data (Thunell *et al.* 1993), this lamina type is deposited during the early winter, and comprises a diverse flora of diatoms. Prevailing summer southerly winds undergo a transition to strong northwesterly winds during November causing the onset of wind-mixing, and upwelling on the mainland coast during winter and spring (Thunell *et al.* 1993). Diatoms that are present in the euphotic zone at this time, taxa such as *Roperia tesselata*, *Actinocyclus* spp., and *Fragilariopsis doliolus* and *Thalassionema nitzschioides* multiply (Sancetta 1993) and frustules are deposited over the summer and early autumn lithogenic lamina. CZCS data show that November through to April is the period of highest surface pigment concentrations across the central Gulf, which encompasses the period of water column stratification breakdown and onset of coastal upwelling (Thunell *et al.* 1993). Zooplankton also increase at this time, to take advantage of the increased phytoplankton biomass (Alvarez-Borrego & Lara-Lara 1991), the increased grazing pressure is probably responsible for the degree of fragmentation of the frustules in this lamina type. Studies of sediment trap fluxes in British Columbian fjords have also ascribed high fluxes of fragmented frustules to grazing by zooplankton (Sancetta 1989). This situation, with diverse diatom species blooming, may persist for up to three months (Thunell *et al.* 1993).

Near-monospecific flora laminae

The near-monospecific flora lamina type has various origins, distinguishable by the diatom species present. The laminae of *Chaetoceros* spp.

(*setae* or resting spores) (Fig. 6) and the more rare *Skeletonema costatum* are deposited either from, or at the end of, the coastal upwelling blooms during the spring, when northwesterly winds are at their strongest (Sancetta 1993). As upwelling plumes spread across the Gulf, diatoms are sedimented. Large fluxes of *Chaetoceros* spp. resting spores are known to settle out of the euphotic zone when nutrients are exhausted at the end of the upwelling season in the eastern Gulf, when the northwesterly winds slacken (Donegan & Schrader 1982). *Chaetoceros* resting spores and *setae* are commonly preserved in the sediment because they are relatively robust, whereas *Chaetoceros* vegetative cells and *Skeletonema costatum* are more weakly silicified, suffer more grazing pressure and are less frequently preserved. Some dissolution of silica may be occurring within the water column and, therefore, a preservational bias may be present in the sediment record (Sancetta 1992). Donegan & Schrader (1982) note that the *Chaetoceros* resting spores are consistently more abundant in the dark, lithogenic laminae. BSEI data do not support this result, showing that the resting spores commonly form discrete laminae below the lithogenic component. The lamina-by-lamina sampling methods of Donegan & Schrader (1982) may have led to the discarding of the majority of this resting spore lamina as they sampled light and dark lamina separately, resulting in a false record. Laminae comprising *Coscinodiscus* spp. or *Rhizosolenia* spp. are the result of sedimentation from blooms of these open oceanic species as the summer thermocline decays (Thunell *et al.* 1993). This is essentially the same mechanism that forms diatom sub-laminae within the top of lithogenic laminae, the difference between the two reflecting the amount of silt and clays present. The lithogenic sub-laminae which occur within the near-monospecific flora lamina type probably represent solitary large storm or shelf mixing events during the early summer. Thunell *et al.* (1993) describe a peak in lithogenic flux recorded in sediment trap data during November and December 1991, and ascribe this to the influence of El Niño. This may also provide a mechanism for the formation of a proportion of the observed lithogenic sub-laminae within the diatomaceous component.

Summary of scanning electron microscope analysis

The particle flux from the water column in the Gulf produces a complicated laminated sediment fabric, recording seasonal variations in the atmospheric and oceanic climate of the region (Byrne & Emery 1960; Calvert 1966). The three-component alternation records the deposition of: (1) mud during the summer and autumn, with occasional blooms of open ocean diatoms as the thermocline collapses in late autumn to early winter; (2) a diverse diatom flora during the early winter partially grazed in the water column by zooplankton, fragmenting frustules; and (3) a near-monospecific diatom flora from the spring coastal upwelling bloom during the period of strongest northwesterly winds with sub-laminae of mud due to early summer storm activity (Fig. 4). The presence of sub-laminae of diatoms and mud, within the lithogenic and near-monospecific flora lamina types respectively, has implications for the interpretation of single lamina composition data (Donegan & Schrader 1982; Kelts & Niemitz 1982). A diatomaceous lamina may have a relatively

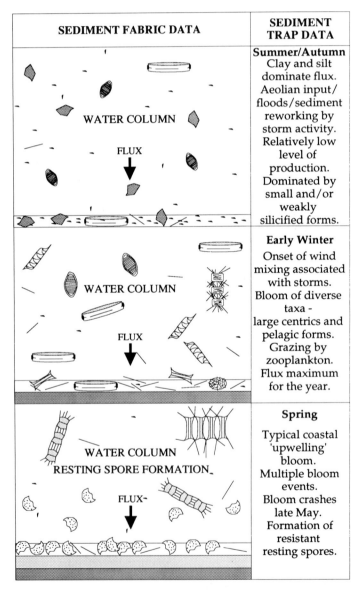

SEDIMENT FABRIC DATA	SEDIMENT TRAP DATA
WATER COLUMN / FLUX	**Summer/Autumn** Clay and silt dominate flux. Aeolian input/floods/sediment reworking by storm activity. Relatively low level of production. Dominated by small and/or weakly silicified forms.
WATER COLUMN / FLUX	**Early Winter** Onset of wind mixing associated with storms. Bloom of diverse taxa - large centrics and pelagic forms. Grazing by zooplankton. Flux maximum for the year.
WATER COLUMN RESTING SPORE FORMATION / FLUX	**Spring** Typical coastal 'upwelling' bloom. Multiple bloom events. Bloom crashes late May. Formation of resistant resting spores.

Fig. 7. Model showing the formation of the three-component sediment fabric in comparison with sediment trap data from the Gulf of California.

high mud content, however, this may be contained entirely within a sub-lamina, representing deposition from one water column event rather than continuous flux over the entire period of lamina deposition.

The two-component alternation represents a 'damped' version of the three-component signal, where deposition from some flux events maybe completely missing due to grazing pressure, or do not occur because of a period of unusually low winter productivity or an anomalously dry rainy season.

Comparisons with sediment trap data

A time series of sediment trap data has been continually compiled, since August 1990, from two traps deployed on the eastern and western side of the central Gulf of California (Thunell *et al.* 1993). Analysis of the sequence of fluxes reveals a dominance of lithogenic material in the summer and autumn and a production maximum during winter and spring. Complications are introduced by high fluxes of mud, or minima in production, late in the winter (Thunell *et al.* 1992). Opaline silica flux peaks in February, and is generally high from November through to February. Sea surface pigment concentrations remain high during the spring, due to coastal upwelling, however, opal flux is relatively low. It is likely that this lower flux is an artifact of grazing and recycling in the euphotic zone, opal still dominates the lithogenic flux (Thunell *et al.* 1993) and diatom-rich sediment continues to be deposited. A general sequence may be discerned in the flux of diatom species. An early winter bloom event occurs throughout the Gulf which is characterized by a diverse range of genera including large *Rhizosolenia* spp., *Coscinodiscus* spp. and other pelagic species such as *Thalassiosira* spp. and *Nitzschia* spp.. A flux of this nature may continue for three months and is followed by a transition to typical coastal bloom taxa such as *Chaetoceros radicans* and *Skeletonema costatum* in the spring. This is followed in the summer and autumn by a period of low diatom production, dominated by weakly silicified taxa such as *Synedra indica* and *Hemiaulus hauckii* (Sancetta 1993).

Flux patterns described from the sediment trap data are reflected in BSEI photomosaics of Holocene sediments from the central Gulf. Sublaminae of *Rhizosolenia* spp. are seen within the top of the lithogenic component and *Coscinodiscus* spp., representing the early winter bloom, are often seen to follow the lithogenic component in sediments from the slopes of the western

Guaymas Basin. The sequence of summer and autumn lithogenic laminae, followed by early winter mixed-flora laminae (comprising centric and pennate diatoms and representing the opal flux maximum for the year) succeeded by spring near-monospecific flora laminae (comprised of *Chaetoceros* spp. resting spores) is also preserved in the sediments (Fig. 4). Lithogenic sublaminae (Fig. 5a) can also be related to late winter periods of high mud flux recorded in the sediment traps. Therefore, the sequence of flux recorded in Holocene sediments of Gulf of California compares favourably with recent sediment trap data (Fig. 7).

Conclusions

Direct comparison of sediment trap with new BSEI data reveals that the present day three component sequence of annual flux has been common throughout much of the period of Holocene sedimentation in Guaymas Basin, Gulf of California. Deposition of diatomaceous material begins as the thermocline collapses in early winter and water column mixing begins. This biogenic flux maxima is followed by deposition from the spring coastal upwelling bloom, and a lithogenic flux maxima occurs during the summer and early autumn. Whereas other sampling methods, such as single lamina sampling and homogenization, may lead to loss of valuable sediment fabric information, BSEI enables investigation of the intact sediment and is a powerful tool for palaeoceanographic and palaeoclimatic reconstructions.

JP wishes to acknowledge the receipt of NERC Research Studentship GT4/92/263/G, and AESK acknowledges the receipt of a NERC Small Grant GR9/203. We are grateful to L. D. Keigwin and R. C. Thunell, chief scientists of the Atlantis II cruise 125/ leg 8 for providing piston and box core material; to C. Sancetta for sharing diatom sediment trap data; and to C. B. Lange for invaluable help with diatom identification. We also thank A. C. Aplin and A. J. Powell for their reviews of this manuscript.

References

ALVAREZ-BORREGO, S. & LARA-LARA, J. R. 1991. The physical environment and primary productivity of the Gulf of California. *In*: DAUPHIN, J. P. & SIMONEIT, B. R. T. (eds) *The Gulf and Peninsular Province of the Californias*. American Association of Petroleum Geologists Memoir, **47**, 555–567.

APLIN, A. C., BISHOP, A. N., CLAYTON, C. J., KEARSLEY, A. T., MOSSMANN, J.-R., PATIENCE, R. L., REES, A. W. G. & ROWLAND, S. J. 1992. A lamina-scale geochemical and sedimentological study of sediments from the Peru Margin (Site 680, ODP Leg 112). *In*: SUMMERHAYES, C. P., PRELL, W. L. & EMEIS, K. C. (eds) *Upwelling Systems: Evolution Since the Early Miocene.* Geological Society, London, Special Publication, **64**, 131–149.

BABA, J., PETERSON, C. D. & SCHRADER, H. J. 1991. Fine-grained terrigenous sediment supply and dispersal in the Gulf of California during the last century. *In*: DAUPHIN, J. P. & SIMONEIT, B. R. T. (eds) *The Gulf and Peninsular Province of the Californias.* American Association of Petroleum Geologists Memoir, **47**, 589–602.

BADAN-DANGON, A., KOBLINSKY, C. J. & BAUMGARTNER, T. 1985. Spring and summer in the Gulf of California: observations of surface thermal patterns. *Oceanologica Acta* 8, 13–22.

BANDY, O. L. 1961. Distribution of foraminifera, radiolaria and diatoms in sediments of the Gulf of California. *Micropaleontology*, **7**, 1–26.

BAUMGARTNER, T. R. 1987. *High Resolution Paleoclimatology from the Varved Sediments of the Gulf of California.* PhD Thesis, Oregon State University.

——, FERREIRA-BARTRINA, V. & MORENO-HENTZ, P. 1991. Varve formation in the central Gulf of California: a reconsideration of the origin of the dark laminae from the 20th Century varve record. *In*: DAUPHIN, J. P. & SIMONEIT, B. R. T. (eds) *The Gulf and Peninsular Province of the Californias.* American Association of Petroleum Geologists Memoir, **47**, 617–635.

BRAY, N. A. 1988a. Thermohaline circulation in the Gulf of California. *Journal of Geophysical Research*, **93**, 4993–5020.

——1988b. Watermass formation in the Gulf of California. *Journal of Geophysical Research*, **93**, 9223–9240.

—— & ROBLES, J. M. 1991. Physical oceanography of the Gulf of California. *In*: DAUPHIN, J. P. & SIMONEIT, B. R. T. (eds) *The Gulf and Peninsular Province of the Californias.* American Association of Petroleum Geologists Memoir, **47**, 511–553.

BRODIE, I. & KEMP, A. E. S. 1994. Variation in biogenic and detrital fluxes and formation of laminae in late Quaternary sediments from the Peruvian coastal upwelling zone. *Marine Geology*, **116**, 385–398.

BYRNE, J. V. & EMERY, K. O. 1960. Sediments of the Gulf of California. *Bulletin of the Geological Society of America*, **71**, 983–1010.

CALVERT, S. E. 1964. Factors affecting distribution of laminated diatomaceous sediments in the Gulf of California. *In*: VAN ANDEL, T. H. & SHOR, G. G. (eds) *Marine Geology of the Gulf of California.* American Association of Petroleum Geologists Memoir, **3**, 311–330.

——1966. Origin of diatom-rich varved sediments from the Gulf of California. *Journal of Geology*, **74**, 546–565.

DAUPHIN, J. P. & SIMONEIT, B. R. T. (eds) 1991. *The Gulf and Peninsular Province of the Californias.* American Association of Petroleum Geologists Memoir, **47**.

DONEGAN, D. & SCHRADER, H. 1982. Biogenic and abiogenic components of laminated hemipelagic sediments in the central Gulf of California. *Marine Geology*, **48**, 215–237.

ENFIELD, D. B. & ALLEN, J. S. 1983. The generation and propagation of sea level variability along the Pacific coast of Mexico. *Journal of Physical Oceanography*, **13**, 1012–1033.

FENSBY, S. S. & GASTIL, R. G. 1991. Geologic-tectonic map of the Gulf of California and surrounding areas. *In*: DAUPHIN, J. P. & SIMONEIT, B. R. T. (eds) *The Gulf and Peninsular Province of the Californias.* American Association of Petroleum Geologists Memoir, **47**, 79–83.

GOLDSTEIN, J. I., NEWBURY, D. E., ECHLIN, P., JOY, D. C., FIORI, C. & LIFSHIN, E. 1981. *Scanning Electron Microscopy and X-ray Microanalysis.* Plenum, New York.

IDSO, S. B. 1976. Chubasco. *Weather*, **31**, 224–226.

JIM, C. Y. 1985. Impregnation of moist and dry unconsolidated clay samples using Spurr resin for microstructural studies. *Journal of Sedimentary Petrology*, **55**, 597–599.

JUILLET-LECLERC, A. & SCHRADER, H. 1987. Variations of upwelling intensity recorded in varved sediment from the Gulf of California during the past 3000 years. *Nature*, **329**, 146–149.

KEIGWIN, L. D. & JONES, G. A. 1990. Deglacial climatic oscillations in the Gulf of California. *Paleoceanography*, **5**, 1009–1023.

KELTS, K. & NIEMITZ, J. 1982. Preliminary sedimentology of late Quaternary diatomaceous muds from Deep Sea Drilling Project Site 480, Guaymas Basin slope, Gulf of California. *In*: CURRAY, J. R., MOORE, D. G. *et al.* (eds) *Initial Reports DSDP, 64.* Washington (US Govt. Printing Press), 1191–1210.

KEMP, A. E. S. 1990. Sedimentary fabrics and variation in lamination style in Peru continental margin upwelling sediments. *In*: SUESS, E., VON HEUNE, R. *et al.* (eds) *Proceedings of ODP, Scientific Results, 112.* Ocean Drilling Program, College Station, TX. 43–58.

—— & BALDAUF, J. G. 1993. Vast Neogene laminated diatom mat deposits from the eastern equatorial Pacific Ocean. *Nature*, **362**, 141–144.

KRINSLEY, D. H., PYE, K. & KEARSLEY, A. T. 1983. Application of backscattered electron microscopy in shale petrology. *Geological Magazine*, **120**, 109–114.

MERRIFIELD, M. A. & WINANT, C. D. 1989. Shelf circulation in the Gulf of California: a description of the variability. *Journal of Geophysical Research*, **94**, 18 133–18 160.

MURPHY, C. P. 1986. *Thin Section Preparation of Soils and Sediments.* A. B. Academic, Berkhampsted.

PATIENCE, R. L., CLAYTON, C. J., KEARSLEY, A. T., ROWLANDS, S. J., BISHOP, A. N., REES, A. W. G., BIBBY, K. G. & HOPPER, A. C. 1990. An integrated biochemical, geochemical, and sedimentological study of organic diagenesis in sediments from Leg 112. *In*: SUESS, E., VON HUENE, R. *et al.* (eds) *Proceedings of ODP, Scientific Results, 112.* College Station, TX (Ocean Drilling Program), 135–153.

PÉWÉ, T. L. (eds) 1981. *Desert Dust: Origin, Characteristics, and Effect on Man.* Geological Society of America, Special Paper **186**.

PIKE, J. & KEMP, A. E. S. 1996. Preparation and analysis techniques for studies of laminated sediments. *This volume.*

POLYSCIENCES, Inc. 1986. Spurr Low-Viscosity Embedding Media. *Polysciences Data Sheet 127.*

PYE, K. & KRINSLEY, D. H. 1984. Petrographic examination of sedimentary rocks in the SEM using backscattered electron detectors. *Journal of Sedimentary Petrology*, **54**, 877–888.

REYES, S. & MEJÍA-TREJO, A. 1991. Tropical perturbations in the eastern Pacific and the precipitation field over north-western Mexico in relation to the ENSO phenomenon. *International Journal of Climatology*, **11**, 515–528.

RINES, J. E. B. & HARGRAVES, P. E. 1988. The Chaetoceros Ehrenberg (Bacillariophyceae) flora of Narragansett Bay, Rhode Island, USA. *Bibliotheca Phycologica*, **79**, 1–196.

ROBLES, J. M. & MARINONE, S. G. 1987. Seasonal and interannual thermohaline variability in the Guaymas Basin of the Gulf of California. *Continental Shelf Research*, **7**, 715–733.

RODEN, G. I. 1964. Oceanographic aspects of Gulf of California. *In*: VAN ANDEL, T. H. & SHOR JR., G. G. (eds) *Marine Geology of the Gulf of California.* American Association of Petroleum Geologists Memoir, 3, 30–58.

RUSNAK, G. A., FISHER, R. L. & SHEPARD, F. P. 1964. Bathymetry and faults of the Gulf of California. *In*: VAN ANDEL, T. H. & SHOR, G. G. (eds.) *Marine Geology of the Gulf of California.* American Association of Petroleum Geologists Memoir, 3, 59–75.

SANCETTA, C. 1989. Spatial and temporal trends of diatom flux in British Columbian fjords. *Journal of Plankton Research*, **11**, 503–520.

——1992. Comparison of phytoplankton in sediment trap time series and surface sediments along a productivity gradient. *Paleoceanography*, **7**, 183–194.

——1993. Seasonal succession of diatoms in the Gulf of California derived from two years of moored sediment traps. *EOS, Transactions, AGU* 74 (1993 Fall Meeting Supplement), 371.

SANTAMARÍA-DEL-ANGEL, E., ALVAREZ-BORREGO, S. & MÜLLER-KARGER, F. E. 1994. Gulf of California biogeographic regions based on coastal zone color scanner imagery. *Journal of Geophysical Research*, **99**, 7411–7421.

SCHRADER, H. & BAUMGARTNER, T. 1983. Decadal variation in upwelling in the central Gulf of California. *In*: THIEDE, J. & SUESS, E. (eds) *Coastal Upwelling, its Sediment Record. Part B: Sedimentary records of ancient coastal upwelling.* NATO Conference Series. iv, Marine Sciences; 10B. Plenum, New York. 247–276.

THUNELL, R., PRIDE, C., TAPPA, E., MULLER-KARGER, F., SANCETTA, C. & MURRAY, D. 1992. Seasonal sediment fluxes and varve formation in the Gulf of California. *In*: SARNTHEIN, M., THIEDE, J. & ZAHN, R. (eds) *Fourth International Conference on Paleoceanography: Short- and Long-Term Global Change.* Kiel, Germany. Program and Abstracts, 280–281.

——, PRIDE, C., TAPPA, E. & MULLER-KARGER, F. 1993. Varve formation in the Gulf of California: Insights from time series sediment trap sampling and remote sensing. *Quaternary Science Reviews*, **12**, 451–464.

VAN ANDEL, T. H. 1964. Recent marine sediments of the Gulf of California. *In*: VAN ANDEL, T. H. & SHOR, G. G. (eds). *Marine Geology of the Gulf of California.* American Association of Petroleum Geologists Memoir, 3, 216–310.

WYRTKI, K. 1962. The oxygen minima in relation to ocean circulation. *Deep-Sea Research*, **9**, 11–23.

——1967. Circulation and water masses in the eastern equatorial Pacific Ocean. *International Journal of Oceanology and Limnology*, **1**, 117–147.

ZEITZSCHEL, B. 1969. Primary productivity of the Gulf of California. *Marine Biology*, **3**, 201–207.

The nature of varved sedimentation in the Cariaco Basin, Venezuela, and its palaeoclimatic significance

KONRAD A. HUGHEN[1], JONATHAN T. OVERPECK[1,2],
LARRY C. PETERSON[2] & ROBERT F. ANDERSON[3,4]

[1] *Institute of Arctic and Alpine Research and Department of Geological Sciences,
University of Colorado, Boulder, Colorado 80309, USA*
[2] *NOAA Paleoclimatology Program, NGDC, Boulder, Colorado 80303, USA*
[3] *Rosenstiel School of Marine and Atmospheric Science, University of Miami,
Miami, Florida 33149, USA*
[4] *Lamont-Doherty Earth Observatory, Columbia University, Palisades,
New York 10964, USA*

Abstract: A laminated sediment record of the last 12 600 radiocarbon years from the anoxic Cariaco Basin provides a rare opportunity to study interannual to millenial-scale climatic change in a marine setting. Sedimentological and radiometric analyses of laminae couplets in the basin sediments indicate that they are annually deposited varves. The varves consist of light and dark laminae which are deposited as the result of the Intertropical Convergence Zone's annual north–south migration over the tropical North Atlantic, and the impact of this migration on regional upwelling and rainfall patterns. Light laminae contain planktonic fossils (predominantly diatoms) and are an indicator of productivity over the basin during the winter–spring upwelling season. Dark laminae contain terrigenous mineral grains and are an indicator of runoff from northern South America during the late summer–fall rainy season. Varve measurements developed using X-ray images and thin sections indicate that a continuous 14 000 calendar year long varve chronology may be generated. Preliminary measurements of thickness of individual light and dark laminae for the period 12.6 to 9 ka BP show similar palaeoclimate signals. Thickness of dark laminae, interpreted as measuring changes in regional runoff, increased rapidly at 10.8 ka BP and decreased rapidly at 9.8 ka BP, an interval coincident with the Younger Dryas cold period. Thickness of light (diatom-rich) laminae, indicative of productivity and possibly upwelling intensity, also show an increase at 10.8 ka BP, and a rapid decrease at 9.8 ka BP, simultaneous with the changes in the dark laminae. Current work is focusing on building long multiple-core varve chronologies and using these for palaeoclimatic reconstructions and geochronological research.

Annually dated and resolved records of climate change are important for understanding short-term (annual to decadal) changes in the global climate system. Of the many timescales of climatic change, the patterns and causes of interannual to millenial-scale variability are among the least well understood (Overpeck 1991). A global network of high-resolution terrestrial and marine palaeoclimate records is necessary to advance our knowledge in this realm (Rea 1987). The lack of precisely dated, high-resolution records is particularly apparent in marine studies. The identification and analysis of annually laminated marine sediments is thus an important aspect of palaeoclimatic research.

The Cariaco Basin, located off the northern coast of Venezuela, is a marine pull-apart basin separated from the open Caribbean Sea by shallow submarine sills. The sills average less than 100 m deep and are interrupted by two channels that reach depths of 120 and 146 m and which control the hydrography of the basin's deep waters. The Cariaco Basin itself actually consists of two 1400 m deep sub-basins separated by a 900 m deep saddle (Fig. 1). The detailed biological and physical oceanography of the basin are described in detail elsewhere (Peterson *et al.* 1991) and will only be summarized briefly here.

The Cariaco Basin is presently anoxic below a water depth of about 300 m. This anoxia is caused by oxidation of the large flux of organic material raining down due to high production rates at the surface. The anoxic conditions of the basin waters are maintained by the shallow sills, which restrict ventilation by oxygenated outside water. The basin has been recognized for several decades as a repository for laminated sediments deposited at a rapid rate (Heezen *et al.* 1959; Kipp & Towner 1975; Overpeck *et al.* 1989). The highest sedimentation rates are found in the eastern sub-basin and flank of the saddle, where, due to the lack of a benthic fauna, a 4–10 m thick sequence of finely laminated sediments is preserved above a deeper, bioturbated unit.

From Kemp, A. E. S. (ed.), 1996, *Palaeoclimatology and Palaeoceanography from Laminated Sediments*,
Geological Society Special Publication No. 116, pp. 171–183.

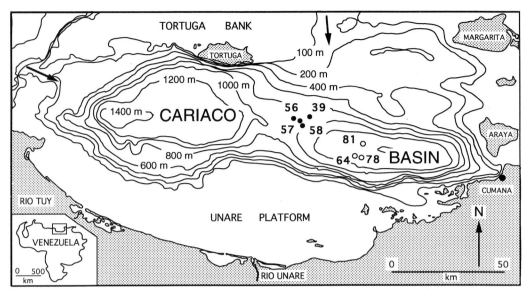

Fig. 1. Location and bathymetry of the Cariaco Basin, Venezuela. The basin consists of two deep sub-basins separated by a saddle. Shallow sills prevent deep Caribbean Sea circulation from entering the basin. Arrows indicate two channels (120 m to the north, 146 m to the west) allowing limited inflow. Locations of cores used in this study are shown by the circles. Solid circles indicate piston cores PL07-39PC, PL07-56PC, PL07-57PC, and PL07-58PC on the eastern slope of the central saddle, used for laminae analysis and generating time series of varve and laminae thickness. Open circles indicate box cores PL07-64BC, PL07-78BC, and PL07-81BC from the floor of the eastern basin, used for ^{210}Pb and upper sediment analysis. Cumana, a site with meteorological data and historical records, is shown on the eastern edge of the basin.

These laminated sediments reflect the shifting sediment production and depositional regimes over the basin, which in turn are controlled by the seasonal climate cycle.

Seasonal climate over northern South America is highly variable and responds to the annual north–south migration of the intertropical convergence zone (ITCZ) over the tropical North Atlantic. During January and February, the ITCZ is at its southernmost position and strong northeasterly trade winds blow along the Venezuelan coast, causing Ekman drift-induced coastal upwelling. Sea surface temperatures within the basin drop as low as 20°C (Fig. 2; Herrera & Febres-Ortega 1975; Aparicio 1986), and primary production rates, measured by chlorophyll A and organic carbon fluxes, are at their maximum during this period (Fig. 3; Ballester 1969). By June or July, the ITCZ has migrated north, trade wind intensity has been reduced, and the upwelling process has slowed or mostly shut down. This causes sea surface temperatures to rise to about 28°C (Fig. 2; Herrera & Febres-Ortega 1975; Aparicio 1986), and primary production rates to fall nearly to zero (Fig. 3; Ballester 1969). In addition to affecting the intensity of upwelling and biological production over the basin, the seasonal change in location of the ITCZ also

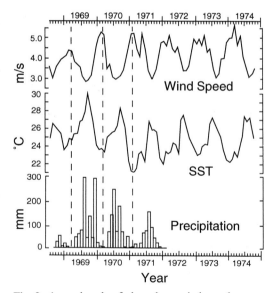

Fig. 2. Annual cycle of alongshore wind speed, sea surface temperature and precipitation near Cumana, at the eastern margin of the Cariaco Basin. Increased wind stress results in increased upwelling, shown here by decreased sea surface temperatures. Note the seasonal antiphase relationship between maximum upwelling and maximum precipitation. Data from Herrera & Febres-Ortega (1975) and Aparicio (1986).

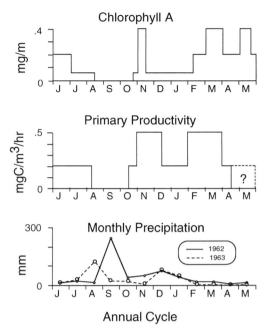

Fig. 3. Annual cycle of surface productivity in the Cariaco Basin, shown with precipitation data from Cumana. Productivity data are sparse and are shown for one year (1968) only. Note that peak productivity occurs during minimum precipitation and vice versa. Seasonal shifts in climate and physical oceanographic parameters, such as intensity of trade wind-induced upwelling and precipitation, result in a seasonal biological response. Data from Ballester (1969).

controls the amount of rainfall over northern South America, and hence river discharge. Annual rainfall in this region also exhibits a distinct seasonality, reaching a maximum in late summer and early fall. At present, the rainy season in the area begins in June or July, increasing clastic input into the basin just as organic production is falling off (Fig. 3).

The purpose of this study has been to examine in detail these laminated sediments and to evaluate their mechanism of formation in light of the pronounced seasonal productivity and hydrographic cycles. The quality and correlatability of the laminated sequences from different cores is also addressed, as well as how they are being used to bridge gaps and compare parameters between cores. We discuss several lines of evidence showing that the laminae couplets are varves and will eventually be used to construct an annual chronology for the basin. Preliminary varve and laminae thickness records are presented, with the ultimate goal of using laminae thickness and composition as palaeoclimate proxy indicators.

Methods

Field work

In June of 1990, a field coring expedition (R/V *Thomas Washington*, Cruise PLUME-07) collected over 100 box, piston and gravity cores from all parts of the basin (Peterson *et al.* 1990). In general, one of each of these types of core was taken at every coring site. Soutar box cores (Soutar & Crill 1977) were used to preserve the fragile sediment–water interface, but these cores were limited to sampling only the uppermost 0.4 to 1 m of sediment. To bring material back to the lab for analysis, we sampled the box cores on board the ship by taking and X-raying unfrozen rectangular sub-cores, and by taking freeze-cores (Shapiro 1958; Fig. 4). We also sampled the box cores by taking unfrozen cylindrical sub-cores. Piston cores typically recovered records

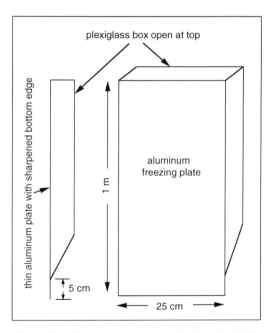

Fig. 4. New freeze-coring method developed by J. T. Overpeck for collecting undisturbed samples of box-cored sediments, including the delicate sediment-water interface. Immediately after box-core recovery, an aluminum and plexiglass sub-corer was pushed gently into the center of the box. The rectangular sub-corer was then filled and refilled continuously with a slurry of crushed dry ice (manufactured on board using CO_2 gas) and methanol until a 3–5 cm thick rind of sediment (and ice above the sediment-water interface) was frozen on the outside of the aluminum freezing plate. This slab was then excavated, carefully removed from the plate by gentle heating, and stored under frozen conditions.

about 10 m long, but the uppermost sediments and sediment–water interface were rarely preserved. Sedimentation rates, based on the depth in each core to the laminated–bioturbated contact, vary widely across the basin. We examined gravity and piston cores with the highest sedimentation rates, looking for continuity of deposition and lack of disturbances. A suite of the highest quality, high-sedimentation rate piston cores was selected for cross-correlation and detailed varve analyses (Table l; Fig. 1). Those X-radiographs and freeze-cores taken from box cores which showed the best preservation of the sediment-water interface in addition to well defined varves were also selected for study (Table l; Fig. 1).

Laminae analysis

High-resolution images at several scales were necessary to correlate sections 4–10 m long of continuous sub-millimetre laminae between different cores. Full-length, colour photomosaics allowed broad comparisons and correlation of sub-centimetre scale features. To correlate individual laminae at the millimetre scale, we used high-resolution X-radiographs. These were obtained by slicing 1 cm-thick slabs from the surface of split core halves off into plexiglass trays, and X-raying these slabs. This allowed imaging of the undisturbed core centre and eliminated the edge distortion caused by the corer's penetration into the soft sediment, which blurred X-radiographs taken of entire core halves. Although our X-radiographs provided sufficiently clear images to count and measure most laminae couplets, they did not provide the resolution necessary to evaluate and identify structures such as micro-turbidites and sub-laminae which appeared to be laminae, but were not in fact deposited during a full upwelling or rainy season. To this end, we sampled a continuous strip of sediment in a sequence of small blocks for embedding in epoxy resin, for subsequent processing into polished

petrographic thin sections. A series of acetone exchanges followed by resin exchanges (Clark 1988; Pike & Kemp 1996) were used to remove the water from interstitial spaces and ensure that the resin penetrated fully into the sediments. For extremely clay-rich layers, pre-drying of samples in a freeze dryer to completely remove interstitial water and/or additional baking of the resin blocks was necessary for adequate curing of the resin.

We analysed the thin sections with a light microscope to determine the internal structure and composition of the laminae, as well as to count and measure accurately the thicknesses of individual light and dark laminae. For more detailed analysis of the internal structure and composition of individual laminae and sub-laminae, we used electron backscatter imaging (Kemp 1990). This technique revealed internal details that would otherwise have been unattainable, such as microbioturbation structures, mineral distributions within laminae, and relative positions and compositions of plankton bloom layers. Digital images of X-radiographs and thin sections were obtained using a video camera and light microscope connected to a computer. Digital electron backscatter electron images (BSEI) were obtained using a JEOL JSM6400 scanning electron microscope and JEOL 8600 electron microprobe. The images were analysed and processed using Dapple Image Prism View and NIH Image, image analysis programs.

Lead-210 analysis

Box cores from the centre of the eastern sub-basin (Table l; Fig. 1) showed distinct laminations at the sediment–water interface and were selected for ^{210}Pb dating (Appleby & Oldfield 1978). The laminated sequence was interrupted near the top by two turbidites, 7 and 3 cm thick, whose presence had been predicted on the basis of the region's earthquake history (C. Schubert,

Table 1. *Core data for piston and box-cores used in this study*

Core ID	Lat. (N)	Long. (W)	Water depth	Core length	Core type
PL07-39PC	10°42.00′	64°56.50′	790 m	993 cm	piston
PL07-56PC	10°41.22′	64°58.07′	810 m	930 cm	piston
PL07-57PC	10°41.21′	64°58.20′	815 m	1096 cm	piston
PL07-58PC	10°40.60′	64°57.70′	820 m	1130 cm	piston
PL07-64BC	10°33.55′	64°47.31′	1376 m	46 cm	box
PL07-78BC	10°33.35′	64°47.01′	1376 m	60 cm	box
PL07-81BC	10°35.33′	64°46.35′	1350 m	66 cm	box

pers. comm.). An unfrozen sub-core from PL07-BC81 was sampled at about 7 mm intervals over the upper 25 cm for ^{210}Pb dating. ^{210}Pb dating was carried out using the procedure outlined by Hughen *et al.* (1996).

Results

Core descriptions

The general stratigraphy of the cores recovered during the PLUME-07 cruise was described by Peterson *et al.* (1990). Piston cores show an abrupt transition from yellowish-brown, bioturbated sediments at depth to greenish-brown, finely laminated sediments above. This transition corresponds to the change from oxic to anoxic conditions within the Cariaco Basin at about 12.6 ka BP (Peterson *et al.* 1991). Within the laminated sections of the piston cores examined, the laminae showed a consistent pattern of change with time. From the onset of anoxia at 12.6 ka BP until about 9.8 ka BP, the deposited laminae were relatively thick and distinct. In the section laid down since 9.8 ka BP, the laminae are thinner and less distinct, although still visible and undisturbed. Ages for events within the anoxic section have been determined from core PL0739PC

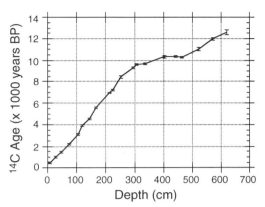

Fig. 5. Radiocarbon age against depth for piston core PL07-PC39. Ages are AMS ^{14}C dates measured on monospecific samples of the planktonic foraminifer, *Globigerina bulloides* (Lin *et al.* in press). Dates are corrected by subtracting 400 years to account for the residence time of carbon in the ocean. The radiocarbon plateau at about 10.0 ka BP was averaged over when calculating ages for varve and laminae thickness time series.

(Lin *et al.* in press; Fig. 5), and are based on 20 AMS ^{14}C dates of monospecific samples of the planktonic foraminifer *Globigerina bulloides*, a species found in high abundance throughout the sediment section. Dates were assigned

Fig. 6. Laminated sediments from piston core PL07-39PC, depths are shown in cm. Note the general lack of disturbances (bent layers in the centre edge were probably formed while coring).

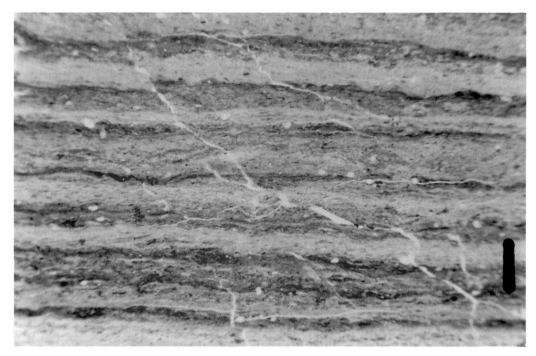

Fig. 7. Optical photomicrograph of laminated sediments in a thin section from piston core PL07-56PC, depth 435 cm. Sediments consist of alternating light and dark laminae. Small voids are planktonic foraminifera, found throughout the sediments. Scale bar = 1 mm.

from PL07-39PC to the other cores in this study on the basis of visual laminae correlations. Piston cores taken from the eastern slope of the central saddle (Fig. 1), including cores PL07-39PC, -56PC, -57PC, and -58PC, have clear, well-defined laminae and are free of major disturbances (Fig. 6). These cores were visually inter-correlated along the entire length of the laminated section and were selected to be the central group for constructing varve chronologies and palaeoclimate records.

Laminae description

This section incorporates detailed observations on the laminated sections of cores PL07-56PC and -57PC. The laminae can be classified as biogenic, after O'Sullivan (1983), consisting of light and dark lamina pairs (Fig. 7) with occasional microturbidites interspersed throughout. The light laminae are composed almost entirely of intact planktonic fossil skeletons – predominantly diatoms, as well as silicoflagellates and coccolith platelets. The materials appear grouped into aggregates (Fig. 8), most likely faecal pellets, surrounded by biogenic

silica. Calcite crystals are fairly common, possibly forming *in situ* following deposition and partial dissolution of calcareous fossils. The dark laminae contain primarily terrigenous minerals, such as quartz and feldspars. Materials in these laminae also seem grouped into aggregates (Fig. 8). Backscatter electron imagery (BSEI) clearly shows the alternating planktonic fossil and terrigenous mineral grain composition, as well as the detailed internal structure, of the laminae (Fig. 9). Detailed BSEI show the distinct compositional differences and clean separation of laminae types (Fig. 10). The plankton-rich and mineral-rich layers are separated into distinct events, with little or no mixing after deposition. High-magnification electron backscatter imagery reveals additional small-scale compositional details of the organic matrix in the plankton-rich layers (Fig. 11). These layers are composed almost entirely of plankton fossils, even in the small void spaces between the larger plankton skeletons.

Both lamina types commonly contain faint sublaminae (Figs 9 & 12). These fine layers probably represent discrete events, plankton blooms or individual storms, that occur within an upwelling or rainy season. Sublaminae, rather than

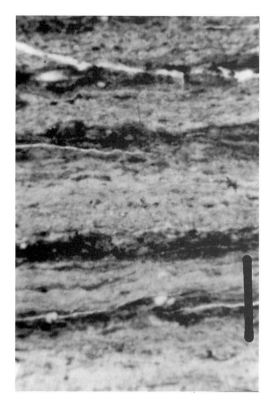

Fig. 8. Close-up optical photomicrograph of laminae from the same thin section as shown in Fig. 7. Note the material aggregated into bundles, giving laminae a distinct textural appearance. Scale bar = 1 mm.

Fig. 9. Electron backscatter mosaic of laminae from a polished thin section from piston core PL07-57PC, depth 749 cm. Because electron backscatter images reflect the atomic number of the material being studied, dense mineral grains appear white, whereas organic material and siliceous fossils appear darker. This is reverse of the relative darkness of laminae in the optical microscope images, where plankton-rich layers are lighter, and terrigenous mineral-rich layers are dark. Note the distinct compositional difference between laminae. Also note the internal structure of the laminae including sub-laminae and minor microbioturbation. Large open ovals and spirals are planktonic forminifera. Scale bar = 1 mm.

occurring occasionally, are actually the building blocks that comprise the laminae themselves. For example, it appears that a light lamina is not a discrete layer of purely fossiliferous material, with no terrigenous material whatsoever, but a layer where there are predominantly more fossiliferous sub-laminae. Similarly, a dark lamina is a layer dominated by a greater number of terrigenous mineral-rich sublaminae. Thus, although the boundaries between sub-laminae are sharp, and there is little, if any, contamination of constituents, the boundaries between laminae themselves are gradational, moving from sections with a higher percentage of light sub-laminae to sections with primarily dark sublaminae. In general, although the occasional dark sub-lamina occurs within a light lamina and vice versa, the core of an individual lamina is usually monotypic, and identification of boundaries is straightforward.

Microbioturbation was recognized by the ubiquitous minor disruption of internal structures such as sub-laminae (Fig. 9). Usually layers were only slightly disturbed and were not homogenized beyond recognition. Occasionally there are sections where microbioturbation has destroyed structure to the point where measurement and counting of varves was impossible. Through correlation with other cores, it should be possible to bridge most of these gaps.

Microturbidites, from 4 mm to less than 0.5 mm thick, appear throughout the sediments (Fig. 13). Under casual inspection, or when using X-radiographs, these can be misidentified as dark or light laminae, which introduces error into varve measurements. However, studying sediments using thin sections made it relatively easy to distinguish between microturbidites and laminae.

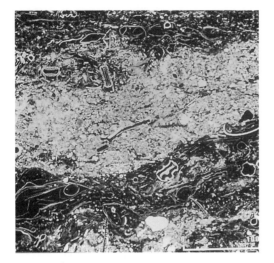

Fig. 10. Close-up electron backscatter image of the same polished thin section as shown in Fig. 9. Note the abrupt boundary separating plankton-rich and mineral-rich layers. Plankton-rich layers contain diverse diatoms, silicoflagellates, coccolith platelets, and other planktonic skeleton fragments. Mineral-rich layers contain terrigenous minerals including quartz and feldspars. Scale bar = 100 μm.

The pelletized structure of the sediments, in addition to the presence of fine sub-laminae, gives undisturbed laminae a distinct textural appearance upon magnification. Microturbidites on the other hand, were identified by their homogenous texture and lack of internal structure, as well as by a colour typically halfway between the light and dark laminae (Fig. 13). Electron backscatter imagery clearly shows the mixing of mineral grains and less dense organic material into a homogenous mass, in contrast

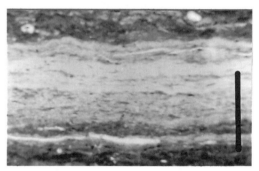

Fig. 12. Optical photomicrograph of laminae in a thin section from piston core PL07-56PC, depth 529 cm. Close-up view of the light lamina shows typical internal structure, particularly the clear sub-laminae. Sublaminae are interpreted as resulting from individual events, rather than from entire upwelling or rainy seasons. Scale bar = 1 mm.

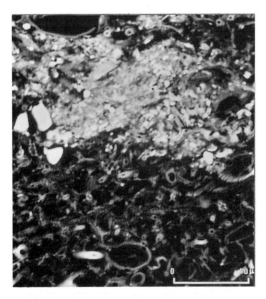

Fig. 11. High-magnification electron backscatter image of the same thin section as shown in Figs 9 and 10 showing a detailed view of the organic matrix between larger plankton skeletons. Note that these layers are composed almost exclusively of planktonic fossils, even at smaller scales. Also note the extremely small size of the clay particles. Scale bar = 10 μm.

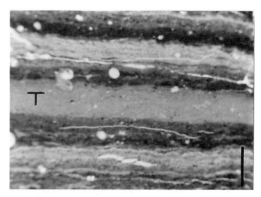

Fig. 13. Optical photomicrograph of laminae in thin section from piston core PL07-56PC, depth 607 cm, showing the appearance of a microturbidite (T) when viewed in thin section. Note the homogenous texture of the turbidite, lacking pelletization, microbioturbation and sublaminae, relative to the complex structure of laminae. Also note somewhat intermediate darkness of turbidite, between that of light and dark laminae. Scale bar = 1 mm.

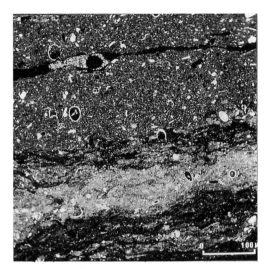

Fig. 14. Electron backscatter image of a microturbidite in thin section from piston core PL07-57PC, depth 747 cm. Note the mixing of dense mineral grains within less-dense organic material in the microturbidite above, contrasted with the clear distinction between mineral grains and planktonic fossils in laminae below. Scale bar = 100 μm.

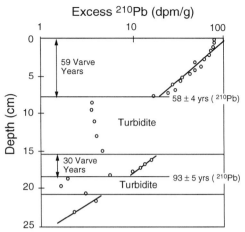

Fig. 15. Results of ^{210}Pb analysis on laminated sediments from box core PL07-81BC. ^{210}Pb dates on the two turbidites in the upper 25 cm yield ages of 58 ± 4 years and 93 ± 5 years before 1990, respectively. Historical records of large earthquakes in and around Cumana give ages of 61 and 90 years before 1990. Varve counts to the turbidites equal 59 and 89 years before 1990, closely matching the independent dating from the other techniques. This confirms that the laminae couplets are in fact annually deposited varves.

to the distinct structure of undisturbed laminae (Fig. 14). Microturbidites generally occur clustered in groups of two to four and seem to be uniformly distributed with depth.

Lead-210 results

Dating by ^{210}Pb of samples from box core PL07-81BC yielded ages for the two turbidites of 58 ± 4 years and 93 ± 5 years before 1990, respectively (Fig. 15). Counts of paired light-dark laminae couplets to the turbidites on X-radiographs of cores PL07-81BC, -78BC, and -64BC equalled 59 and 30 pairs, which agrees with the ^{210}Pb results and supports the annual interpretation for the laminae couplets (Fig. 15). The ^{210}Pb and varve ages are supported further by historical accounts of two earthquakes, one epicentered near Cumana at the eastern edge of the basin (Fig. 1) and one directly under the basin itself. The earthquakes, which very probably triggered turbidity currents, occurred in 1929 and 1900, within one or two years of the dates indicated by ^{210}Pb and varve counting (Paige 1930; Fiedler 1961, 1972). The ^{210}Pb curve (Fig. 15) shows that the cores used for dating came from an area where turbidites are deposited without erosive scouring of the underlying sediments. The equal ^{210}Pb values above and below the upper turbidite show

that no material is missing or disturbed. The offset of values above and below the lower turbidite and shifting of the bracketing points toward lower values was caused by contamination of the bracketing ^{210}Pb samples with some turbidite material.

Palaeoclimate records

Because an absolute varve-based chronology for the Cariaco Basin is not yet complete, we used radiocarbon to assign ages used in the discussion of average varve thickness time series from cores PL07-56PC and -57PC (Fig. 16). The values shown by the dotted line in Fig. 16 were measured from X-radiographs of core PL07-56PC, whereas the values shown by the solid line were measured from thin sections of cores PL07-56PC and -57PC and spliced together to form a single record. The dotted varve thickness values represent averages calculated from measurements over 2–3 centimetres, sampling at approximately 10 cm intervals. The measurements were taken from X-radiographs because thin sections were not yet available for these sediments. Embedding and processing unconsolidated sediments into petrographic thin sections is laborious and has only been completed for the deeper sections of cores PL07-56PC and -57PC. Because the upper

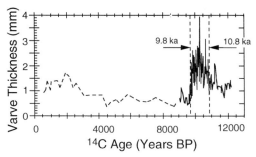

Fig. 16. Average varve thickness record for piston cores PL07-56PC and -57PC. Dotted line from 9 ka BP to present is based on measurements from X-radiographs of core PL07-56PC and is considered preliminary. Solid line from 12.6 to 9 ka BP is based on higher resolution measurements from thin sections of cores PL07-56PC and -57PC. Sections measured from the different cores were spliced together to form a single record. Note the large increase in varve thickness from 10.8 to 9.8 ka BP, concurrent with the Younger Dryas cold period.

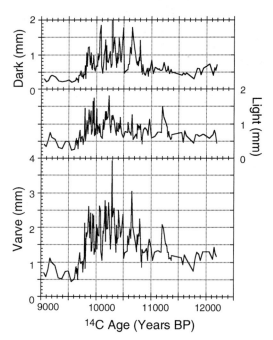

Fig 17. Independently measured dark and light laminae and total varve thickness records for 12.6 to 9 ka BP. Dark laminae thickness is a record of terrestrial sediment carried by runoff, and may be a proxy for precipitation. Light laminae thickness is a record of primary productivity in Cariaco Basin surface waters and may record trade wind-induced upwelling intensity. Further work is needed to clarify these interpretations. Note the abrupt changes in both light and dark laminae at the onset and termination of the Younger Dryas.

(<9 ka BP) varves were fainter and thinner than the older ones below, and the measurements were made using X-radiographs, the identification of microturbidites and disturbances, and the discrimination between laminae and sub-laminae was difficult. For these reasons, the varve thickness measurements for the period 9 ka BP to the present must be considered preliminary. The values shown by the solid line in Fig. 16 were also calculated from varve thicknesses averaged over 2 cm, but were sampled at a much higher resolution, about every 2 cm. The sedimentation rates in this interval are higher, and the measurements were taken from thin sections, allowing greater accuracy in identifying microturbidites and laminae boundaries. Therefore, most of the discussion which follows will focus on this older interval, from 12.6 to 9 ka BP.

The average varve thickness curve for PL07-56/57PC shows great variability, but there are some clear trends in the data. From the onset of anoxia at 12.6 ka BP to a depth in the core dated about 10.8 ka BP, the varves average 1.00 mm thick. Above that depth, they rapidly increase in thickness, reaching a maximum of 3.25 mm and averaging 1.75 mm until a depth dated at approximately 9.8 ka BP. At that point, the varves abruptly become thinner and fainter. The measurements available indicate that from 9.8 to 9 ka BP, the varves average 0.89 mm thick. Low-resolution, preliminary data suggest that average varve thickness remains at this value until 2.6 ka BP. At 2.6 ka BP, there is a gradual increase to an average thickness of 1.3 mm, which remains until the present. The higher resolution record

from 12.6 to 9 ka BP shows a series of large oscillations. In particular, the period of maximum varve thickness between 10.8 and 9.8 ka BP seems to coincide in time with the Younger Dryas cold period of the circum-North Atlantic region (Broecker *et al.* 1988; Wright 1989).

The clarity of images obtained from thin sections enabled us to measure and construct separate thickness records for light and dark laminae (Fig. 17). The separate measurement of these signals allowed us to independently address changes in the different seasonal climate systems. In general, the two records show the same trends. The period of increased sedimentation rate coincident with the Younger Dryas is driven by changes in both types of laminae (Fig. 17). Light and dark laminae increase in thickness abruptly at 10.8 ka BP, followed by an abrupt decrease at 9.8 ka BP. A shorter period of increased thickness from 11.4 to 11.1 ka BP is reflected in both records, as is a brief pulse of decreased thickness during the Younger Dryas at 10.5 ka BP. Despite these similarities, there are some significant differences between the details of the two records. For example, the light laminae show a series of three spikes of increased thickness from 12 to 11 ka BP that are absent or reduced in the dark laminae. In addition, the dark laminae record two spikes of increased thickness during the mid-Younger Dryas (10.3 to 10.0 ka BP) that are not seen in the light laminae. Most of these measurements were made on thin sections from single cores. It is difficult to say whether the detailed observations reported here will be seen in the completed multiple-core laminae thickness records. Regardless, the general pattern of periods of increased thickness leading up to and culminating in the Younger Dryas event occurs in both light and dark laminae records and appears to be robust.

Discussion and conclusions

Laminae analysis

Both independent dating and the clear match of laminae composition with the seasonal sediment production within the basin confirm that the laminae couplets are annually deposited varves and thus useable as a geochronological tool. Our methods of investigating laminae structure and composition also show that individual laminae can be readily identified and measured. The number and quality of cores recovered from the basin, and the correlations of individual laminae between cores suggest that we will be able to bridge disturbances and gaps within single cores. This will make possible the construction of a varve chronology for the Cariaco Basin, which will provide annual dating control for climatic change studies on multiple proxies from the basin. In addition to dating control for palaeoclimate records, the varve chronology will be useful for carbon cycle research. The high abundance of planktonic foraminifera found in the sediments provides the means to create a high-resolution series of AMS ^{14}C dates. By comparing this to an independent varve chronology, we will examine secular ^{14}C variations back to 12 600 radiocarbon years BP. This will give insight into ocean circulation changes and will provide an independent calibration of the marine ^{14}C timescale during this time. This is an important period spanning the Younger Dryas and much of the deglaciation, and our work will contribute needed additional information to the ongoing efforts to confirm the ^{14}C calibration curve beyond about 10 000 radiocarbon years BP (Stuiver & Reimer 1993).

Palaeoclimate records

Using thin sections, we were able to measure light and dark laminae separately, creating independent palaeoenvironmental records. This separated the productivity and runoff records preserved in the sediments, which makes it easier to interpret these signals. It is unclear whether these records are in fact physically independent of each other, as runoff can potentially affect primary productivity, but it is still valuable to distinguish the two signals. Clastic sediment input into the Cariaco Basin from river runoff, recorded by dark laminae thickness, increased sharply during the Younger Dryas cold period and may be the result of increased regional precipitation. Lake sediments from Panama seem to show a similar increase in precipitation during the same period (Bush et al. 1992), but the temporal resolution of this record is coarse, and the correlation is speculative. Additional evidence comes from Lake Valencia, a nearby Venezuelan lake in a valley which drains toward the Cariaco Basin. Palaeoclimatic investigations based on pollen, ostracods, diatoms and sedimentological analyses showed that the region was arid during glacial times but experienced an increase in precipitation around 10.5 ka BP (Bradbury et al. 1981). There did not, however, appear to be a 1000-year pulse of increased precipitation, followed by a return to previous, more arid conditions. Thus, our interpretation of an increase in dark laminae thickness as due to an increase in precipitation is uncertain. Other factors, including sea level change and its effect on the locations of river deltas feeding sediment into the basin, need to be investigated.

The large increase in light (diatom-rich) laminae thickness during the Younger Dryas records an increase in primary productivity. There are several mechanisms that could result in increased productivity in the Cariaco Basin. A leading possibility would be an increase in the intensity of trade wind-induced upwelling. Primary productivity in the basin today is controlled by seasonal changes in the strength of trade winds. However, other factors besides trade wind intensity may influence productivity within the Cariaco Basin. Changes in the nutrient content of the subsurface 'feedstock' waters, for example, could also affect productivity. If the nutrient content of the subsurface water mass being upwelled changes through shifts in ocean circulation, the productivity of the basin will change, even though the intensity of the upwelling remains the same. River discharge also may potentially play a role in affecting nutrient levels. Although the direct influence of changes in the distant Amazon and

Orinoco Rivers may be minimal, the affect of the smaller local rivers that empty directly into the Cariaco Basin could be large. Changes in foraminiferal assemblages in Cariaco Basin sediment cores beginning at 12.6 ka BP have been previously interpreted as an increase in upwelling from greater trade wind intensity (Peterson *et al.* 1991). Continuing studies of additional palaeoclimate proxies from these cores, as well as other nearby records (e.g. corals; Cole 1993) will be required to resolve these issues.

The presence of a potential Younger Dryas-correlative signal in the Cariaco Basin implies that the basin is sensitive to more than local climate changes. It also suggests that the ITCZ in the tropical Atlantic responded sensitively to circum-North Atlantic changes during the Younger Dryas and deglaciation. Another result of interest, although preliminary, is the increase in varve thickness at about 2.6 ka BP in the record shown in Fig. 16. Varves deposited after 2.6 ka BP are nearly as thick as those deposited during the Younger Dryas, during a time of drastically different climate. These X-radiograph-based measurements need to be confirmed through continued high-resolution analysis of thin sections. The Cariaco Basin record will provide an important palaeoclimate record by serving as the annually dated low-latitude counterpart to annually dated Greenland ice cores.

We thank the Republic of Venezuela for permission to do research in the Cariaco Basin (clearance document no. 296, Ministerio de Relaciones Exteriores, May 14, 1990). We specifically thank Lic. Santos Valero of M.R.E., who accompanied us on our cruise as an official observer, and Dr Carlos Schubert for sharing his considerable knowledge of the region. We acknowledge the official sponsorship of our cruise by the Instituto Venezolano de Investigaciones Cientificas (Caracas), and the support of members of the Instituto Oceanografico of the Universidad de Oriente in Cumana. The cruise would not have been successful without the efforts of Captain Thomas Desjardins, the mates, and crew of the R/V Thomas Washington. We also thank the curators and staff at the Lamont-Doherty Earth Observatory core repository for their support and help. This work was supported by the US National Science Foundation and US National Oceanic and Atmospheric Administration.

References

APARICIO, R. 1986. *Upwelling along the Southern Coastal Boundary of the Caribbean Sea: Physical Characterization, Variability, and Regional Implications.* MS thesis, Florida Institute of Technology, Melbourne.

APPLEBY, P. G. & OLDFIELD, F. 1978. The calculation of lead-210 dates assuming a constant rate of supply of unsupported ^{210}Pb to the sediment. *Catena*, **5**, 1–8.

BALLESTER, A. 1969. Periodicidad en la distribution de nutrients en la Fosa de Cariaco. *Memorias de la Societe Ciencias Naturales La Salle*, **29**, 122–141.

BRADBURY, J. P., LEYDEN, B., SALGADO-LABOURIAU, M., LEWIS, W. M. JR, SCHUBERT, C. BINFORD, M. W., FREY, D. G., WHITEHEAD, D. R. & WEIBEZAHN, F. H. 1981. Late Quaternary environmental history of Lake Valencia, Venezuela. *Science*, **214**, 1299–1305.

BROECKER, W. S., ANDREE, M., WOLFI, W., OESCHGER, H., BONANI, G., KENNETT, J. P. & PETEET, D. 1988. The chronology of the last deglaciation: Implications for the cause of the Younger Dryas event, *Paleoceanography*, **3**, 1–19.

BUSH, M. B., PIPERNO, D. R., COLINVAUX, P. A., DEOLIVEIRA, P. E., KRISSEK, L. A., MILLER, M. C. & ROWE, W. E. 1992. A 14300-YR paleoecological profile of a lowland tropical lake in Panama. *Ecological Monographs*, **62**, 251–275.

CLARK, J. S. 1988. Stratigraphic charcoal analysis on petrographic thin sections: application to fire history in northwestern Minnesota. *Quaternary Research*, **30**, 81–91.

COLE, J. E. 1993. Recent upwelling variability in the Cariaco Basin (Venezuela) recorded by coral isotopic records. *Eos Transactions American Geophysical Union*, Oct. 26, Washington DC.

FIEDLER, G. 1961. Areas afectadas por terremotos en Venezuela. *Memoria III Congreso Geologico Venezolano Caracas*, Tomo IV, 1791–1810.

——1972. La liberacion de energia sismica en Venezuela, volumenes sismicos y mapa de isosistas. *Memoria IV Congreso Geologico Venezolano Boletin de Geologia*, Publicacion Especial, **5**, 4, 2441–2462.

HERRERA, L. E. & FEBRES-ORTEGA, G. 1975. Procesos de surgencia y de renovacion de aguas en la Fosa de Cariaco, Mar Caribe. *Instituto de Oceanografia Universidad Oriente, Bolivia*, **14**, 31–44.

HEEZEN, B. C., MENZIES, R. J., BROECKER, W. S. & EWING, M. 1959. Stagnation of the Cariaco Trench. *In*: SEARS, M. (ed.) *Internal Oceanography Congress Preprints*, American Association for the Advancement of Science, Washington DC, 99–102.

HUGHEN, K. A., OVERPECK, J. T., ANDERSON, R. F. & WILLIAMS, K. M. 1996. The potential for palaeoclimate records from varved Arctic lake sediments: Baffin Island, Eastern Canadian Arctic. *This volume.*

KEMP, A. E. S. 1990. Sedimentary fabrics and variation in lamination style in Peru continental margin upwelling sediments. *In*: SUESS, E. & VON HUENE, R. (eds) *Proceedings of the Ocean Drilling Program Scientific Results*, **112**, 43–58.

KIPP, N. G. & TOWNER, D. P. 1975. The last millenium of climate: foraminiferal records from coastal basin sediments. *Proceedings of the WHO/ IAMAP Symposium on Long-Term Climatic Fluctuations*, Report 421, World Meteorological Organization, Geneva, 119–126.

LIN, H. L., PETERSON, L. C., OVERPECK, J. T., TRUMBORE, S. & MURRAY, D. W. in press. Late Quaternary climate change from $\delta^{18}O$ records of multiple species of planktonic foraminfera: high-resolution records from the anoxic Cariaco Basin (Venezuela). *Paleoceanography*.

O'SULLIVAN, P. E. 1983. Annually-laminated lake sediments and the study of Quaternary environmental changes—a review. *Quaternary Science Reviews*, 1, 245–313.

OVERPECK, J. T. 1991. Century-to-millenium-scale climatic variability during the Late Quaternary. *In*: BRADLEY, R. (ed.) *Global Changes of the Past*. UCAR/Office for Interdisciplinary Earth Studies, Boulder, CO, 139–172.

——, PETERSON, L. C., KIPP, N. G., IMBRIE, J. & RIND, D. 1989. Climate change in the circum-North Atlantic region during the last deglaciation, *Nature*, 38, 553–557.

PAIGE, S. 1930. The earthquake at Cumana, Venezuela, January 17, 1929. *Seismic Society of America Bulletin*, 20, 1–10.

PETERSON, L. C., OVERPECK, J. T. & MURRAY, D. W. 1990. A high-resolution paleoenvironmental study of the Cariaco Basin, Venezuela: Late Quaternary to present. *Preliminary Report on R/V Thomas Washington Cruise Plume-07*, June 5–26, 1990.

——, OVERPECK, J. T., KIPP, N. G. & IMBRIE, J. 1991. A high resolution late Quaternary upwelling record from the anoxic Cariaco Basin, Venezuela. *Paleoceanography*, 6, 99–119.

PIKE, J. & KEMP, A. E. S. 1996. Preparation and analysis techniques for studies of laminated sediments. *This volume*.

REA, D. K. 1987. *National Science Foundation Climate Dynamics Workshop on Paleoclimatic Data-model Interaction*. The Woods, Hedgeville, W.Va., April 8–10, 1987, NSF, Washington DC.

SHAPIRO, J. 1958. The core-freezer-a new sampler for lake sediments. *Ecology*, 39, 748.

SOUTAR, A. & CRILL, P. A. 1977. Sedimentation and climatic patterns in the Santa Barbara Basin during the 19th and 20th centuries. *Geological Society of America Bulletin*, 88, 1161–1172.

STUIVER, M. & REIMER, P. J. 1993. Extended ^{14}C Data Base and Revised 3.0 ^{14}C Age Calibration Program. *Radiocarbon*, 35, 215.

WRIGHT, H. E. 1989. The amphi-Atlantic distribution of the Younger Dryas paleoclimatic oscillation. *Quaternary Science Reviews*, 8, 295–306.

Laminated sediments from the oxygen-minimum zone of the northeastern Arabian Sea

H. SCHULZ, U. VON RAD & U. VON STACKELBERG

Federal Institute for Geosciences and Natural Resources,
Stilleweg 2 30655 Hannover, Germany

Abstract: Holocene and late Pleistocene hemipelagic muds with fine, mm-scale laminations were discovered during the International Indian Ocean Expedition (IIOE) 30 years ago in the northwestern Arabian Sea, where an expanded mid-water oxygen-minimum zone (OMZ) impinges on the continental slope and prevents bioturbation. In 1993, R.V. *Sonne* revisited Pakistani waters in order to study in more detail the complex interactions between surface and deep water hydrography, productivity and terrigenous input which form and preserve organic-rich sediments. Laminated sediments frequently occur along the Indian–Pakistani continental margin between 300 and 900 m water depth. The zone of high organic carbon contents roughly coincides with the depth of the present OMZ (200–1200 m water depth). However, resedimentation processes influence the character of sediment laminae and the distribution of organic carbon in the slope sediments. Microscopic and SEM studies suggest that the lamination may be due to the seasonal changes of surface productivity and to the lateral supply of fine-grained sediment. We infer that the OMZ must have varied considerably in space and time. This is documented by the distinct, possibly cyclic alternations of dark-coloured, organic-carbon-rich, laminated intervals with light-coloured, carbonate-rich, bioturbated sedimentary sequences, reflecting variations between anoxic to suboxic and oxic bottom water conditions.

High-resolution studies from the marine sediment record have tremendously improved our knowledge of the timing, causes and consequences of oceanic and climatic change at local, regional or even global scales. However, only a few examples of laminated sediments from open marine environments have been investigated in detail. In this study, we report on laminated sediment cores from the northeastern Arabian Sea, where high productivity and subsequent bacterial decay of organic matter, as well as other oceanographic factors, lead to extremely low oxygen concentrations in the water column. Off Pakistan, this forms a stable mid-ocean oxygen-minimum zone (OMZ) between 200 and 1200 m water depth (von Rad *et al.* 1995). In this area, high surface ocean productivity and a strong seasonal variability in sediment fluxes are closely related to monsoonal circulation changes and associated upwelling intensity (Prell *et al.* 1991; Haake *et al.* 1993).

The lack of bioturbation due to severely reduced or even stagnant bottom water ventilation is currently seen to be the most important factor for the preservation of sediment laminae in marine and lacustrine sediments. Most organic-rich laminated sediments occur in areas of high oceanic productivity that favour strong oxygen depletion in the water column and an enhanced (seasonal) flux of organic matter. This has lead to the ongoing debate of what controls the accumulation of organic matter, polarized in terms of 'anoxia' (von Stackelberg 1972; Thiede & van Andel 1977; Demaison 1991; Dean *et al.* 1994) versus 'productivity' (Pedersen & Calvert 1990; Shimmield *et al.* 1990; Pedersen *et al.* 1992; Calvert & Pedersen 1992). In a recent evaluation, Canfield (1994) has stated that both groups might be right: at high rates of deposition, such as near continental margins, little difference in preservation is found with varying bottom-water oxygen levels, since most carbon is decomposed by anaerobic pathways; at low rates of deposition, euxinic sediments remain anaerobic and the relative efficiencies of aerobic versus anaerobic decomposition could enhance organic matter preservation in low-oxygen environments. However, the issue is far from being resolved. Therefore, laminated sediments which provide a unique opportunity for estimating sediment accumulation rates and carbon burial should be studied with refined sedimentological and geochemical techniques to address these problems.

Another important factor for the formation of sediment laminae may be seen not only in the high biogenic fluxes, but also in the enhanced deposition of terrigenous material. Many laminated sediments are reported either from semi-enclosed ocean environments, such as the California Borderland basins (e.g. the Santa Barbara Basin: Soutar & Crill 1977; Schimmelmann &

From Kemp, A. E. S. (ed.), 1996, *Palaeoclimatology and Palaeoceanography from Laminated Sediments*, Geological Society Special Publication No. 116, pp. 185–207.

Lange 1996; Kennett & Ingram 1995), or from restricted marginal basins, such as the Cariaco Basin (Peterson *et al.* 1991; Hughen *et al.* 1996), the Black Sea (Anderson *et al.* 1994) or the Mediterranean sub-basins (Cita *et al.* 1991). These sites may be regarded as global sinks of marine and terrigenous organic carbon, as well as of terrigenous material supplied by rivers, winds and currents from the continents. Redeposition of fine-grained lithic material may also contribute to the formation of sediment laminae originating from small-scale, low-velocity turbidity currents on the sea floor or from gravitative turbid-layer flows within the water column. Such short-term depositional events may be triggered by earthquakes, heavy storms or by rapid changes in the dynamic depth of water mass boundaries. A specific type of downslope resedimentation of fine-grained material from shallow water depths in the northeastern Arabian Sea has been described by von Stackelberg (1972): this 'turbid-layer transport' may play an important role in the accumulation of organic matter and

the formation of sediment laminae in the northeastern Arabian Sea, as well as in other continental slope environments (Gökcen 1987).

We will discuss some of the early results of the IIOE cruises in the light of new geochemical and sedimentological data and ideas based on the material from the *Sonne-90* cruise in 1993 (Project PAKOMIN, Pakistan Oxygen Minimum), which had the objective of studying the distribution and formation of laminated sediments in the northern Arabian Sea and the temporal and spatial variability of the OMZ (von Rad *et al.* 1995). The northern Arabian Sea may be a key test site to compare organic matter accumulation and the formation of sediment laminae under different depositional regimes (Slater & Kroopnick 1984; Paropkari *et al.* 1993; Calvert *et al.* 1995; van der Weijden *et al.* in press). The results of this study summarize the combined efforts of sedimentological, mineralogical, paleontological and geochemical investigations that have been started by the IIOE and expanded by the PAKOMIN scientific party after the *Sonne-90* cruise (Fig. 1).

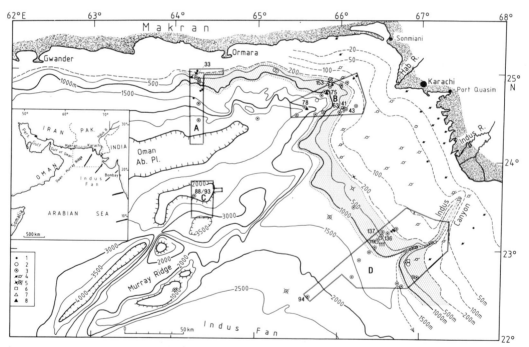

Fig. 1. Area of investigation off Pakistan with location of *Sonne-90* survey areas and sampling stations. Numbers denote some important sediment stations mentioned in the text. The five sampling transects of IIOE (von Stackelberg 1972) off Karachi, the Kathiawar peninsula, Bombay, Goa and Cochin are shown as bars in the inset map of the Arabian Sea. 1, box core; 2, gravity/piston core; 3, box and piston core; 4, IIOE stations (by R.V. *Machhera*); 5, IIOE stations (by R.V. *Meteor*); 6, Kasten (long box) core; 7, TV-guided grab sampler; 8, sediment trap station. Stippled area indicates oxygen-minimum zone (OMZ) impinging on the continental slope.

However, the PAKOMIN research is still in progress and many details will be addressed in more detail elsewhere.

Geological setting and methods

During the International Indian Ocean Expedition (1965) German scientists studied the oceanography and marine geology of the Indian and Pakistani continental margins on R.V. *Meteor* and M.V. *Machhera* (Dietrich *et al.* 1966; Schott *et al.* 1970; von Stackelberg 1972; Marchig 1972; Dietrich 1973; Mattiat *et al.* 1973; Zobel 1973).

Samples from 68 bottom grabs, 33 box cores, 34 gravity cores and 14 piston cores were collected along the more than 2000 km long continental margin between Cochin (7° N) and Karachi (25° N).

On the Indian–Pakistani continental margin six types of surface sediment facies were recognized (Fig. 2). This zonation runs more or less parallel to the coast and to the shelf edge. By more detailed sampling a better differentiation of facies types was possible in the shallow waters off the Indus River (see inset map of Fig. 2). From the coast to the shelf edge, we distinguished the zones of terrigenous sand, micaceous mud, pteropod-rich mud and calcareous relict sand. The deeper

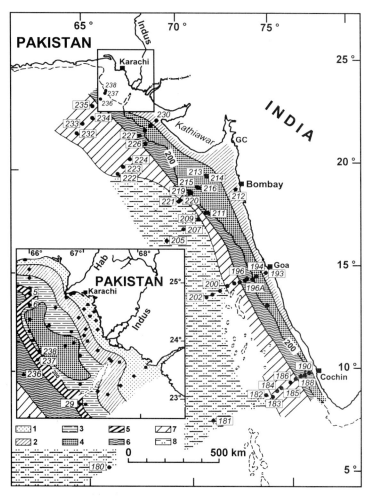

Fig. 2. Facies distribution of surface sediments from the Indian–Pakistan continental margin (modified from von Stackelberg 1972). 1, terrigenous sand; 2, micaceous mud; 3, pteropod-rich mud; 4, calcareous relict sand; 5, laminated silty clay; 6, olive-gray silty clay; 7, dark-brown foraminiferal nannofossil marl; 8, light-brown foraminiferal nannofossil ooze. Inset map: Numbers of Pakistan margin stations where laminated sediments have been first described; ●, sediment stations.

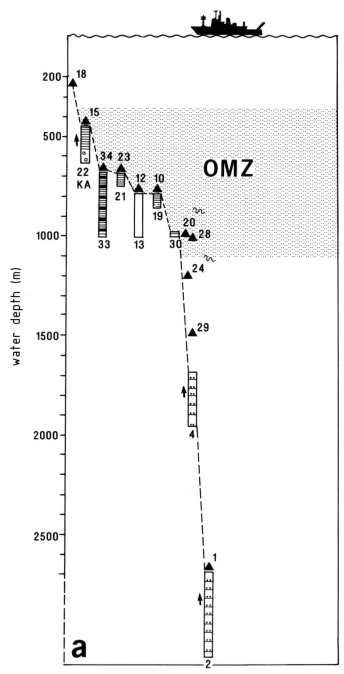

Fig. 3. Lithofacies distribution in PAKOMIN cores across the continental slope off Pakistan. For explanation see legend on Fig. 3c (black bands indicate distinct colour banding). (**a**) Area A Makran transect; (**b**) Areas B/C, Karachi transect/Murray Ridge; (**c**) Area D Indus transect.

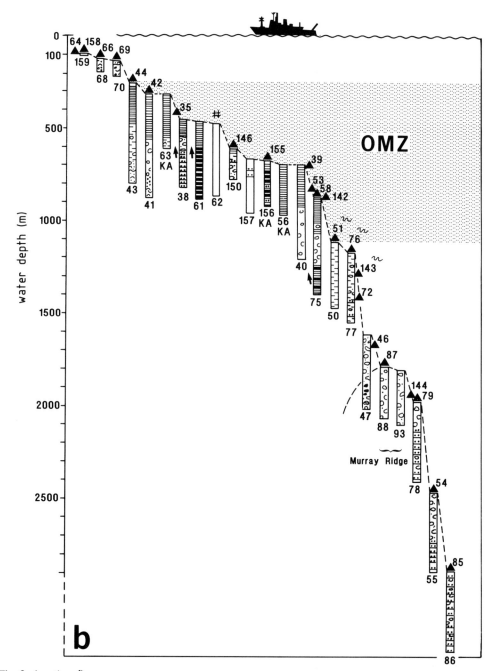

Fig. 3. (*continued*)

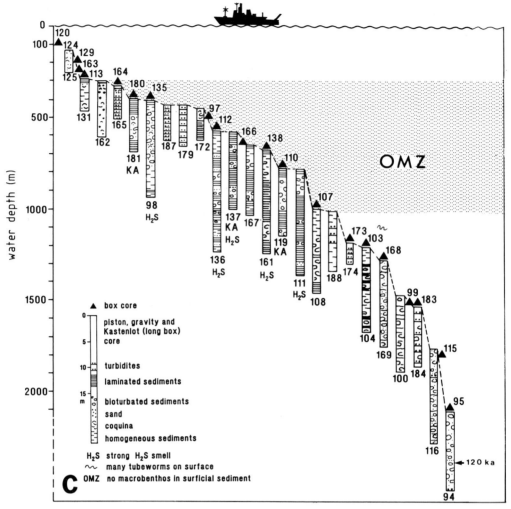

Fig. 3. (*continued*)

slope is characterized by dark-brown foraminiferal nannofossil marl, and the more or less flat deep-sea floor and Indus Fan by a light-brown foraminiferal nannofossil ooze. In contrast, the upper slope is covered by olive-gray foraminiferal silty clay or by a laminated silty clay ('Bänderschlick'). Distinctly laminated sediments had been discovered in the uppermost sections of the IIOE cores 5, 6, 29 and 237 from water depths between 300 and 420 m (Fig. 2). Indistinctly laminated muds were also found in deeper sections of cores from the upper OMZ (e.g. cores 210 and 219) along the 1600 km stretch between Cochin and the Kathiawar peninsula (von Stackelberg 1972). The IIOE cores are still in a good state of preservation and will be used for further studies.

Nearly 30 years later, R.V. Sonne revisited the continental margin off Pakistan in 1993 in order to study four areas within the OMZ in more detail (von Rad and Scientific Shipboard Party 1994; von Rad *et al.* 1995). Altogether, we took 61 box cores, 51 piston/gravity cores (up to 16.5 m long) and 8 (up to 10 m long) kasten cores with 30 × 30 m diameter (total length of sediments recovered 460 m). These cores cover three depth transects across the Pakistan continental margin (Fig. 3). They come from three genetically different depositional environments. Area A (Fig. 3a, Makran Transect) is on the active Makran continental margin where the Arabian Plate is subducted below the Eurasian Plate. The Makran margin is characterized by a narrow shelf and an extremely steep continental slope interrupted by ponded slope basins. In Area B (Fig. 3b, Hab Transect) two subareas were investigated: one belonging to the steep Makran margin (Arabian Plate) and one covering the less steep slope to the southeast 'Sindh passive margin', belonging to the Indian subcontinent. The Sindh margin is characterized by a gentle slope and a broad shelf forming the catchment area for most of the fine-grained continental debris of the Indus River delta. In Area D (Fig. 3c, Indus transect) two core transects were taken NW and SE of the deeply incised Indus Canyon. Here, the gentle continental slope and a broad shelf is much less affected by sediment redeposition. The Indus Canyon and its seaward channel-levee system is the major conduit for funnelling riverine detritus from the shelf down to the large Indus Fan. We took cores also from the northeastern extension of the Murray Ridge (Area C) from below the present OMZ in order to obtain pelagic reference sections, unaffected by downslope reworking.

During the *Sonne* cruise, two sediment trap systems were deployed in order to study the seasonality of sedimentary fluxes (Fig. 1). The Western PAKOMIN trap (WPT) and eastern PAKOMIN trap (EPT) were located in water depths around 2000 m (WPT) and 1000 m (EPT). Trap depths are 500 and 1500 m (WPT) and 500 m (EPT), respectively. These traps have been retrieved by R.V. *Meteor* Leg 32–2 in May 1995. At some stations within the OMZ, additional surface sediments were taken with a multicorer.

Bulk sediment samples were processed for weight% of $CaCO_3$, Corg, coarse fraction and for grain size analyses. Sample preparation and coarse fraction analysis of the PAKOMIN samples were modified from von Stackelberg (1972) in order to quantify the absolute (specimens per gram bulk sediment) and relative abundances of planktic foraminiferal species, benthic foraminifera, pteropods, planktic foraminiferal debris and other lithogenic and biogenic components. We also determined lithic grains, megafossils, siliceous components (diatoms, radiolaria), and fecal pellets. For time series studies, age control comes from radiocarbon dating and from $\delta^{18}O$-stratigraphy using the planktic foraminifer species *Globigerinoides ruber*. Stable isotope ratios were measured with a Finnigan MAT 251 mass spectrometer, attached to a Carbo-Kiel automated CO_2 preparation system (H. Erlenkeuser, ^{14}C-laboratory of the Institut für Kernphysik, Kiel).

For analyzing the varve-type laminations we applied the following methods: core photography and X-radiography, colour density logs and discrete gray-value measurements (1 to 10 cm spacing), digital imaging by gray-value scanning with a resolution of 0.1 mm (Ruhr-Universität Bochum, Schaaf & Thurow 1994), microscopic analysis (impregnated thin-sections) and scanning electron microscopy with energy-dispersive X-ray micro-analysis (EDX). In some cores, individual varve-type laminae were counted. Although our preliminary results agree well with the ages deduced from the chronostratigraphic matchpoints, these time-consuming studies will have to be validated by other methods, such as high-resolution AMS-^{14}C and ^{210}Pb-dating.

Oceanographic environment

Generally, the northeastern Arabian Sea may be characterized as a warm, eutrophic oceanic environment, semi-enclosed by NE Africa, Arabia and the Indian peninsula. In terms of basin and open ocean, the northeastern Arabian Sea may represent a coastal marine environment along the upper slope of the Indian–Pakistani continental margin.

The surface circulation in the northern Arabian Sea is mainly controlled by the seasonal reversal of monsoonal winds. During summer, there is a strong anticyclonic circulation driven by the SW monsoon. Due to the configuration of the coast, the surface currents flow eastward along the Makran Coast (Wyrtki 1971). Southeast of the Indus delta the currents are deflected to SE, then flow parallel to the coast of the Indian subcontinent. A complete reversal of surface currents during winter, driven by the NE-monsoon, occurs in the western Arabian Sea (for example off Oman). In the northeastern Arabian Sea, off Pakistan and northern India, seasonal current trends are more scattered. Off Pakistan, summer sea-surface temperatures generally show little spatial variability ranging between 26 and 29°C (Qasim 1982). This is in clear contrast to the broad coastal and open ocean upwelling areas with temperatures below 23°C that develop on the western side of the Arabian Sea during the summer SW-monsoon in August to September (Prell & Streeter 1982). However, there is a distinct upslope migration of the mixed-layer depth from around 100 m (February) to 40–50 m in July–October which clearly extends over the time of strong SW-monsoonal winds (August–September, Banse, 1984). A typical temperature profile of the upper 600 m is shown in Fig. 4a. Off Pakistan cool temperatures down to 20°C are restricted to the very shallow coastal waters and are much less pronounced offshore. Generally, coldest SST's

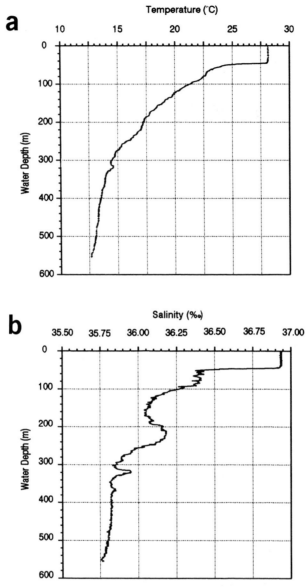

Fig. 4. Oceanographic parameters of the water column (examples from CTD-water sampling of September 1993);
(a) temperature and (b) salinity from the upper continental slope off Karachi (Area D SO90–132MS,
T. Jennerjahn in von Rad and Scientific Shipboard Party 1994).

are linked to the cool NE-monsoonal winds during winter (Banse 1984).

High-salinity intermediate water masses originating from the Persian Gulf (and possibly from the Red Sea) contribute to the stratification of the OMZ waters. A salinity maximum at intermediate depths between 200 and 350 m (Fig. 4b) is attributed to the eastern extension of Persian Gulf Water (PGW). Also the Red Sea Water (RSW)

may be detected in the eastern Arabian Sea at water depths above 800 m. However, PGW and RSW are only minor contributors to the main water mass at intermediate depths in the northeastern Arabian Sea, i.e. the North Indian Central Water which originates from the low latitudes of both hemispheres (You & Tomczak 1993). The high-salinity tongue at 200–350 m roughly coincides with the upper edge of the

OMZ at approximately 200 m and may suppress vertical convection and deep mixing. Recent tracer studies using a transient anthropogenic trace gas (Freon 11) by Olson *et al.* (1993), however, demonstrated that the residence time for water in the OMZ is not abnormally long (about 10 years) and oxygen consumption is not exceptionally high. This may imply that low-oxygen waters are preformed to some degree outside the OMZ, and are advected laterally, mostly from the low latitudes of the southeast Arabian Sea (Swallow 1984). Following Olson *et al.* (1993), the near-zero oxygen concentration is maintained by waters with initially low oxygen concentrations that pass through the layer at moderate speed. Finally, the low solubility of O_2 in the warm surface waters off Pakistan (with an annual mean of up to 25.5°C at the surface, Qasim 1982) may also contribute to low oxygen levels of sub-surface waters and hence to the maintenance of the OMZ.

A high stock of biomass and subsequent bacterial oxygen consumption by the decay of organic matter is another reason for the strong oxygen depletion and hence for the formation of the OMZ that is clearly seen in the CTD oxygen profiles and on the ocean floor (Fig. 5). Generally, surface water productivity, deduced from satellite observations and integrated chlorophyll concentrations, is very high during the summer monsoon in the northern Arabian Sea (Banse 1984; Goddard Space Flight Center 1989). However, nutrient concentrations, such as phosphate or nitrate, are much lower off Pakistan than in the

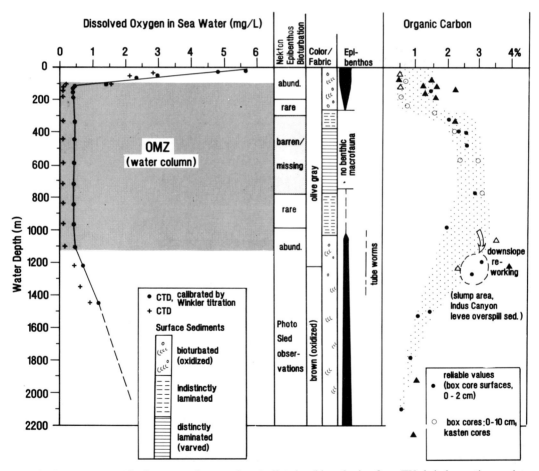

Fig. 5. Oxygen contents in the water column, nekton/epibenthos/bioturbation from TV-sled observations, colour and sediment facies of box core-tops, and organic carbon contents from Sonne-90 surface samples of the Indus transect (modified from von Rad *et al.* 1995). Triangles indicate C_{org} values of surface samples from the Karachi and Kathiawar transects from IIOE cores after von Stackelberg (1972).

western upwelling areas off Oman (Wyrtki 1971). Further evidence for the lack of upwelling off Pakistan comes from the nearly uniform distribution of $\delta^{18}O$ ratios preserved in the shells of the near surface-dwelling planktic foraminifer *Globigerinoides ruber* (Schulz & Erlenkeuser, unpublished data) from the PAKOMIN sediment surface samples. Provided that this species is most frequent during the upwelling season (i.e. the time when most food is available, Curry *et al.* 1992), light $\delta^{18}O$ values of $< -2‰$ may indicate generally warm temperatures and only minor coastal upwelling off Pakistan. In contrast, in the western Arabian Sea a clear upwelling signal is found in the $\delta^{18}O$ values of planktic foraminifera (Prell & Curry 1981; Duplessy 1982).

We conclude that the high productivity in the surface waters of the eastern Arabian Sea may be temporally and spatially linked to the summer upwelling intensity off Oman. Possibly, a very productive surface water mass with a high standing stock of biomass drifts roughly northeastward from Oman to the Pakistani margin, forced by the strong summer SW-monsoonal winds. There are several factors leading to the maintenance of an expanded oxygen minimum zone between 200 and 1200 m water depth in the northeastern Arabian Sea. The most important are: (1) the somewhat 'land-locked' coastal configuration with generally warm sea surface temperatures and reduced vertical mixing due to highly saline subsurface waters originating from the Persian Gulf and the Red Sea; (2) lateral advection of already preformed low-oxygen middepth waters and (3) the high surface ocean productivity leading to strong oxygen consumption by bacterial decay of organic matter. However, the oceanography of the northeastern Arabian Sea is complex and many details are still poorly understood, due to the lack of nearshore seasonal hydrographic and biological data.

Along the Indian–Pakistani continental margin the OMZ increases in thickness and intensity from south to north (Dietrich *et al.* 1966): off Cochin O_2-values below 1.4 mg/l occur between 130 and 1050 m (below 0.7 mg/l between 140 and 750 m) with a minimum value of 0.3 mg/l. Off Karachi, minimum values of 0.3 mg/l are found

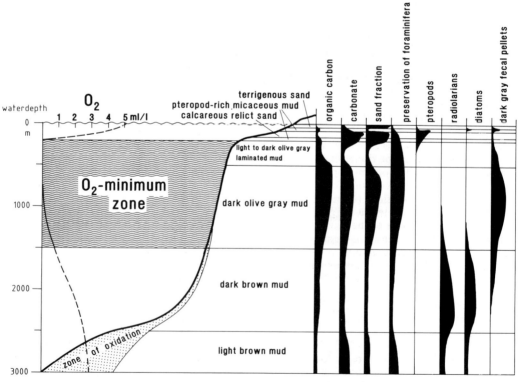

Fig. 6. Schematic vertical section across the Indian-Pakistan continental margin with oxygen contents in the water column (left), sediment facies (centre), C_{org} contents and relative abundances of selected coarse fraction components (right). IIOE results (modified from von Stackelberg 1972). Note that the lower boundary of the OMZ is here put at around 1500 m water depth, somewhat deeper than in Fig. 5.

between 180 and 1600 m and 0.7 mg/l between 190 and 1400 m respectively (Fig. 6). CTD data calibrated by Winkler titration (T. Jennerjahn, pers. comm.) from the PAKOMIN cruise clearly exhibit a uniform OMZ with continuously low oxygen concentrations <0.5 mg/l between 200 and 1200 m water depth. Based on the recent data, we define the lower boundary of the OMZ at around 1200 m water depth (Fig. 5). Below that depth, values exceed 0.5 mg/l.

Surface sediments

Eighty-nine surface sediment samples taken in transects across the continental margin were analyzed quantitatively and semi-quantitatively by von Stackelberg (1972) in order to delineate the zonation of sedimentary facies changing with water depth and hence with oxygenation related to the OMZ (Fig. 6). Additionally, 55 surface samples from *Sonne-90* box cores were processed for $CaCO_3$ and C_{org} and coarse fraction analysis. Generally, the percentage of the coarse fraction (>63 μm) is extremely low at the central slope within the OMZ and rarely exceeds 5%. This is due to the strong dilution by fine-grained terrigenous material, as seen in the X-radiographs from the sediment profiles, where calcitic tests of planktic and benthic foraminifera can be identified as small, dark dots. On the shelf, the coarse fraction and $CaCO_3$ contents may reach 70% in areas where biogenic relict carbonate sands are deposited. On the slope the carbonate contents also closely correlate with the percentage of the sand fraction. Generally, $CaCO_3$ contents on the slope range from 10 and 20% and increase with water depth, i.e. with the decreasing dilution by terrigenous debris. The optimum preservation of calcareous remains (coccoliths, foraminifera) is observed on the upper slope (Cepek 1973; Zobel 1973). Pteropods are only found on the shelf and upper slope. This is due to the elevated position of the aragonite compensation depth (ACD). It has been shown by several authors (e.g. Berger 1978, Ganssen & Lutze 1982) that the ACD is raised along continental margins with high productivity and high accumulation rates of organic carbon.

The distribution of siliceous organisms shows a pattern generally opposite to the calcareous components and organic matter (Fig. 6). Radiolarians and diatoms are rare or totally absent in the upper slope sediments off Pakistan but show a maximum in water depths between 1500 and 2500 m. This is surprising, because the enhanced surface water productivity should result in diatom blooms and a strong flux of biogenic opal to the ocean floor, as has been reported from other areas with laminated sediments (e.g. the Gulf of California; Calvert 1966). In fact, abundant living diatoms and radiolarians were found in plankton nets taken from the uppermost 40 m of waters overlying the upper continental slope (shipboard observations and H. Oberhänsli, pers. comm. 1995). The lack of diatoms in the sediments suggests strong silica dissolution in the rapidly accumulating, organic-rich slope sediments. This conflicts with the general observation that the downcore depth of the opal dissolution zone is positively correlated with the rate of sedimentation (Schrader 1972). We speculate that this dissolution may be caused by the strong dilution by detrital material, leading to an undersaturation of silica in the pore water.

In area D the facies zonation on the sea-floor (observed during the Sonne-90 cruise by underwater TV profiles and on the well preserved box core tops) correlates well with the CTD/titration values of the oxygen distribution in the water column (Fig. 5, Indus transect). In the zone of lowest oxygen concentrations, surface sediments are distinctly or indistinctly laminated with no or very scarce macrobenthic life at the surface. Distinctly laminated sediments occur in the center of the OMZ, between 300 and 900 m water depth. Indistinctly laminated sediments with rare benthos and slight bioturbation (dysaerobic conditions) are preserved above and below this central zone. Organic carbon analyses of surface sediments indicate a 2–3-fold increase between about 180 and 1000 m water depth, as compared to the outer shelf and deeper slope samples. The broad concentration maximum corresponds very well with the depth where the OMZ impinges on the continental slope. We should expect a maximum in the C_{org} concentrations on the uppermost slope (200–500 m), because of the highest bioproductivity in that areas (Banse 1984). However, the distribution of organic-rich sediments on the upper slope is also controlled by the redeposition of organic carbon-poor material from the shelf. This is documented by a surface sample from 340 m water depth, taken on the top of an isolated, 600 m high elevation on the uppermost slope at about 14° N which surmounts the zone of turbid-layer flow. At that position a maximum value of 9.1% C_{org} was measured (IIOE Station 196A, see Fig. 2), the highest concentration of organic matter we have ever found in our sediment samples.

During the *Sonne-90* cruise at the end of the period of the SW-monsoon (September–October), we observed orange–brown bacterial films within the OMZ on the well preserved

sediment surfaces of several box cores and
by TV-sled surveys (von Rad and Scientific
Shipboard Party 1994). Similar thin bacterial
carpets were not found on the coretops taken by
Meteor M 32–2 in the intermonsoonal period
(April–May). Possibly, the bacterial films at the
sediment–water interface during the summer
monsoon reflect the seasonally enhanced productivity of the surface waters.

Preliminary geochemical (Rock-Eval, C/N-
ratios, $\delta^{13}C_{org}$) and microscopic data (A.
Lückge, S. Schulte, pers. comm.) indicate that
about 90% of the (mostly amorphous) organic
material is of marine origin. Surprisingly, we did
not find any indication for Indus water discharge, either in the distribution of stable
isotopes, lacking a freshwater signal derived
from the Indus river (area D), or in changes of
the nature of the organic matter.

Regional and temporal variations of sedimentation patterns

The PAKOMIN sediment cores were taken from
a wide range of different depositional environments along the Pakistani continental margin
intersected by the present OMZ (Fig. 4). However, in most cores from the OMZ, we observe a
strong alternation between dark-coloured, laminated, organic-rich/carbonate-poor (2–4% C_{org},
10–25% $CaCO_3$) silty clays, and homogeneous,
light greenish to whitish intervals which contain
less C_{org} (<1%) and more $CaCO_3$ (>25–65%). In
the laminated sections, both planktonic (e.g.
Globigerina bulloides) and benthic foraminifera
reach highest relative abundances, indicating high
surface productivity and high nutrient flux to the
sea floor (Schulz *et al.* 1995). Only a low-diversity
benthic fauna, mainly consisting of the foraminiferal genera *Globobulimina*, *Brizalina* and
also *Uvigerina* (Zobel 1973) appears to be adapted
to live under nearly anoxic conditions within the
uppermost mm below the sediment surface. The
light-coloured, carbonate-rich intervals generally

exhibit strong bioturbation. In contrast to the
laminated sections, they contain a more diverse
fauna of planktonic and arenaceous/calcitic benthic foraminiferal species and macrobenthos.
Weakened surface productivity (also indicated
by maximum abundances of pteropods) and/or
stronger deep ventilation results in higher oxygen
levels supporting benthic life and bioturbation at
the sea floor.

Core SO90–43KL was taken in 255 m water
depth near the upper boundary of the OMZ
(Fig. 7). Distinct to indistinct lamination occurs
from the surface to a sub-bottom depth of 5.3 m.
Below that depth the dark olive-gray muds are
mainly homogeneous (or very faintly laminated).
A maximum in the sand fraction around 11.5 m
core depth is mainly due to pteropods and to
reduced terrigenous input, as indicated by the
highest $CaCO_3$ contents. An upward-fining section between 12 and 11.3 m core depth may
indicate the rapid transgression and deepening
at this site. A shift in the $\delta^{18}O$-record at around
5.3 m (S) coincides with the transition between
the laminated and the homogeneous (to faintly
laminated) interval. A very similar Holocene
$\delta^{18}O$-pattern has been recently described by
Sirocko (1995) in a core off the entrance of the
Persian Gulf. We interpret this mid-Holocene
shift at around ?5–6 ka BP to reflect decreased
surface salinity during the early Holocene,
marking the end of the early Holocene humid
interval (van Campo *et al.* 1982). Comparing the
Holocene (0–9.0 m (E5) core depth) and late
Pleistocene (9.0 m (E5)– 13.5 m (E1) core depth),
sedimentation rates are much higher during the
Holocene than during the late Pleistocene. Age
control for these calculations comes from stratigraphic $\delta^{18}O$ matchpoints (E1 to E6) in this
core, by analogy to an AMS-dated record of
Sirocko *et al.* (1993) from the Arabian Sea.

Kasten core SO90–137KA from the Indus
Transect in area D (Fig. 8) is also from the OMZ
at 573 m water depth. In this core, the isotopic
termination is documented in more detail than
in core SO90–43KL. The early glacial termination is also marked by a discrete peak in the sand

---

Fig. 7. Core SO 90–43KL from the upper continental slope off Karachi (Sindh), Area B (235 m water depth).
(**a**) Lithology, (**b**) % of coarse fraction (>63 μm), (**c**) Corg concentrations (1 cm intervals, estimated from
sediment lightness measurements, Schulz *et al.* 1995), (**d**) $\delta^{18}O$ vs. PDB of the planktic foraminifer *Globigerinoides
ruber*. E1 to E6 denote deglacial $\delta^{18}O$-events. Following Sirocko *et al.* (1993), the age of these events is E1,
17.80 cal-ka BP; E2, 17.00; E3, 16.06; E4, 11.45; E5, 9.70 and E6, 8.05 cal-ka BP, respectively. L marks the onset of
lamination-style sedimentation at 6.3 m core depth, coincident with a shift in the organic carbon contents; above
5.3 m core depth, coherent with a slight shift in $\delta^{18}O$ (S) indistinct to distinct lamination. Symbols of lithology
column: 1, bioturbated/homogeneous; 2, indistinctly laminated; 3, distinctly laminated; 4, light-colored
homogeneous; 5, dark-coloured sandy-silty clay; 6, dark-coloured clayey-silty sand; 7, light-coloured mollusc-rich
sandy marl (shelf facies); dashes indicate turbidites (<1 cm thick). Full line, C-turbidites; heavy dashes,
D-turbidites; fine dahes, E-turbidites; +C, closely adjacent or combined C turbidites with (D) or (E).

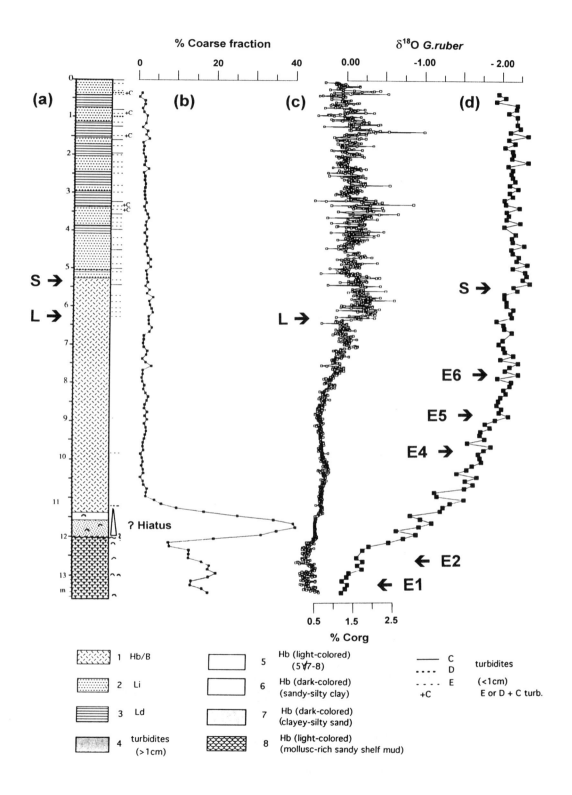

SO90 - 137KA (573 m W.D.)

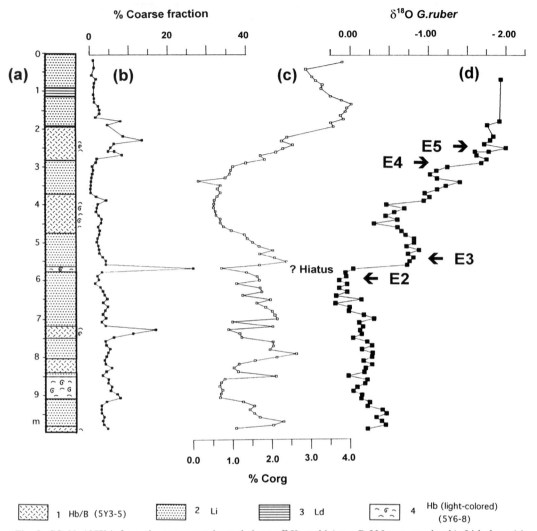

Fig. 8. SO 90–137KA from the upper continental slope off Karachi (area B 235 m water depth). Lithology (**a**), $\delta^{18}O$ against PDB of the planktic foraminifer *G. ruber* (**b**), % of coarse fraction (**c**), and C_{org} concentrations (**d**). Deglacial events (E2–E5) and symbols of lithology see caption of Fig. 7.

fraction at 5.7 m core depth in core SO90–137KA. This light-coloured, bioturbated interval may indicate a hiatus in the sedimentary record, coinciding with a strong shift in the $\delta^{18}O$ record. However, there are numerous alternations of dark-coloured and frequently laminated intervals with light-coloured, bioturbated ones that are rich in $CaCO_3$ and poor in C_{org} which are not associated with shifts in the $\delta^{18}O$ record. Independent of the facies changes, the $\delta^{18}O$ curve exhibits a rather uniform pattern with only

minor fluctuations below 5.7 m core depth. The Holocene section of 0 to 2.0 m core depth (E5) in core SO90–137KA is considerably shorter, indicating much lower Holocene sedimentation rates off the Indus than in Areas A and B.

Most cores in area B exhibit a hiatus between the uppermost Holocene and late Pleistocene sections. Based on $\delta^{18}O$ stratigraphy, only two cores placed off the eastern (Sindh) side in area B may document the whole glacial/interglacial cycle (e.g. cores SO90–41KL and SO90–43KL).

From the tectonic and geological settings, it is clear that downslope redeposition forming allochthonous fine-grained clastic deposits is active along the Indian–Pakistani continental slope. As shown by von Stackelberg (1972), sediment redeposition may take place by turbid-layer transport and by low-density turbidity currents. There are transitions between these two processes which are still poorly understood. For turbid-layer transport, we assume a pulsating flow of material stirred up mainly from the relict zone of the shelf. The sediment particles are kept in suspension forming a turbid layer which slowly moves across the shelf edge and down the upper slope, where they are being deposited. Turbid-layer sediments were first described by Moore (1969) from the California Continental Borderland. They consist of indistinct flasers or lenses of sand- to silt-sized sediment material intercalated in hemipelagic sediments (Fig. 9a). The lenticular to phacoidal sedimentary fabric does, however, not reflect annual (monsoonal) rhythms, but episodic gravitational redeposition events, triggered by major storms, cyclones or earthquakes, or mobilization of shelf deposits due to rapid sea-level changes. The polymodal grain size distribution of the turbid layers is caused by the mixture of hemipelagic and allochthonous material, mainly silt to fine sand. Normal grading was never observed in these turbid-layer deposits (e.g. IIOE core 219; von Stackelberg 1972). Sometimes a slight upcore decrease in grain size was noticed, suggesting a longer-term upwards-fining sedimentation cycle as known from turbidites. The fabric is characterized by very indistinct bedding caused by intercalations of several layers of allochthonous silt to fine sand.

Origin of sediment laminae

Sedimentation rates are highest in areas A and B on the steep slope off Makran. In these areas, most of the PAKOMIN sediment records from the OMZ are of Holocene age and are continuously laminated (Figs 3a–c). However, gradual changes between distinct and indistinct laminations occur, for instance in the deeper core sections (as in cores SO90–43KL and SO90–137KA), slightly disturbed by burrow mottling. 'Indistinct lamination' is caused by a subtle disturbance of fine laminae by 'micro-bioturbation', i.e. limited horizontal burrowing by tiny sediment-ingesting polychaete worms or benthic foraminifera that does not reach deep enough to destroy the lamination completely.

Based on X-radiographs, microscopic analyses of impregnated thin-sections and scanning electron microscopy with EDX (Figs 9b–d), we attempted a classification of the sediment laminae inferred from variations of thickness, colour, texture and fabric. We distinguish the following lithotypes which we interpret to be predominantly of hemipelagic (A, B and Hb) or allochthonous (C-G) origin. These results are summarized in Figs 10a and b.

Lithotype A

The light-coloured hemipelagic A-laminae are well sorted and contain almost exclusively land-derived clay minerals, mica, chlorite, quartz, and very few coccoliths.

Lithotype B

Concentrations of amorphous organic matter (apparently fecal pellets or aggregates) are concentrated mostly in the dark-coloured hemipelagic B-laminae which contain slightly more $CaCO_3$ and C_{org} than the light-coloured laminae (von Stackelberg 1972), although the differences are very small. The B-laminae are about 0.4–0.8 mm thick, somewhat thicker than the lighter-coloured detrital A-laminae. Generally the dark-coloured laminae are poorly sorted, contain abundant benthic and planktonic foraminifera (often filled by pyrite and with a geopetal fabric) and coccoliths.

Lithotype Bw

A subtype of dark laminae consists of foraminiferal- and quartz-rich laminae which can tentatively be explained by the winnowing activity of tidal or thermohaline bottom currents active at the slope.

Lithotype Hb

'Normal' (homogeneous-bioturbated) hemipelagic sediments.

Lithotype C

Most abundant are light-gray, homogeneous (rarely indistinctly graded) clayey silt turbidites which contain mainly clay minerals, mica, chlorite, quartz and some reworked carbonates, all of terrigenous origin. The source area might be the mica-rich clayey silt belt (Fig. 2) at the middle part of the Pakistani shelf (von Stackelberg 1972). These layers are typically 1–3 mm thick, but may

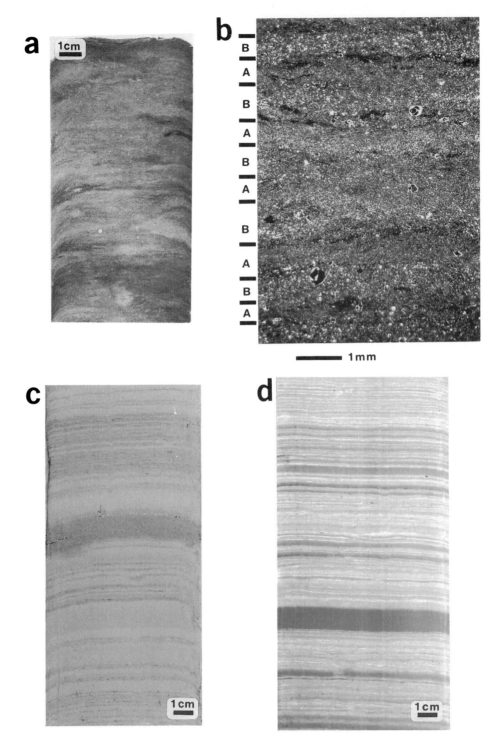

Fig. 9a–d. (a) X-radiograph of turbid-layer sediments with fine phacoidal bedding (fecal pellets concentrated in the 'dark' zones, Core IIOE 210KK (189–200 m); (b) section of sediment core SO90–33KL (93.5 cm core depth) with sub-millimetric varve-type lamination (A, light-coloured; B, dark-coloured lamina); (c) photography of OMZ core SO90–33KL (Area A, 661 m water depth, section 116–134 cm) with laminated hemipelagic sediments and abundant light-gray C-turbidites, and a few D-turbidites, (d) X-radiograph of core SO90–33KL (section 95–109 cm); 'dark' layers are C-turbidites.

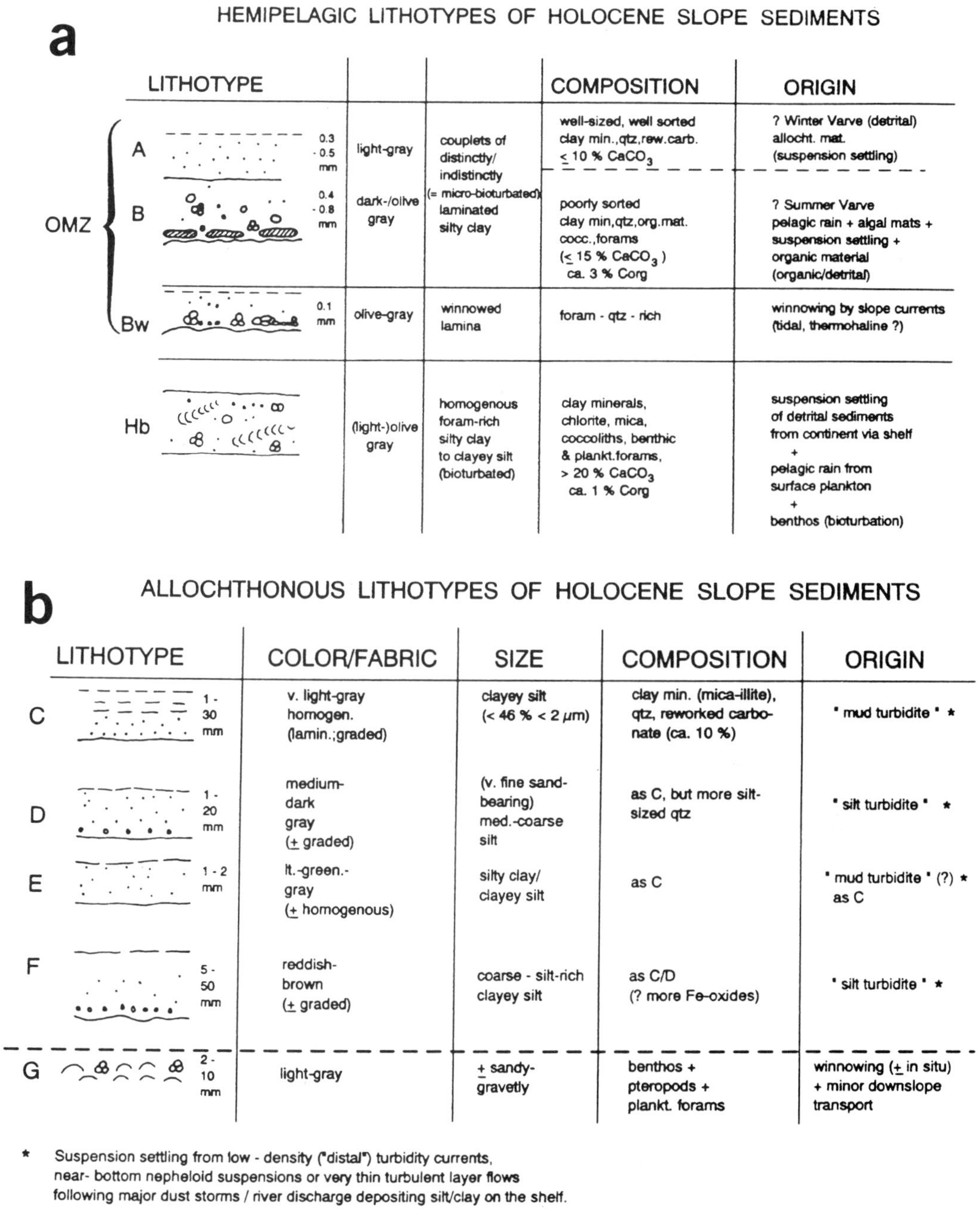

a

HEMIPELAGIC LITHOTYPES OF HOLOCENE SLOPE SEDIMENTS

LITHOTYPE				COMPOSITION	ORIGIN
OMZ {	A	0.3 - 0.5 mm	light-gray	couplets of distinctly/ indistinctly (= micro-bioturbated) laminated silty clay	well-sized, well sorted clay min.,qtz,rew.carb. ≤ 10 % CaCO₃ → $\leq 10\% \; CaCO_3$	? Winter Varve (detrital) alloct. mat. (suspension settling)
	B	0.4 - 0.8 mm	dark-/olive gray		poorly sorted clay min,qtz,org.mat. cocc.,forams ($\leq$ 15 % $CaCO_3$) ca. 3 % Corg	? Summer Varve pelagic rain + algal mats + suspension settling + organic material (organic/detrital)
Bw		0.1 mm	olive-gray	winnowed lamina	foram - qtz - rich	winnowing by slope currents (tidal, thermohaline ?)
Hb			(light-)olive gray	homogenous foram-rich silty clay to clayey silt (bioturbated)	clay minerals, chlorite, mica, coccoliths, benthic & plankt.forams, > 20 % $CaCO_3$ ca. 1 % Corg	suspension settling of detrital sediments from continent via shelf + pelagic rain from surface plankton + benthos (bioturbation)

b

ALLOCHTHONOUS LITHOTYPES OF HOLOCENE SLOPE SEDIMENTS

LITHOTYPE		COLOR/FABRIC	SIZE	COMPOSITION	ORIGIN
C	1 - 30 mm	v. light-gray homogen. (lamin.;graded)	clayey silt (< 46 % < 2 μm)	clay min. (mica-illite), qtz, reworked carbonate (ca. 10 %)	" mud turbidite " *
D	1 - 20 mm	medium-dark gray (± graded)	(v. fine sand-bearing) med.-coarse silt	as C, but more silt-sized qtz	" silt turbidite " *
E	1 - 2 mm	lt.-green.-gray (± homogenous)	silty clay/ clayey silt	as C	" mud turbidite " (?) * as C
F	5 - 50 mm	reddish-brown (± graded)	coarse - silt-rich clayey silt	as C/D (? more Fe-oxides)	" silt turbidite " *
G	2 - 10 mm	light-gray	± sandy-gravelly	benthos + pteropods + plankt. forams	winnowing (± in situ) + minor downslope transport

* Suspension settling from low - density ("distal") turbidity currents, near- bottom nepheloid suspensions or very thin turbulent layer flows following major dust storms / river discharge depositing silt/clay on the shelf.

Fig. 10. (a) Sediment facies, colour, fabric, composition and origin of hemipelagic lithotypes (A, B Hb); **(b)** Sediment facies, colour, fabric, grain size composition and origin of allochthonous lithotypes of Holocene slope sediments (C–G). Note that 'turbid-layer sediments' were omitted in this diagram (lithotype C grades into turbid-layer type sediments).

also reach up to 5 cm in some places. Most of the apparently ungraded light-gray laminae found in the cores from areas A and B can be attributed to this mud turbidite type.

Lithotype D

Medium to dark-gray silt turbidites occur either at the base of a graded C-turbidite or as a distinct silt layer. They have distinct lower boundaries, are mostly graded and contain very fine terrigenous sand or medium to coarse silt.

Lithotype E

Light greenish-gray turbidites are very similar to type C.

Lithotype F

Reddish-brown silt turbidites (Type F) were only found in the late Holocene sections of area B forming distinct marker horizons that could be correlated in some cores.

Lithotype G

Coarse-grained foraminiferal pteropod-rich layers (G) can probably be explained by *in situ* winnowing by near bottom currents.

Although these lithotypes of sediment laminae were found in all three PAKOMIN areas, it is obvious that the tectonic and sedimentary differences between these areas are reflected in the style and frequency of sediment laminations. Figure 11 shows an X-radiograph and a gray-value scan (digital imaging) of Holocene sediment laminations of the long box core SO90–56KA from the OMZ off Karachi. In this area, laminated sediments were discovered for the first time in four IIOE cores (Fig. 2, von Stackelberg 1972). In core SO90–56KA, light/dark-colored couplets of laminae are clearly preserved from 0 to 6.5 m core depth (5750 uncorrected ^{14}C-years BP). This core currently has only three ^{14}C-ages. However, it was possible to count these varve-type couplets at 16 discrete intervals (each about 10–20 cm long) using the X-radiographs. Assuming that one dark/light couplet was deposited within one year, we inferred sedimentation rates of

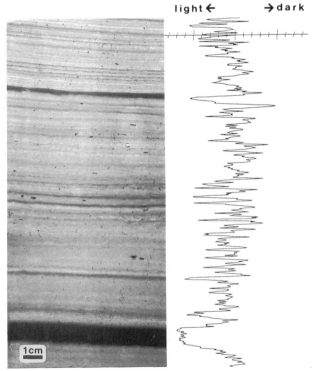

Fig. 11. X-radiograph and gray-scale scan of OMZ core SO90–56KA (Area B, 695 m water depth, section 69–84.5 cm). Note mm-thick, sometimes graded light-coloured C-turbidites.

0.8–1.3 mm/year for those intervals. These data correlate well with the sedimentation rates resulting from our preliminary ^{14}C-ages (0.9 to 1.5 mm/year) and confirm that the dark/light-coloured couplets are indeed varves. They can be interpreted to represent changes in the deposition of organic carbon, calcareous plankton and fine detrital components forced by the monsoonal cycle.

A seasonal model of varve formation on the steep Makran continental slope where laminae are most distinct is shown in Fig. 12. During the summer monsoon (Fig. 12a), onshore winds lead to rather stratified waters on the shelf and to an elevated thermocline causing coastal plankton blooms and high pelagic rain of organic and calcitic material. In contrast, during winter (Fig. 12b), cool offshore winds (NE-monsoon and northwesterlies) enhance turbulent mixing and reworking on the shelf. Suspended sediment is transported across the shelf edge by thin near-bottom nepheloid layers and episodic turbidity currents. The low content of organic carbon in the winter laminae may be explained by decreased surface ocean productivity during the NE-monsoon and by the admixture of eolian dust, originating from the Baluchistan and Thar deserts in the north and northeast (Sirocko & Sarnthein 1989).

Conclusions and open questions

1. Laminated sediments have been frequently found along the Indian-Pakistani continental margin in water depths roughly between 250 and 800 m coinciding with the core of the OMZ, which exhibits nearly constantly low oxygen levels between 200 and 1200 m water depth. Examination and counting of varve-type laminae of a presumably continuously laminated Holocene record covering the last 6000 years should be possible (e.g. cores SO90–56KA, SO90–33KL). Preliminary results suggest that the sediment laminae are of seasonal origin. Their formation may be strongly related to seasonal changes in the flux of organic matter in phase with the atmospheric monsoonal circulation changes.

2. The accumulation of organic matter in the OMZ sediments is controlled by a number of factors, of which we address four: (1) intensity and zonation of productivity belts and coastal plankton blooms, (2) admixture of allochthonous material on the upper slope, (3) downslope reworking,

leading to a deep maximum of organic carbon accumulation within the lower part of the OMZ and beyond , and (4) bottom water anoxia. Up to now these factors have not been sufficiently evaluated. The hypotheses on the preferential accumulation and preservation of organic matter at continental margins within a well developed OMZ have to be supported by quantitative data of the organic matter and other particle fluxes to the sea floor. Detailed stratigraphic work and high-resolution dating by AMS[14]C and ^{210}Pb methods is needed to calculate reliable accumulation rates for the different environments. The high sedimentation rates and the undisturbed sedimentation record of the laminated Holocene (and some Pleistocene) sections, however, offer a unique opportunity to address these problems.

3. Longer time-series in our sediment cores covering at least the past 30 ka clearly depict gradual changes in the intensity of the OMZ through time, as indicated by facies changes between distinctly to faintly laminated (organic-rich) and bioturbated (oxidized) sections with a more diverse benthic fauna, mainly of benthic foraminifera and macrofaunal remains. Generally, the light-coloured, bioturbated intervals are poor in organic carbon and rich in pteropod debris. These layers do not contain more reworked material than the laminated sections and may represent an *in situ* hemipelagic facies. We consider the intercalation between laminated and bioturbated intervals to reflect major changes in bottom water chemistry (such as the aragonite compensation depth), ventilation (indicated by bioturbation intensity), and changes in surface ocean productivity. Distinct intervals with abundant pteropod debris were also found in the pelagic realm (e.g. on the isolated top of the Murray Ridge). This further suggests that these changes are not a local phenomenon of redeposition on the continental margin.

4. Apparently, the supply of allochthonous material from shallower to deeper waters by turbid-layer flow, turbidity currents and slumping strongly influences the C_{org} content of the sediments. No doubt, the primary production is of great importance for both, the formation of the OMZ and the organic carbon concentration on the sea floor. The process of deposition by suspension settling either from very thin low-density turbidity currents or from near-bottom nepheloid suspensions is still

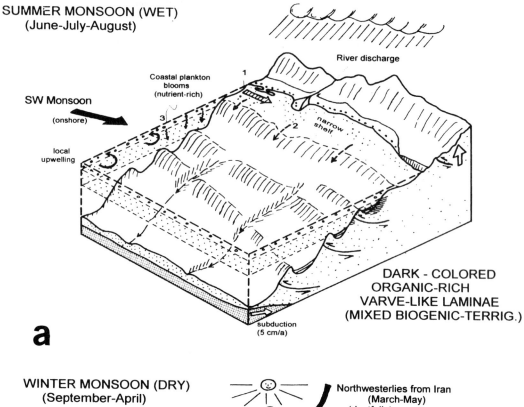

SUMMER MONSOON (WET)
(June-July-August)

River discharge

Coastal plankton blooms (nutrient-rich)

SW Monsoon

(onshore)

narrow shelf

local upwelling

DARK - COLORED ORGANIC-RICH VARVE-LIKE LAMINAE (MIXED BIOGENIC-TERRIG.)

subduction (5 cm/a)

a

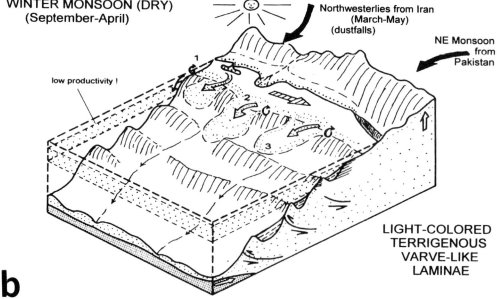

WINTER MONSOON (DRY)
(September-April)

Northwesterlies from Iran (March-May) (dustfalls)

NE Monsoon from Pakistan

low productivity !

LIGHT-COLORED TERRIGENOUS VARVE-LIKE LAMINAE

b

Fig. 12a, b. Descriptive model of varve formation along the Pakistan continental margin (see text). (**a**) during summer (SW-) monsoon, (**b**) during winter (NE-) monsoon.

obscure. These mechanisms are poorly studied, but probably very active and frequent at the upper slopes of continental margins with high rates of coastal uplift, narrow shelves, and rapid rates of terrigenous deposition, such as at the Makran and Sindh margins.

5. The reason why almost no siliceous plankton was preserved on the sea-floor and in the sediment cores within the OMZ is still obscure. The obvious lack of siliceous microfossils in the sediments as tracers of environmental change limits the potential of our laminated sediments for the reconstruction of seasonal to interannual oceanographic changes. However, calcareous remains such as coccoliths and benthic and planktic foraminifera are exceptionally well preserved. Time-series studies of flux rates of the different lithogenic and biogenic components from the PAKOMIN sediment traps will help to refine our model of varve formation and to verify the seasonal origin of the sediment laminae in the northeastern Arabian Sea.

6. From the distinct pattern of alternations between dark-coloured, anoxic to suboxic (laminated) and light-colored, oxic (bioturbated) sedimentary sequences (e.g. core SO90–137KA) we infer that the OMZ must have varied considerably in space and time during the past 30 ka (von Rad *et al.* 1995). Similar alternations have been recently reported from laminated sediments cored in ODP Site 893A in the Santa Barbara Basin (Kennett & Ingram 1995). This pattern of temporal variations of the OMZ might reflect short-term fluctuations, such as the Dansgaard–Oeschger Events described from high-resolution Greenland ice cores via deep-water and/or atmospheric circulation changes, or by rapid sea-level fluctuations. We suggest that the sedimentary environment of the OMZ is extremely sensitive to such changes, since the Arabian Sea is situated at the terminus of the oceanic conveyor belt in the northern Indian Ocean.

This work benefitted from the help and discussions with U. Berner, M. Geyh, V. Marchig, J. Merkt, V. Riech, U. Röhl and M. Weber (all Hannover). M. Schaaf, M. Rivas-Koslowski (both University of Bochum), S. Schulte (University of Oldenburg), A. Lückge (KFA Jülich) and T. Jennerjahn (University of Hamburg) provided unpublished data (gray-value scanning, organic petrography, and oceanographic data). H. Erlenkeuser (University of Kiel) provided stable isotope data. We thank M. Szurlies, B. Stenschke, M. Schmidtke, H. Karman for laboratory assistence, H. Kawohl and R. Goergens for obtaining high-quality sediment cores and Captain H. Papenhagen (*Sonne-90*) and Captain M. Kull (*Meteor 32–2*) and their skilled crews for help and good collaboration on board. This study was generously funded by the German Federal Ministry of Education and Research (BMBF), Projects 03G0090A and 03F0137C.

References

ANDERSON, D. M. & PRELL, W. L. 1993. A 300 KYR record of upwelling off Oman during the Late Quaternary: Evidence of the Asian Southwest Monsoon. *Paleoceanography*, **8/2**, 193–208.

ANDERSON, R. F., LYONS, T. W. & COWIE, G. L. 1994. Sedimentary record of a shoaling of the oxic/anoxic interface in the Black Sea. *Marine Geology*, **116**, 373–384.

BANSE, K. 1984. An overview of the hydrography and associated biological phenomenea in the Arabian Sea, off Pakistan. *In*: HAQ, B. U. & MILLIMAN, J. D. (eds) *Marine Geology and Oceanography of the Arabian Sea and Coastal Pakistan*. Van Nostrand Reinhold, New York, 271–304.

BERGER, W. H. 1972. Deep-sea carbonate: pteropod distribution and the aragonite compensation depth. *Deep-Sea Research*, **25**, 447–452.

CALVERT, S. E. 1966. Origin of diatom-rich, varved sediments in the Gulf of California. *Journal of Geology*, **74**, 546–565.

—— & PEDERSEN, T. F. 1992. Organic carbon accumulation and preservation in marine sediments: How important is anoxia? *In*: WHELAN, J. K. & FARRINGTON, J. W. (eds) *Productivity, Accumulation and Preservation of Organic Matter in Recent and Ancient Sediments*. Columbia University Press, New York, 231–263.

——, PEDERSEN, T. F., NAIDU, P. D. & VON STACKELBERG, U. 1995. On the organic carbon maximum on the continental slope of the eastern Arabian Sea. *Journal of Marine Research*, **53**, 269–296.

CANFIELD, D. E. 1994. Factors influencing organic carbon preservation in marine sediments. *Chemical Geology*, **114**, 315–329.

CEPEK, P. 1973. Die Art *Pontosphaera indooceanica* n.sp. und ihre Bedeutung für die Stratigraphie der jüngsten Sedimente des Indischen Ozeans. "Meteor"-Forsch.- Ergebnisse, Reihe C 12: 1–8.

CITA, M. B., DELANGE, G. J. & OLAUSSON, E. (eds) 1991. Anoxic Basins and sapropel deposition in the eastern Mediterranean: past and present. *Marine Geology*, **100**, 1–182.

CURRY, W. B, OSTERMANN, D. R., GUPTHA, M. V. S. & ITTEKKOT, V. 1992. Foraminiferal production and monssonal upwelling in the Arabian Sea: evidence from sediment traps. *In*: SUMMERHAYES, C. P., PRELL, W. L. & EMEIS, K. C. (eds) *Evolution of Upwelling Systems since the Early Miocene*. Geological Society, London, Special Publication, **64**, 93–106.

DEAN, W. E., GARDNER, J. V. & ANDERSON, R. Y. 1994. Geochemical evidence for enhanced preservation of organic matter in the oxygen minimum zone of the continental margin of northern California during the late Pleistocene. *Paleoceanography*, **9**, 47–61.

DEMAISON, G. 1991. Anoxia vs. productivity: What controls the formation of organic-carbon rich sediments and sedimentary rocks? *American APG Bulletin*, **75**, 499.

DIETRICH, G. 1973. The unique situation in the environment of the Indian Ocean. *In*: ZEITZSCHEL, B. & GERLACH, S. A. (eds) *The Biology of the Indian Ocean*. Springer, Berlin, 1–6.

——, DÜING, W., GRASSHOFF, K. & KOSKE, P. H. 1966, Physikalische und chemische Daten nach Beobachtungen des Forschungsschiffes "Meteor" im Indischen Ozean 1964/1965. *'Meteor' Forsch.-Ergebnisse, Reihe A*, **2**, 1–5, Tabellen.

DUPLESSY, J. C. 1982. Glacial to interglacial contrasts in the northern Indian Ocean. *Nature*, **295**, 494–498.

GANSSEN, G. & LUTZE, G. F. 1982. The aragonite compensation depth of the northeastern Atlantic continental margin. *'Meteor' Forsch.-Ergebnisse, Reihe C*, **36**, 57–59.

GODDARD SPACE FLIGHT CENTER 1989. Mission to Planet Earth. *Aviation Week and Space Technology*, March, 13 1989, 34–41.

GÖKCEN, S. L. 1987. Effects of turbidity currents on the modes of vertical and horizontal particle flux in the sea. *Mitt. Geol.-Paläont. Inst. Univ. Hamburg, SCOPE/UNEP Sonderband*, **62**, 269–278.

HAAKE, B., ITTEKKOT, V., RIXEN, T., RAMASWAMY, V., NAIR, R. R. & CURRY, W. B. 1993. Seasonality and interannual variability of particle fluxes to the deep Arabian sea. *Deep-Sea Research*, **40**, 1323–1344.

HUGHEN, K., OVERPECK, J. T., PETERSON, L. C. & ANDERSON, R. 1996. Varve analysis and Palaeoclimate from Sediments of the Cariaco Basin, Venezuela. *This volume*.

KENNETT, J. P. & INGRAM, B. L. 1995. Paleoclimatic evolution of Santa Barbara Basin during the last 20 k.y. *Proceedings of the ODP, Scientific Results*, **146** (Pt2), 309–325.

MARCHIG, V. 1972. Zur Geochemie rezenter Sedimente des Indischen Ozeans. *"Meteor" Forsch.-Ergebnisse, Reihe C*, **11**, 1–104.

MATTIAT, B., PETERS, J. & ECKHARDT, F. J. 1973. Ergebnisse petrographischer Untersuchungen an Sedimenten des indisch-pakistanischen Kontinentalrandes (Arabische See). *"Meteor" Forsch.-Ergebnisse, Reihe C*, **14**, 1–50.

MOORE, D. G. 1969. Reflection profiling studies on the California Borderland: structure and Quaternary turbidite basins. Geological Society of America, Special Paper, **107**, 142pp.

OLSON, D. B., HITCHCOCK, G. L., FINE, R. A. & WARREN B. A. 1993. Maintenance of the low-oxygen layer in the central Arabian Sea. *Deep-Sea Research*, **40**, 673–685.

PAROPKARI, A. L., BABU, C. P. & MASCARENHAS, A. 1993. New evidence for enhanced preservation of organic carbon in contact with oxygen minimum zone on the western continental slope of India. *Marine Geology*, **111**, 7–13.

PEDERSEN, T. F. & CALVERT, S. E. 1990. Anoxia versus productivity: what controls the formation of organic-carbon-rich sediments and sedimentary rocks? *AAPG Bulletin*, **74**, 454–466.

——, SHIMMIELD, G. B. & PRICE, N. B. 1992. Lack of enhanced preservation of organic matter in sediments under the oxygen minimum on the Oman margin. *Geochimica et Cosmochimica Acta*, **56**, 545–551.

——, OVERPECK, J. T., KIPP, N. G. & IMBRIE, J. 1991. A high-resolution late Quaternary upwelling record from the anoxic Cariaco Basin, Venezuela. *Paleoceanography*, **6**, 99–119.

PRELL, W. L. & CURRY, W. B. 1981. Faunal and isotopic indices of monsoonal upwelling: Western Arabian Sea. *Oceanologica Acta*, **4/1**, 91–98.

—— & STREETER, H. F. 1982. Temporal and spatial patterns of monsoonal upwelling along Arabia: a modern analogue for the interpretation of Quaternary SST anomalies. *Deep-Sea Research*, **40**, 143–155.

——, NIITSUMA, N., EMEIS, K.-C. et al. 1991. *Proceedings of the Ocean Drilling Program, Scientific Results*, **117**, College Station, Texas.

QASIM, S. Z. 1982. Oceanography of the northern Arabian Sea. *Deep-Sea Research*, **29**, 1041–1068.

SCHAAF, M. & THUROW, J. 1994. A fast and easy method to derive highest-resolution time-series data sets from drill cores and rock samples. *Sedimentary Geology*, **94**, 1–10.

SCHIMMELMANN, A. & LANGE, C. B. 1996. A review of the Holocene Santa Barbara Basin laminated sediment record. *This volume*.

SCHOTT, W., von STACKELBERG, U., ECKHARDT, F.-J., MATTIAT, B., PETERS, J. & ZOBEL, B. 1970. Geologische Untersuchungen an Sedimenten des indisch-pakistanischen Kontinentalrandes (Arabisches Meer). *Geologische Rundschau*, **60**, 246–275.

SCHRADER, H. J. 1972. Kieselsäure-Skelette in Sedimenten des ibero-marokkanischen Kontinentalrandes und angrenzender Tiefsee-Ebene. *'Meteor' Forsch.-Ergebnisse, Reihe C*, **8**, 10–36.

SCHULZ, H., VON RAD, U., BERNER, U., & ERLENKEUSER, H. 1995. Holocene to Pleistocene sediment color cycles, stable isotope stratigraphy, and organic carbon accumulation in the Northeastern Arabian Sea. Abstracts, European Union of Geoscientists (EUG), 8th meeting, Strasbourg 9–13 April 1995, Terra Nova, 7: 215.

SHIMMIELD, G. B., PRICE, N. B. & PEDERSEN, T. F 1990, The influence of hydrography, bathymetry, and productivity on the sediment type and composition of the Oman margin and in the northwest Arabian Sea. *In*: ROBERTSON, A. H. F. et al. (eds) *The Geology and Tectonics of the Oman Region*. Geological Society, London, Special Publication, **49**, 759–769.

SIROCKO, F. 1995. Abrupt change in monsoonal climate: evidence from the geochemical composition of Arabian Sea sediments. Habilitation Thesis, University of Kiel.

—— & SARNTHEIN, M. 1989. Wind-borne deposits in the northwestern Indian Ocean: record of Holocene sediments versus modern satellite data. *In*: LEINEN, M. & SARNTHEIN, M. (eds) *Paleoclimatology and Paleometerology: Modern and Past Patterns of Global Atmospheric Transport*. NATO ASI Series C 282: 401–322.

——, ——, ERLENKEUSER, H., LANGE, H., ARNOLD, M. & DUPLESSY, J. C. 1993. Century-scale events in monsoonal climate over the past 24000 years. *Nature*, **364**, 322–324.

SLATER, R. D. & KROOPNICK, P. 1984. Controls on dissolved oxygen distribution and organic carbon deposition in the Arabian Sea. *In*: HAQ, B. U. & MILLIMAN, J. D. (eds) *Marine Geology and Oceanography of the Arabian Sea and Coastal Pakistan*. Van Nostrand Reinhold, New York, 305–313.

SOUTAR, A. & CRILL, P. A. 1977. Sedimentation and climatic patterns in the Santa Barbara Basin during the 19th and 20th centuries. *Geological Society of America Bulletin*, **88**, 1161–1172.

SWALLOW, J. C. 1984. Some aspects of the physical oceanography of the Indian Ocean. *Deep-Sea Research*, **31**, 639–650.

THIEDE, J. & VAN ANDEL, T. H. 1977. The paleoenvironment of anaerobic sediments in the Late Mesozoic South Atlantic Ocean. *Earth & Planetary Science Letters*, **33**, 301–309.

VAN CAMPO, E., DUPLESSY, J. C. & ROSSIGNOL-STRICK, M. 1982. Climatic conditions deduced from a 150-kyr oxygen isotope-pollen record from the Arabian Sea. *Nature*, **296**, 56–59.

VAN DER WEIJDEN, C. H., REICHART, G. J. & VISSER, H. J. in press. Evidence for enhanced preservation of organic matter in sediments from the oxygen minimum zone in the northeastern Arabian Sea. *Geochimica et Cosmochimica. Acta*.

VON RAD, U., SCHULZ, H. & SONNE-90 SCIENTIFIC PARTY 1995. Sampling the oxygen minimum zone off Pakistan: glacial-interglacial variations of anoxia and productivity (preliminary results). *Marine Geology*, **124**, 7–19.

—— & SCIENTIFIC SHIPBOARD PARTY 1994. PAKOMIN-Influence of the Oxygen Minimum Zone on the sedimentation at the upper continental slope off Pakistan (NE Arabian Sea), Research Cruise SO-90 with R. V. Sonne. Unpublished cruise report, BGR Hannover.

VON STACKELBERG, U. 1972. Faziesverteilung in Sedimenten des indisch-pakistanischen Kontinentalrandes (Arabisches Meer). *"Meteor" Forsch.-Ergebnisse, Reihe C*, **9**, 1–73.

WYRTKI, K. 1971. *Oceanographic Atlas of the International Indian Ocean Expedition*. Washington, National Science Foundation.

YOU, Y. & TOMCZAK, M. 1993. Thermocline circulation and ventilation in the Indian Ocean derived from water mass analysis. *Deep-Sea Research*, **40**, 13–56.

ZOBEL, B. 1973. Biostratigraphische Untersuchungen an Sedimenten des indisch-pakistanischen Kontinentalrandes (Arabisches Meer). *"Meteor" Forsch. Ergebnisse, Reihe C*, **12**, 9–73.

Oligocene laminated limestones as a high-resolution correlator of palaeoseismicity, Polish Carpathians

GRZEGORZ HACZEWSKI

Institute of Geography, Kraków Pedagogical University,
ul. Podchorążych 2, 30-084 Kraków, Poland

Abstract: Thin intercalations of laminated coccolith limestones in a series of flysch and black shales define a regional marker horizon in the Oligocene strata of the Outer Carpathians. Distinctive sequences of laminae in the coccolith limestones have been correlated over distances of up to 240 km and across major facies zones. The correlated succession of laminae provided the basis for correlation of intercalated turbidites. The successions of laminae are disturbed at all studied localities by erosional gaps, intercalations of non-laminated limestone and deformation of some intervals of the laminated sequence. The disturbances themselves have been precisely correlated between three major nappes of the Outer Carpathians. The characteristics of the disturbances and their long-distance continuity indicate their seismic origin. The coccolith ooze at the top of the sediment column was firm and its simultaneous disturbance over large areas required earthquakes of magnitudes above 6.5. Laminated pelagic limestones provide a precise chronostratigraphic framework for long-distance correlation of seismic and other events recorded within their sequence.

Study of the temporal and spatial patterns of ancient seismic activity requires a precise areal chronostratigraphical framework. Seasonal varves in some pelagic sediments may provide such framework. Time resolution of the varve record is close to one year, and individual laminae or their distinctive successions may be traced over distances of up to hundreds of kilometres. This paper presents an attempt at recognizing the record of syn-sedimentary seismic events in the Oligocene Paratethys by the study of lamination in pelagic limestones.

Geological setting

The Oligocene laminated limestones described in this paper occur in the Outer Carpathians fold-and-thrust belt. The Outer Carpathians are a pile of nappes thrust northwards at least 40 km on the margin of the European platform (Książkiewicz 1956; Royden & Karner 1984). Sedimentary sequences in separate nappes correspond roughly to major facies zones.

The major nappes of the Polish sector of the Outer Carpathians include, from internal outwards: the Magura, Dukla, Silesian, Subsilesian and Skole nappes (Fig. 1). The nappes consist mostly of flysch sediments, up to 7000 m thick, of earliest Jurassic through Early Miocene age. Pelagic sediments are important in the Subsilesian nappe, comprising sediments laid down on a submarine high between the Skole and Silesian troughs supplied by gravity mass flows from

surrounding continental areas and from intra-basinal sources. The youngest part of the sequence, Oligocene-Early Miocene in age, consists of two informal lithostratigraphical divisions: the Menilite Beds and the overlying Krosno Beds.

The Menilite and Krosno Beds form relatively small outcrops in the western part of the Polish Carpathians; they become widespread eastwards, where they crop out over large areas (Fig. 1). The Menilite and Krosno Beds are thickest and most widespread in the Skole and Silesian nappes. They are thinner in the Subsilesian nappe and only locally present in the Dukla nappe. Oligocene deposits similar to the Menilite and Krosno Beds occur also in the more internal Magura nappe.

The distinctive lithology of the Menilite Beds are organic-rich, dark brown shales, mostly carbonate-free, at some horizons siliceous. Sandstones occur as thin, cross-laminated layers or lenses, and as thick fluxoturbidite sequences. The thickness of the Menilite Beds varies from 100 metres in sections dominated by shales, to more than 500 metres in the sections with thick fluxoturbidites. The Krosno Beds are up to 3500 m thick, changing up-section from sandy, to normal, to shaly flysch. Their upper boundary is erosional.

Fossils in the Menilite–Krosno series have not provided a basis for detailed biostratigraphical zonation. Fish faunas are abundant in some layers (Jerzmańska 1960), and six regional ichtyofaunal zones have been distinguished in the

From Kemp, A. E. S. (ed.), 1996, *Palaeoclimatology and Palaeoceanography from Laminated Sediments,*
Geological Society Special Publication No. 116, pp. 209–220.

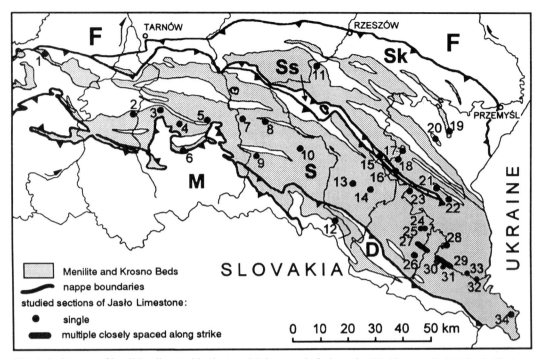

Fig. 1. Index map of localities discussed in the text. Main tectonic-facies units: M, Magura; D, Dukla; S, Silesian; Ss, Subsilesian; Sk, Skole; F, foredeep. For details of locality names and location see Haczewski (1989) and Koszarski & Żytko (1961).

Menilite Beds of the Skole nappe by Kotlarczyk & Jerzmańska (1976). Foraminifers are rare, usually poorly preserved and of low stratigraphical value (Olszewska 1984b). Diatoms also have low stratigraphical value, although they are abundant in some layers (Kotlarczyk & Kaczmarska 1987). Coccolith assemblages consist mainly of long-ranging taxa (Gaździcka, pers. comm. 1985), and have not been hitherto studied in the whole sequence.

Thin white layers of coccolith limestones occur in the Menilite and Krosno Beds. They are restricted to narrow stratigraphic intervals, are very persistent laterally, and are easily discernible by their lithology and colour among the host rocks. The limestone-bearing intervals are referred to as limestone horizons and are used as regional markers, named (from bottom to top): Tylawa, Jasło and Zagórz horizons (Haczewski 1989). All the limestone horizons occur beneath the inferred position of the Oligocene/Miocene boundary. The markers demonstrate that the boundary between the Menilite and Krosno Beds is strongly diachronous, younging northwards (Koszarski & Żytko 1961; Jucha 1969); the same is true for the major facies boundaries within the Krosno Beds (Fig. 2).

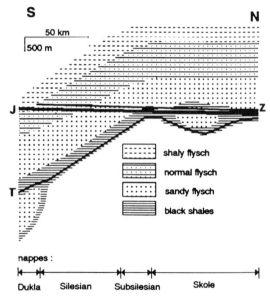

Fig. 2. Schematic cross-section through unfolded Menilite and Krosno Beds showing positions of limestone horizons. Scales approximate. T, Tylawa; J, Jasło; Z, Zagórz.

The base of the Menilite Beds is isochronous, lowermost Oligocene in age (Van Couvering *et al.* 1981; Olszewska 1984*a*). The shales of the Upper Krosno Beds in the south of the Silesian nappe may be as young as early Miocene (Olszewska & Szymakowska 1984), while the bulk of the Krosno Beds in the northern part of the Skole nappe is attributed to the Miocene (Nowak *et al.* 1985). Nevertheless, the position of the Oligocene/Miocene boundary has yet to be precisely defined. The Jasło Limestone horizon is tentatively dated at the upper part of the NP 23 or 24 zone.

Jasło Limestone horizon

The Jasło Limestone horizon is widespread in the Silesian, Subsilesian and Skole nappes. It lies in the middle part of the Menilite Beds in the external part of the Skole nappe and its position in the lithostratigraphic column gradually rises southwards, to the upper part of the Krosno Beds in the southern part of the Silesian nappe and in the Dukla nappe (Fig. 2).

The changing host facies determine the aspect of the limestone horizon. It consists of one layer, about 30 cm thick, near the eastward termination of the Subsilesian nappe, in a facies of hemipelagic black shales laid down on a submarine high. At all other localities the limestone is split by clastic layers whose lithology, thickness and number depend on the host facies at each locality. Figure 3 shows examples of the relationship between the Jasło Limestone and various host facies.

Numerous outcrops of the Jasło horizon are situated at close intervals along strike in the meandering course of the San river in the eastern part of the Silesian nappe (localities 29-33 in Fig. 1). They permit bed-by-bed correlation of the intercalated turbidites (Haczewski 1981). Some turbidites have been traced downcurrent to their terminations while others proved very persistent laterally. Turbidites in the Skole and Silesian nappes belong to different turbidite systems, supplied from different sources, hence they differ in petrographic composition, palaeocurrent patterns and internal structures. In the Silesian nappe, the pattern of lateral change in thickness and sedimentary structures (Haczewski 1981) indicates their transport from NW through a system of elongated depressions oriented approximately parallel with the present-day structural strike.

The maximum total thickness of the pelagic limestone layers within the Jasło Limestone horizon is about 35 cm. At the base the limestone is non-laminated, white-grey, with bluish or violet shade. Thin, horizontal lamination appears about 6 cm above the bottom and the limestone is laminated upwards for about 26 cm. The laminae are very fine, about 25 white-and-dark couplets per centimetre. The white laminae vary in thickness and distinctive patterns are discernible in their vertical sequence, especially in the upper half of the horizon. Towards the top, the white laminae are less pronounced, and the limestone passes gradually upwards to dark brown marl, rich in organic matter and containing sub-millimetre laminae of white micrite for several centimetres up-section.

The structure of the limestone, as viewed in thin section, is best described as micronodular laminated; it is a microscale equivalent of the nodular laminated structure described by Garrison & Kennedy (1977) from the English Chalk. The lamination is due to alternation of layers varying in structure rather than composition. The laminae are built of micrite nodules separated by wavy, often bifurcating, dark sheaths composed of organic matter, pyrite framboids and clay minerals. The layers built of greater nodules, up to 600 μm long and 300 μm thick, appear macroscopically as white laminae. The dark laminae have the sheaths more densely spaced, so that they delineate smaller nodules, up to 50 μm long and about 15 μm thick.

The micrite consists of coccoliths, which are recrystallized to a varying degree. Numerous partly dissolved, often collapsed, tests of planktonic foraminifers are scattered within the micrite nodules. Parting surfaces parallel to lamination expose numerous perfectly preserved fish remains, occasional seaweed and leaves of terrestrial plants, and very rare insects. The preservation of the organic remains is often perfect.

Bioturbation and benthic fossils are absent, except for threadlike *Trichichnus* burrows in the upper half of the horizon at locality 16 in the Subsilesian nappe.

The laminae in the Jasło Limestone, although irregular and diffuse at close view, are very persistent laterally. The distinctive patterns of laminae, especially in the upper part of the horizon, have been correlated between most sections of the Jasło Limestone, over distances up to 240 km along the structural strike, and across the boundaries of three major nappes – Skole, Subsilesian and Silesian (Haczewski 1984). The laminae were correlated on polished surfaces or on their enlarged photographs. The limestone becomes fissile upon weathering (it was described as Jasło Shale in older literature) and discoloured. For this reason, many, or even all laminated limestone layers were unusable for correlation of laminae in some outcrops where detailed sections were measured. The sequences of laminae slightly change laterally, so that the smaller the distance, the better is the match of the compared sequences. The most distant sections may need the use of intermediate sections for convincing correlation.

Therefore, although the Jasło Limestone has been identified in the Vineţisu section in the southern part of the East Carpathians in Romania (550 km from the type locality in

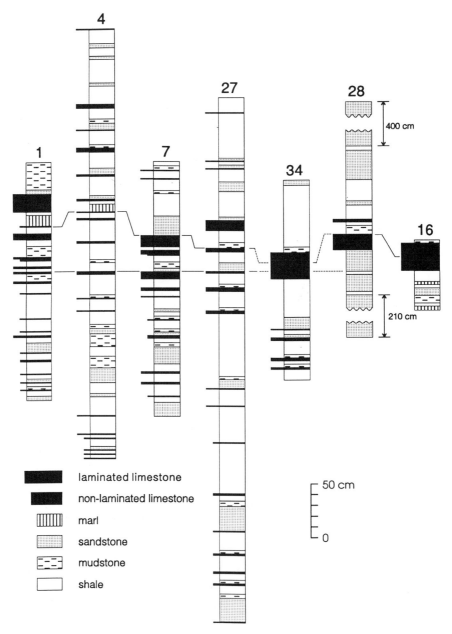

Fig. 3. Selected logs of Jasło Limestone horizon showing its relation to various host facies. Left part of each column shows limestone layers alone. See Fig. 1 for location of the sections identified by the numbers at top.

Jasło; Haczewski 1989), no convincing match was achieved by the comparison of laminae in complete sections of all limestone layers. Nevertheless, laminae in the overlying Zagórz limestone horizon have been correlated (Haczewski 1989).

The correlation of laminae in the Jasło Limestone has been unsuccessful in the southern part of the Silesian nappe and in the Dukla nappe where the Jasło Limestone is contaminated with abundant fine clastic material, present as fine clayey laminae and as clayey admixture in micrite. Numerous redeposited Cretaceous and older Paleogene foraminifers (Olszewska 1984*b*) and coccoliths (Gaździcka, pers. comm. 1985) were found in the samples of the Jasło Limestone from this area.

Despite these limitations, the succession of laminae in the Jasło Limestone provides a precise regional chronostratigraphic framework whose resolution is accepted as annual (Haczewski 1989). Laminae in analogous modern coccolith ooze in the Black Sea have been shown to represent annual varves (Degens *et al.* 1978; Hay & Honjo 1989). Turbidites and other phenomena recorded within the Jasło Limestone horizon can be precisely located within the sequence of laminae so that their temporal and spatial relationships may be studied.

Disturbances of the Jasło Limestone laminae

A complete continuous succession of laminae in the Jasło Limestone has been not found at any locality. It is disrupted by (i) erosional gaps, (ii) intercalations of non-laminated limestones, (iii) soft-sediment deformation structures. Though the disturbances are so common, the part of the laminated succession that remains undisturbed was used for precise location and areal correlation of the disturbances themselves. The position of the disturbances was determined by correlation of the directly overlying undisturbed laminae.

Erosional gaps. The total amount of erosional removal of the laminated limestone at any locality is determined by the difference between the maximum thickness of the laminated limestone (26 cm) found elsewhere and the total thickness of laminated layers at that locality. At some localities only a few centimetres are left and at some perfectly exposed sections the Jasło Limestone is completely absent, especially within the facies of thick bedded sandstones of the lower Krosno Beds, e.g. for several kilometres NW of locality 23 (Fig. 1). The limestone horizon usually reappears within a few kilometres along strike.

The erosional surfaces are mostly concordant with the planes of lamination, only exceptionally are they incised discordantly. The concordant erosional gaps occur beneath thin- and medium-bedded turbidites, especially those with Bouma division c at the base. The correlation of laminae beneath the distal parts of turbidites with the laminae beyond the terminations of these turbidites shows that the extent of erosion beneath these layers is limited to the depth of sole marks. Lateral continuity of the concordant erosional gaps varies; some gaps are filled up within a few kilometres, others extend for tens of kilometres.

The discordant erosional scours into the laminated limestone, visible at localities 7, 27, 29 and 33, all occur at the same position, at the base of the unique turbidite layer which includes clasts of the laminated limestone (Figs 4 & 5). The thickness and the grain size of this turbidite are the same as those of the turbidites which are not incised into their limestone substrate and bear no limestone clasts. At one of the exposures marked collectively as locality 27 in Fig. 1 a 22 mm layer of laminated limestone beneath this turbidite is gently thinned out by erosion until its complete removal (Fig. 4). The gentle scour sharply deepens where it reaches to the underlying shale. At locality 33 (Fig. 1) loosened sheets of limestone are imbricated over one another, perpendicular to the palaeoflow direction in the overlying turbidite, shown by cross-lamination and sole marks (Figs 4 & 5). This indicates that the disruption and sliding of the clasts preceded deposition of the turbidite. At this and neighbouring localities, the 22 mm layer of limestone displays straight parallel anticlinal folds about 3 cm wide, about 5–10 m from one another. An erosional surface is present at this position in all studied sections.

Non-laminated limestone layers. These are up to 4 cm thick. They overlie siliciclastic turbidites or laminated limestone and are everywhere overlain by laminated limestone, except for locality 1. Pure limestone layers lack any distinguishable internal structure, except for a slight upward grading in shade from greyish to white. Their tops and bottoms are horizontal and smooth; only in a few cases, when the thin micritic layers overlie a slump or in situ distorted laminated sediment, do they fill the relief at their base.

Some of the thicker, 1.5–4 cm non-laminated carbonate layers consist of two parts: basal dark marl, overlain by homogenous white limestone. The boundaries are sharp, but irregular. The white micrite thickens locally and is sunken into the lower, darker marly sediment (Fig. 6). This deformation is manifest as winding elongated

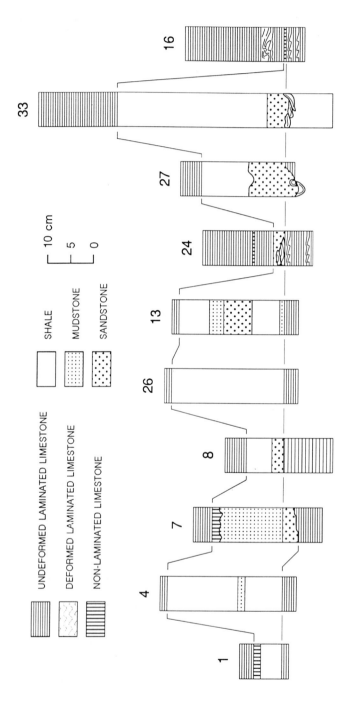

Fig. 4. Details of the lower disturbed horizon described in the text. Correlation lines show the base of the intervening clastic layer (datum) and the base of the overlying undisturbed laminae which were the basis for correlation. Numbers above columns refer to their locations in Fig. 1.

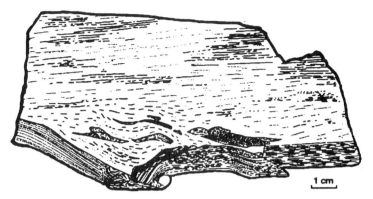

Fig. 5. Imbricated slabs in disrupted layer of laminated limestone in locality 29 (Fig. 1), thrust on one another from right to left. The overlying sandstone turbidite displays a palaeocurrent direction towards the viewer.

Fig. 6. Non-laminated limestone and dark marl at locality 7 covered with the same undisturbed laminae as the slump fold at locality 16, shown in Fig. 7.

ridges on the soles of the non-laminated limestone layers. Outcrop conditions rarely permit observation of sole surfaces but in one case (locality 7) the ridges clearly displayed superposition of some ridges on others, indicative of their non-simultaneous origin, much like superposed animal traces. The variable and diffuse forms of the ridges do not match any trace fossil taxon and no remains of possible trace makers were found. These dubious bioturbations cover the sole completely but they do not extend to neighbouring exposures, even a few kilometres apart.

The non-laminated limestone layers have been traced for tens of kilometres. They thin gradually in the same directions as the accompanying turbidites. They are especially numerous and thick at locality 1.

Deformed laminae. The deformational structures include a slightly wavy fabric with aggregation of white micrite into bigger nodules, and fine wrinkles, which pass upwards to folds, sometimes overturned (Fig. 7). Laminae in the most intensely deformed fragments are completely obliterated. This deformation dies out

downwards, towards a well defined horizontal base. In most cases the deformation does not involve lateral transport of sediment. The folds, wrinkles and thrusted slices (locality 35) are strongly elongated in one direction at every locality where they occur.

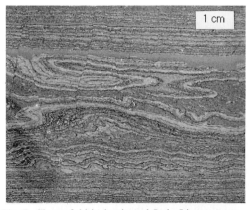

Fig. 7. Slump fold in laminated Jasło Limestone at locality 16 (Fig. 1), covered with the same undisturbed laminae as the non-laminated limestone at locality 7 shown in Fig. 6.

Another type of deformation is present within intervals up to about 1 cm thick. Laminae remain generally horizontal but they are sheared: slightly wavy, discontinuous, forming lozenge-like bundles. Deformation structures of this sort are very limited in areal extent, they have never been correlated between various outcrops, and tend to reappear at various positions in the same section, especially at locality 31.

Correlation of disturbed horizons

The most intense disturbances have been correlated over large parts of the studied area. Two horizons are especially well manifest. One includes the exceptional turbidite with erosional scours at base and limestone clasts inside (Fig. 4). Wrinkles, convolution and partial obliteration of laminae occur at this horizon at locality 16 in the Subsilesian nappe (Fig. 7). Imbricated slabs and wrinkles are present at localities 27 and 29 (Fig. 5). A sandstone turbidite is present at this horizon at most localities. It has discordant erosional scours at its base (localities 7, 27, 29, 33) and includes clasts of laminated limestone (localities 7, 24, 29, 33), features not observed in other correlated layers within the Jasło Limestone horizon. Clasts of Jasło Limestone in sandstone were reported from other localities by Koszarski & Żytko (1961) and Jucha (1969). These occurrences have not been precisely located in the section and possibly belonged to the same layer.

Besides the localities mentioned the sandstone turbidite at the same position is present at localities 2, 5, 13, 30, 31, 32 and mudstone or shale at localities 1 and 34 (locations in Fig. 1). Taking into account the palaeoflow of the turbidity current from NW it is important to note that at least 6 cm of laminated limestone is missing beneath a 16 mm thick shale layer at locality 1 which lies farthest northwest. Section 16 is the only correlated section where no distinct siliciclastic layer is present in this position, but there is a discontinuous sub-millimetre clayey lamina. A 2 mm layer of non-laminated limestone is present at approximately the same position at locality 20. At other localities precise identification of this horizon was not achieved because the limestone layers were weathered or too thin for correlation, or the sections were incompletely exposed.

The other widespread disturbed horizon is situated higher in the section. The greatest preserved thickness of the laminated limestone between these levels is 38 mm at locality 24. Only there and at locality 16 (Subsilesian nappe) is

there no clastic layer in this interval, but at locality 16 this interval is severely slump-deformed (Fig. 7) and its thickness (60 mm) seems to be increased by slumping. A shale or mudstone layer a few centimetres thick is present at the same position at most localities. A horizon of non-laminated limestone, sometimes marly or accompanied by marl, overlies it at localities 1 (10 cm of marl with 5 mm of limestone at base), 3, 4 (65 mm), 7 and 8 (10–16 mm, with the dubious organic traces; Fig. 6), 16, 17 (20 mm), 21, 22, 24 (4 mm), 27 (1.5 mm). A sandstone turbidite is present at the same position in the Skole nappe at locality 20. The laminated limestone layer beneath it is slightly distorted down to the depth of about 20 mm beneath its base. The sandstone is medium-grained and distinct from the sandstones of the Krosno Beds.

Another extensive disturbed horizon is discernible at some localities beneath the lower horizon, but it is usually obliterated by the latter, hence it cannot be described in detail.

Both of the disturbed horizons described here were identified in all sections where correlation of laminae permitted it. They are probably represented by clayey laminae at locality 34, by apparent downcurrent terminations of sandstone turbidites present at localities 27, 29, 33 and 32. However, the corresponding part of the weathered 24 cm thick limestone layer at locality 34 was unusable for precise correlation.

Origin of the disturbances

Erosion, disruption and deformation in the laminated limestone

A surface layer of laminated coccolith mud has been removed beneath many turbidites. The lack of limestone clasts in the turbidites (all but one) and smooth erosional boundaries indicate that erosion proceeded by sand-blasting rather than by tearing off greater portions of cohesive sediment.

The coccolith mud was more cohesive then the clastic sediments intercalated with it. Its firm consistency is proven by the steep-walled erosional scours and by clasts in turbidites, plastically deformed and abraded during transport, but preserving their integrity and internal structure. Clastic dikes are very common in the Oligocene strata of the Outer Carpathians but none has been reported to cut through a layer of laminated coccolith limestone.

The only case of a different relation between the turbidite and underlying laminated limestone is the turbidite which includes clasts of laminated limestone and has sharp basal erosional scours. The features of the underlying limestone layer beneath this peculiar turbidite indicate that the calcareous mud was weakened, broken and deformed before the turbidity current overrode it. The thickness and the grain size of this turbidite do not depart from those of the other turbidites intercalated with the Jasło Limestone. The exceptional erosive action of this particular turbidity current is best explained by the state of the underlying sediment which had been weakened, folded, fractured, and even slumped before the arrival of the turbidity current. It was deformed in similar way even beyond the extent of the turbidite.

Another turbidite was laid down at the same position in the Skole part of the basin. Its thickness also does not exceed that of the others in the same section, but the laminated limestone layer beneath it is slightly distorted.

The bottom sediment became deformed at the same time (within the time resolution of the seasonal lamination) over the area of the basin floor between sections 7, 20, 33, regardless of the local relief and position in the turbidite depositional system. The deformation was shortly followed by the descent of at least two turbidity currents, in distant parts of the basin, and from discrete source areas. The characteristics of the disturbances are best explained by their earthquake origin (cf. Roep & Everts 1992). The turbidity currents are ruled out as the causative agents because their areal distributions differ from those of the deformation structures and the strain patterns are incompatible with the palaeoflow directions.

The affected area is 100 km long and 35 km wide. The source area of the turbidity current in the Silesian Unit was located outside this area, northwest of section 7 and possibly even of section 1. The original width of this area on the basin floor may be estimated at c. 100 km, based on the palinspastic reconstruction of the basin along a section running near localities 21 and 25 (Roure et al. 1993). The deformation of the laminated limestone is strongest in the eastern part of the area, and the most probable location of the epicentre was near the boundary of the Silesian and Subsilesian units, between localities 14 and 16, near an area where the Jasło Limestone is locally absent.

Attempts at quantification of palaeoearthquake magnitudes from the soft-sediment deformation structures are based mainly on liquefaction phenomena in terrestrial environments (Allen 1986; Vittori et al. 1991). Such constrains as water content of sediment and slope angle can be approximated only very crudely for ancient marine sediments, leaving a wide margin of uncertainty. For the lower of the observed deformed horizons, the data on the areal extent are the best constraints available. Data presented by Allen (1986) and Vittori et al. (1991) indicate that earthquakes of magnitudes close to 7 produce liquefaction at distances up to 50 km from the epicentre. The disruption of the laminated ooze probably required stronger shaking than liquefaction of water-saturated sand. Liquefaction features have not been observed in underlying sandstones. Their potential for liquefaction was low due to the high clay and mud content. The magnitude of the earthquake that caused the lower of the two observed disturbed horizons is thus estimated as 6.5–7 or higher. The higher disturbed horizon was apparently caused by a less strong earthquake which caused deformation of limestone over a smaller area, though turbidity currents were generated at distant localities.

A comparison of the seismically triggered turbidites with the other turbidites in the host sequence indicates that the former do not differ markedly in grain-size and composition from the latter. An important difference lies in the evidence of failure or distortion of the underlying firm sediment before the arrival of the turbidity current. Seismically triggered turbidites in the Hellenic Trench also do not differ from typical turbidites in their host sequence (Anastasakis & Piper 1991).

Non-laminated limestone layers

The origin of the non-laminated limestone intercalations is less clear. Rapid deposition (within one year) is demonstrated for all of these layers whose position within the sequence of correlated laminae could be traced laterally beyond their extent. The nearly homogenous internal structure, with only a slight vertical grading, as well as the position atop siliciclastic turbidites, suggest deposition from suspension. A seismic tremor itself does not seem capable of raising firm calcareous mud into suspension.

Venting of fine sediment from fractures in the basin floor during strong earthquakes could create a cloud of suspended sediment moving downslope and settling down for many days or weeks. However, clastic intrusions, abundant in the Menilite Beds and common in the Krosno Beds, are always built of sand or pebbly sand. Examples of sandstone dikes venting to the basin floor, described by Dżułyński & Radomski (1956) and another one observed by the author, indicate that the clastic intrusions could erupt sediment to the water column. There is no explanation, however, for the calcareous composition of sediment in such a cloud. The coccolith ooze was then present as a thin layer at the surface and a few thin layers below. These layers would be more likely to break to coarse fragments and could hardly feed a cloud of suspended calcareous sediment.

The numerous erosional gaps prove that the laminated coccolith ooze, though firm, was susceptible to scouring by violent currents. There is, however, no correlation between the erosional gaps and the occurrence of the non-laminated limestone layers. Erosion in the form of steep-walled scours produced cohesive clasts. Turbidity currents could not rise the clouds of pure carbonate suspension as they were themselves clouds of suspended siliciclastic sediment. Violent bottom water currents would produce the effect we are looking for if they attained velocities above threshold erosion velocity. The threshold erosion velocity for this type of sediment needs to be evaluated intuitively as no close modern analogue is known for the Jasło Limestone. Figures derived experimentally for abyssal red clay (Lonsdale & Southard 1974) and calcareous ooze (Southard et al. 1971) indicate that a bottom current velocity well above 50 cm/s, perhaps 70 cm/s or more, would be required for erosion and entrainment of the firm laminated coccolith ooze of relatively low water content. Bottom roughness, caused by folding, fracturing and sliding of surface layer, could decrease the threshold velocity.

Strong, pulsating bottom currents may accompany an advance of a tsunami. Kastens & Cita (1981) provide a formula that links the maximum height of a tsunami wave with the water depth and maximum bottom current velocity:

$$u_{\max} = \frac{H}{2d}\sqrt{gd}$$

where H is the wave height, $u_{\max}$ is the maximum water particle velocity in horizontal direction, d is the water depth and g is the acceleration due to gravity. Using this relation to find the height of surface wave that would result in a bottom current of maximum velocity of 70 cm/s at a depth of 200 m (a conservative estimate for the intrabasinal elevations, e.g. in the Subsilesian realm):

$$H = \frac{u_{\max} \cdot 2d}{\sqrt{gd}}$$

we find a wave height of 6.3 m; a figure implying a tsunami related to a large displacement at the seafloor. The bottom current caused by a tsunami is pulsating and the period of pulsation is shorter than time needed for suspended fine sediment to settle down. So the pulsating current may raise a successively growing cloud of suspended fine sediment that could take many days or weeks to settle on the bottom. The suspended sediment would move by gravity, following the local relief of the basin floor. This process provides a plausible explanation for the features of the homogenous limestone layers. The mixed layers probably may be attributed to the same process with siliciclastic mud entrained in addition to the calcareous ooze and then settling faster due to greater size and density of its particles.

Concluding remarks

Annual lamination in Oligocene pelagic Jasło Limestone has been correlated laterally over distances of up to 240 km. The laminated sediments represent a time interval of about 800 years. A record of two exceptionally strong earthquakes was recognized in these limestones and correlated areally over many tens of kilometres. No record of volcanic activity accompanies the palaeoseismicity record. The epicentres were probably located at a zone roughly corresponding to the boundary between the Subsilesian and Silesian realms and may be related to early movements along the leading thrust of the future Silesian nappe.

Precise time calibration, coupled with areal correlation of the calibrated sediments is rarely acquired in ancient sediments. The long-distance continuity of laminae demonstrated in the Jasło Limestone is likely to be present in other similar sediments. Thin bands of coccolith limestone have been described from various sedimentary sequences with organic-rich sediments (e.g. Best & Müller 1972; Busson & Noel 1972; Bréhéret 1983; Gallois & Medd 1979; Müller & Blaschke 1969, 1971; Nagymarosy 1983; Voronina & Popov 1984) and some of these bands proved very persistent laterally and finely laminated (Bréhéret 1983; Gallois & Medd 1979; Voronina & Popov 1984). The potential of such sediments for precise long-distance correlation of enclosed events offers a clear return for the investment of time in the lateral tracing of their lamination.

The study on sedimentary record of palaeoseismicity was supported by the KBN grant 6 6298 91 02 to the author, registered at the Institute of Geological Sciences, Polish Academy of Sciences. Participation in the conference 'Palaeoceanography and Palaeoclimatology from Laminated Sediments' was facilitated by Dr Tadeusz Ziętara, the head of the Physical Geography Division in the Institute of Geography,

Kraków Pedagogical University (WSP). The Institute of Geography provided facilities and covered the cost of manuscript preparation. Helpful discussions with Jarosław Tyszka and Kurt Grimm are gratefully acknowledged. I thank D. I. M. Macdonald for his helpful review of the manuscript.

References

ALLEN, J. R. L. 1986. Earthquake magnitude-frequency, epicentral distance, and soft-sediment deformation in sedimentary basins. *Sedimentary Geology*, **46**, 67–75.

ANASTASAKIS, G. C. & PIPER, D. J. W. 1991. The character of seismo-turbidites in the S-1 sapropel, Zakinthos and Strofadhes basins, Greece. *Sedimentology*, **38**, 717–733.

BEST, G. & MÜLLER, C. 1972. Nannoplankton-Lagen im Unter-Miozän von Frankfurt am Main. *Senckenbergiana Lethaea*, **53**, 103–117.

BRÉHÉRET, J.-G. 1983. Sur des niveaux de black shales dans l'Albien inférieur et moyen du domaine vocontien (sud-est de la France): étude de nannofacies et signification des paleoenvironments. *Bulletin de Musée National d'Histoire Naturelle Paris*, **4e sér. 5**, section C, 113–159 [with English abstract].

BUSSON, M. G. & NOËL, D. 1972. Sur la constitution et la genèse de divers sédiments finement feuilletés ('laminites'), ê alternances de calcaire et de matière organique ou argileuse. *Comptes Rendus d'Academie des Sciences Paris*, **sér. D. 274**, 3172–3175 [with English abstract].

DEGENS, E. T., STOFFERS, P., GOLUBIĆ, S. & DICKMAN, M. D. 1978. Varve chronology: estimated rates of sedimentation in the Black Sea deep basin. *In: Initial Reports of the Deep Sea Drilling Project*, **42**, pt 2, 499–508.

DŻUŁYŃSKI, S. & RADOMSKI, A. 1956. Clastic dykes in the Carpathian flysch. *Annales Societé géologique Pologne*, **26**, 225–264.

GALLOIS, R. W. & MEDD, A. W. 1979. Coccolith-rich marker bands in the English Kimmeridge Clay. *Geological Magazine*, **116**, 247–334.

GARRISON, R. E. & KENNEDY, W. J. 1977. Origin of solution seams and flaser structure in Upper Cretaceous chalks of southern England. *Sedimentary Geology*, **19**, 107–137.

HACZEWSKI, G. 1981. Extent and lateral variation of individual turbidites in flysch, horizons with Jasło Limestones, Krosno Beds, Polish Carpathians. *Studia Geologica Polonica*, **68**, 13–27.

—— 1984. Korelacja lamin w chronohoryzontach wapienia jasielskiego i wapienia z Zagórza (Karpaty Zewnętrzne). *Kwartalnik Geologiczny*, **28**, 675–688 [with English abstract].

—— 1989. Poziomy wapieni kokkolitowych w serii menilito-wokrosnieniej – rozróżnianie, korelacja i geneza. *Annales Societatis Geologorum Poloniae*, **59**, 435–523 [with English summary].

HAY, B. J. & HONJO, S. 1989. Particle deposition in the present and Holocene Black Sea. *Oceanography*, **2**, 26–31.

JERZMAŃSKA, A. 1960. Ichtiofauna łupków jasielskich z Sobniowa. *Acta Palaeontologica Polonica*, **5**, 367–419 [with English abstract].

JUCHA, S. 1969. Łupki jasielskie, ich znaczenie dla stratygrafii i sedymentologii serii menilitowo-krośnieńskiej (Karpaty fliszowe). *Prace Geologiczne, Polska Akademia Nauk, Oddział w Krakowie*, **52**.

KASTENS, K. A. & CITA, M. B. 1981. Tsunami-induced sediment transport in the abyssal Mediterranean Sea. *Geological Society of America Bulletin*, **92**, 845–857.

KOSZARSKI, L. & ŻYTKO, K. 1961. Łupki jasielskie w serii menilitowo-krośnieńskiej w Karpatach Srodkowych. *Instytut Geologiczny, Biuletyn*, **166**, 87–232 [with English abstract].

KOTLARCZYK, J. & JERZMAŃSKA, A. 1976. Biostratigraphy of Menilite Beds of Skole Unit from the Polish Flysch Carpathians. *Bulletin d'Academie Polonaise des Sciences, Serie des Siences de la Terre*, **24**, 55–62.

—— & KACZMARSKA, I. 1987. Two diatom horizons in the Oligocene and (?) Lower Miocene of the Polish Outer Carpathians. *Annales Societatis Geologorum Poloniae*, **57**, 143–188.

KSIĄŻKIEWICZ, M. 1956. Geology of the Northern Carpathians. *Geologische Rundshau*, **45**, 369–411.

LONSDALE, P. & SOUTHARD, J. B. 1974. Experimental erosion of North Pacific red clay. *Marine Geology*, **17**, M51–M60.

MÜLLER, G. & BLASCHKE, R. 1969. Zur Entstehung des Posidonienschiefers (Lias ξ). *Die Naturwissenschaften*, **56**, 635.

—— 1971. Coccoliths: important rock-forming elements in bituminous shales of central Europe. *Sedimentology*, **17**, 119–124.

NOWAK, W., GEROCH, S. & GASIŃSKI, A. 1985. Oligocene/Miocene boundary in the Carpathians. *In: VIIIth Congress of the Regional Committee on Mediterranean Neogene Stratigraphy. 12–22 September 1985, Budapest. Abstracts*. Hungarian Geological Survey, Budapest, 427–429.

NAGYMAROSY, A. 1983. Mono- and duospecific nannofloras in Early Oligocene sediments of Hungary. *Proceedings of the Koniglijske Nederlandskoe Akademie van Wetenschappen*, **ser. B, 86**, 273–283.

OLSZEWSKA, B. 1984a. Benthonic foraminifera of the SubMenilite Globigerina Marls of the Polish Outer Carpathians. *Prace Instytutu Geologicznego*, **110**.

—— 1984b. Kilka uwag o zespolach otwornic towarzyszacych wapieniom jasielskim w polskich Karpatach Zewnętrznych. *Kwartalnik Geologiczny*, **28**, 689–700 [with English abstract].

—— & SZYMAKOWSKA, F. 1984. Olistostroma w Kołaczycach koło Jasła (Karpaty Sródkowe) i czas jej powstania w swietle nowych badan mikropaleontologicznych. *Biuletyn Instytutu Geologicznego*, **346**, 117–145 [with English abstract].

ROEP, T. B. & EVERTS, A. J. 1992. Pillow-beds: a new type of seismites? An example from an Oligocene turbidite fan complex, Alicante, Spain. *Sedimentology*, **39**, 711–724.

ROYDEN, L. & KARNER, G. D. 1984. Flexure of lithosphere beneath Apennine and Carpathian foredeep basins: Evidence for an insufficient topographic load. *AAPG Bulletin*, **68**, 704–712.

ROURE, F., ROCA, E. & SASSI, W. 1993. The Neogene evolution of the Outer Carpathian flysch units (Poland, Ukraine and Romania): kinematics of a foreland/fold-and-thrust belt system. *Sedimentary Geology*, **86**, 177–201.

SOUTHARD, J. B., YOUNG, R. A. & HOLLISTER, C. D. 1971. Experimental erosion of fine abyssal sediments. *Journal of Geophysical Research*, **76**, 5903–5909.

VAN COUVERING, J. A., AUBRY, M.-P., BERGGREN, W. A., BUJAK, J. P., NAESER, C. W. & WIESER, T. 1981. The terminal Eocene event and the Polish connection. *Palaeogeography, Palaeoclimatology, Palaeoecology*, **36**, 321–362.

VITTORI, E., SYLOS LABINI, S. & SERVA, L. 1991. Palaeoseismology: review of the state of the art. *Tectonophysics*, **193**, 9–32.

VORONINA, A. A. & POPOV, S. V. 1984. Solenovskiy gorizon vostochnogo Paratetysa. [The Solenovskiy horizon of the Eastern Paratethys]. *Izvyestiya Akademii Nauk SSSR, Seriya Geologicheskaya*, 1984, 41–53 [In Russian].

High-resolution sedimentology and micropalaeontology of laminated diatomaceous sediments from the eastern equatorial Pacific Ocean (Leg 138)

R. B. PEARCE[1], A. E. S. KEMP[1], J. G. BALDAUF[2] & S. C. KING[1]

[1] *Department of Oceanography, University of Southampton,*
Southampton Oceanography Centre, Waterfront Campus, European Way,
Southampton, SO14 3ZH, UK
[2] *Department of Oceanography and Ocean Drilling Program, Texas A&M University,*
College Station, TX 77840, USA

Abstract: Scanning electron microscope (SEM)-based analyses of the laminated diatom oozes encountered during Leg 138 reveal three major laminae types. The first lamina type is composed of multiple layers of approximately 20 μm thick diatom mats, which form laminae dominated by assemblages of the pennate diatom, *Thalassiothrix longissima*. More than one variety/subspecies of *T. longissima* occurs within these laminae (referred to as the *T. longissima* Group). The second lamina type is composed of a mixed-assemblage of several species of diatoms (centric and pennate varieties), calcareous nannofossils, and subordinate quantities of radiolarians, silicoflagellates and foraminifera. The third lamina type is dominated by an assemblage of nannofossils and minor amounts of those fossil components mentioned above. This last form of lamination is compositionally similar to the background sediment type, foraminifer-nannofossil ooze (F-NO). Two lamina associations occur within the laminated intervals; the first comprises of alternations of *T. longissima* Group and mixed-assemblage laminae (average thickness is approximately 6 mm) and the second is composed of *T. longissima* and nannofossil-rich laminae (average thickness is approximately 3.5 mm). The arrangement of laminae probably originates from the deposition of multiple layers of 20 μm thick mats from one mat-flux episode. The much thinner nannofossil-rich laminae are interpreted to represent periods of more 'normal' deposition between mat-flux episodes. The occurrence of several varieties/subspecies of *T. longissima* within individual mat layers is consistent with observations of *Rhizosolenia* diatom mats in the modern world ocean.

Laminated diatomaceous marine sediments can provide unique information about the record of interannual and seasonal variability in marine systems, for example in the California Borderland Basins (Gorsline 1984), within the Gulf of California (Baumgartner *et al.* 1985) and off Peru (Kemp 1990). However, the distribution of such laminated sediments hitherto has been thought to be restricted to relatively isolated anoxic basins or continental margins, where an intense oxygen minimum zone intersects the continental shelf/slope. The discovery of widespread deposits of deep-sea laminated diatom ooze in the oxygenated deep waters of the eastern equatorial Pacific Ocean (Kemp & Baldauf 1993) provides the first opportunity to identify ancient individual flux events in the deep sea and relate these to short timescale oceanographical processes.

Recent research has shown that only backscattered electron imagery (BSEI) of resin-impregnated sediment has sufficient resolution to identify the individual components of laminated fine-grained sediments (Kemp 1990; Grimm 1992; Kemp & Baldauf 1993; Brodie & Kemp 1994). The purpose of this paper is to report detailed sedimentological investigations of laminated diatomaceous sediments recovered during Ocean Drilling Program (ODP) Leg 138 drilling and to relate these using electron microscope techniques to micropalaeontological observations.

Methodology

Electron microscopy

The sediments were prepared for SEM examination using the method described by Kemp (1990). In brief, samples were taken using standard ODP palaeomagnetic sample cubes or u-channels, which minimize deformation of the sediment. The sediments were refrigerated prior to preparation to avoid drying them out. The material was

Previously published in: Pisias, N. G., Mayer, L. A., Janecek, T. R., Palmer-Julson, A. & van Andel, T. H. (eds) 1995. *Proceedings Ocean Drilling Program, Scientific Results*, **138**, College Station, TX (Ocean Drilling Program), 647–663.

From Kemp, A. E. S. (ed.), 1996, *Palaeoclimatology and Palaeoceanography from Laminated Sediments,* Geological Society Special Publication No. 116, pp. 221–241.

prepared for sectioning by carefully cutting subsamples from each sample using scalpel and forceps. The subsamples were placed in plastic boxes and, while still damp, removed to a Logitech vacuum impregnator. The impregnator was evacuated to an internal pressure of about 10^{-6} mb prior to the introduction of pre-evacuated low-viscosity epoxy resin. Because each sample effectively dried out during the evacuation process, the pump-down time ranged up to 36 hours. Impregnated samples were mounted on frosted glass slides and polished using a Logitech lapping machine. The preparation process was finished with a cloth impregnated with aluminium powder that maximized the polish. Polished slides were carbon-coated and analysed with a JEOL JSM 6400 scanning electron microscope (SEM) fitted with a Tracor Series 2 energy dispersive system and a solid-state backscattered electron detector. Goldstein *et al.* (1981) discussed the use of BSEI in SEM work, while geological applications were outlined in Krinsley *et al.* (1983) and references therein.

Diatoms

Samples were processed (using HCl and H_2O_2), and slides were prepared for examination following a procedure similar to that discussed by Boden (1991). Each glass slide was examined using $\times 750$ and $\times 1250$ magnifications with a Zeiss transmitted light microscope. For the data presented in Fig. 10, relative abundance of specific species were based on the number of specimens observed per field of view at $\times 750$ magnification. The abundance of the *Thalassiothrix longissima* Group (lengths greater than 70 μm), and *Denticulopsis* are defined as follows: barren (B), no specimens observed; rare (R), less than one specimen observed per horizontal transverse; few (F), one to five specimens observed per horizontal transverse; common (C), one to five specimens per field of view; and abundant (A), greater than five specimens per field of view. In each sample, the abundance of diatom species (shown in Fig. 10 as 'Other Diatoms'), other than specimens of the *T. longissima* Group, but including *Denticulopsis* specimens, is presented as the average number (in percent) of specimens observed per field of view.

Diatom preservation was assessed qualitatively by (1) comparing the relative abundance of finely silicified forms (e.g. specific species of *Thalassiosira* and *Nitzschia*) to robust forms (e.g. specific species of *Coscinodiscus*, *Craspedodiscus* and *Denticulopsis*), (2) examining the degree of dissolution (pitting, etching, etc.) of individual frustules, and (3) analysing the degree

of fragmentation of individual frustules (specimens of specific species, other than those from the *T. longissima* Group).

Poorly preserved samples typically contain rare specimens of several species (generally *Coscinodiscus*, *Rossiella* and *Craspedodiscus*) that exhibit pitting and etching. *Craspedodiscus coscinodiscus* specimens, if present, are fragmented, with only the central portion of each valve face preserved. In addition, *Denticulopsis* specimens, when present, typically contain only the girdle portion of the valves. Specimens of *Nitzschia* or *Thalassiosira*, as well as silicoflagellates, are extremely rare if present at all.

Moderately preserved samples contain few specimens exhibiting pitting or etching. Specimens of *C. coscinodiscus* generally exhibit the valve face and a portion of the mantle. The number of *Denticulopsis* girdle bands was seen to be reduced and generally low, when compared with the abundance of complete valves of the species. Silicoflagellates, and specimens of *Nitzschia* and *Thalassiosira* are rare, but generally present.

Samples containing a well preserved diatom assemblage consist of diatom frustules without pitting or etching. Specimens are generally complete, with the exception of those from the *T. longissima* Group.

Foraminifera and radiolarians

The absolute abundances of planktonic foraminifera, benthic foraminifera, and radiolarians were estimated for several intervals from the $\times 20$ magnification BSEI photomosaics. For each interval, the total number of foraminifera and radiolarians was counted, and the number per square centimetre of sediment calculated. The abundances of planktonic foraminifera, benthic foraminifera, and radiolarians were defined as follows: barren (B), no specimens observed; rare (R), less than one specimen per square centimetre; common (C), one to five specimens per square centimetre; abundant (A), more than five specimens per field of view.

Occurrence of the laminated diatom ooze intervals

Laminated diatom ooze (LDO) of middle Miocene through early Pliocene age was recovered from ODP sites 844 and 847 (located north and west of the Galapagos Islands respectively), and 849 through 851 (Fig. 1) which form part of the 110° W north–south equatorial transect. LDO

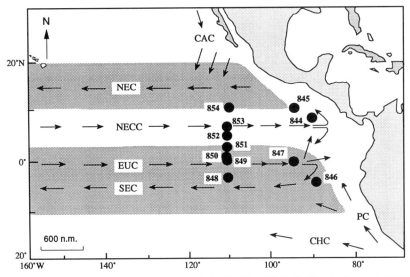

Fig. 1. Location map of Leg 138 drilling sites containing LDOs. CAC, California Current; NEC, North Equatorial Current; NECC, North Equatorial Counter Current; EUC, Equatorial Undercurrent; SEC, South Equatorial Current; CHC, Chile Current; and PC, Peru Current.

also occurs at DSDP sites 572 through 574 (133° W) (Kemp & Baldauf 1993) and DSDP Site 158 (85° W) (11.1 Ma). Thus, deposition of these sediments has occurred across as much as 3000 km of the eastern equatorial Pacific Ocean, in an east–west direction. The north–south extent of the LDOs is more restricted. Plate tectonic backtracking of the Leg 138 site positions shows that LDOs were probably deposited in two latitudinally separate regions: (1) within a narrow band immediately south of the equator between 0° and 2° S; and (2) north of the equator, centred around the early to mid Miocene positions of Sites 844 and 845 (4–6° N) (see Kemp et al. 1995). The tight latitudinal control on LDO between 0° and 2° S, is well illustrated at Sites 850 and 851 by the gamma-ray attenuation porosity evaluator (GRAPE) record across the youngest mat interval (4.4 Ma) (see Fig. 6 in Kemp et al. 1995). The LDOs are concentrated within the time intervals: 12.8–12.5, 11.8–11.5, 11.1, 10–9.5, 6.3–6.1, 5.8–5.1, and 4.4 Ma, with less widespread intervals at 14.6, 8.2, 7.35–7, 6.75, and 5.1–4.5 Ma. The episodes of mat deposition at each site have been plotted against depth (metres below sea floor, mbsf) and age (Ma) (Figs 2A and B respectively). Individual LDO deposits can be correlated between sites using biostratigraphy, and continuous GRAPE-data (see Kemp et al. 1995; Bloomer et al. 1995).

Preservation of the laminated diatom ooze intervals

(i) Carbonate dissolution. Extensive SEM investigations of the LDO intervals reveals no substantive evidence that carbonate dissolution occurred preferentially in the *T. longissima* mat deposits. Planktonic and benthic foraminifera tests from *T. longissima*-rich laminae show no signs of dissolution (Figs 8 & 9) and both intact and deeply pitted/etched coccoliths occur within individual laminae. Further details on silica/carbonate variation in the sediments of the eastern equatorial Pacific Ocean can be found in Kemp et al. (1995) and Kemp (1995).

(ii) Bioturbation. Bioturbation of the laminated intervals occurs preferentially at the upper boundary with the overlying F-NO (e.g. 4.4 Ma interval, Fig. 3). The lower boundary of the laminated intervals is characterized by well laminated and rarely bioturbated sediment (Fig. 6). The preservation of the diatomaceous laminated intervals is thought to relate to the physical subjugation of the benthos by the *T. longissima* meshwork (Kemp & Baldauf 1993; King et al. 1995), rather than a decrease in dissolved oxygen concentration. Burrows within the laminated intervals are most frequently horizontal and

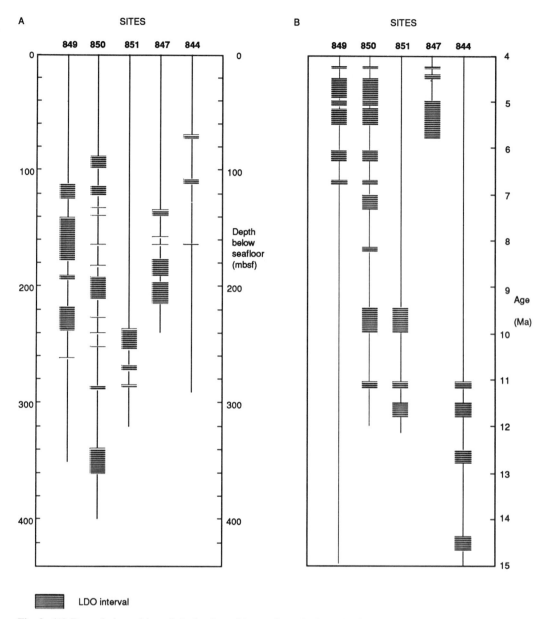

Fig. 2. (A) Down-hole position of the laminated intervals at the Leg 138 sites. (B) Episodes of LDO deposition using the Berggren *et al.* (1985) timescale. The resolution of the age dating varies down-core, such that the number of LDO intervals between (A) and (B) are not equal.

show little or no sign of disrupting the overlying laminae (Figs 8A & 9A). In some instances, the diatom mats have been parted by the action of the organism. The composition of the burrows resemble that of the laminae in which they occur, but the frustules within them are very fragmented (see later section on 'Synthesis of the sedimentology and micropalaeontology').

(iii) Sedimentation rates. The LDO intervals of greatest mat production/flux have abnormally high bulk sedimentation rates. For example, at Site 850, accumulation rates across the time intervals 6.3–6.1, and 5.1–4.66 Ma are 122 and 97.73 m/Ma respectively (Mayer, *et al.* 1992). At other sites there is a similar record; sedimentation rates of 84.1 m/Ma are recorded

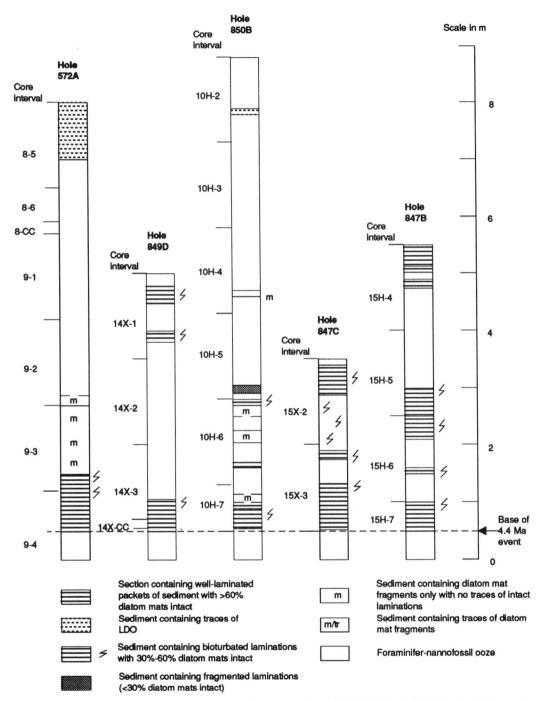

Fig. 3. The sedimentary record across the 4.4 Ma interval at sites 847, 850, 849 and 572. The distribution and preservation of LDO packets is variable.

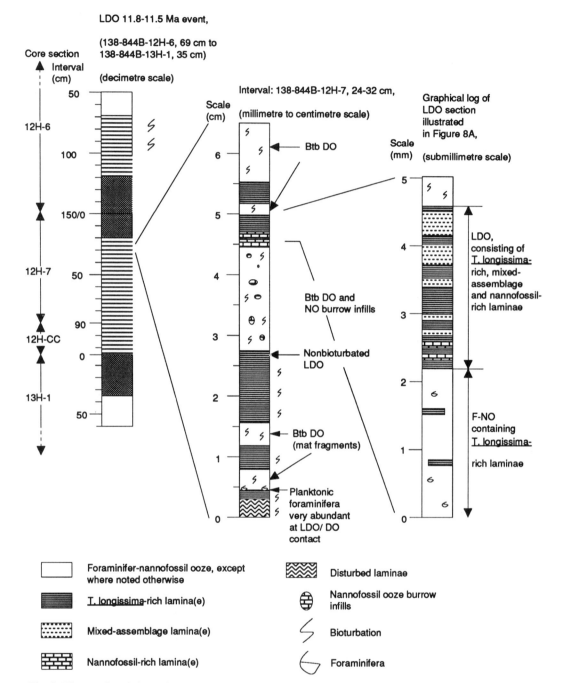

Fig. 4. Three scales of alternation are seen between LDO and the background sediment type (F-NO). At Site 844, the 11.8–11.5 Ma interval consists of several extensive LDO sections, one of which is illustrated to show the submillimetre and the millimetre to centimetre scale alternation. The sedimentary interval corresponding to the submillimetre scale alternation is shown in Fig. 8(A).

from Site 847 (Hole 847B) at 4.4 Ma, and at Site 849 (Hole 849D) 92.2–108 m/Ma are recorded between 4.98–4.5 Ma (Mayer *et al.* 1992). New calculations based on the orbitally tuned time-scale of Shackleton *et al.* (1995) give rates through the 6.3–6.1 Ma interval at Site 850 as high as 160 m/Ma.

Lamination style

Scales of lamination

The LDO intervals most frequently occur as packets of sediment between 10 and 20 m thick in the time intervals discussed above. The 4.4 Ma interval (Fig. 3) is a typical example. Normally within these packets, three scales of alternation can be seen between LDO and the background sediment type, F-NO. These alternations are at the submillimetre, the millimetre to centimetre, and the decimetre scale (Figs 4 & 5). The submillimetre lamination is discussed later. The millimetre to centimetre scale alternation is between nannofossil-bearing diatom ooze (NDO) and purer, diatom ooze (DO) (Fig. 5B). These laminae vary in colour from white or pale cream to green. Under an optical microscope (up to ×20 magnification), the white laminae appear either slightly translucent, with a layered texture, or homogeneous and nontranslucent. The translucent layered laminae are composed of successive, high-porosity, diatom mats. The non-translucent white laminae are composed of bioturbated diatom-rich laminae and non-bioturbated NDO laminae. Some non-bioturbated NDO laminae also are green, as are most bioturbated sediments (burrowed laminae and pellets). Pyrite-bearing sediment is normally dark green. F-NO appears cream to beige in colour.

On a decimetre scale, beds of unlaminated F-NO alternate with thinner beds of LDO and/or bioturbated DO. Occasionally, laminated beds more than 0.5 m thick can occur; for example, the sedimentary record across the 11.8 to 11.5 Ma interval at Site 844 is continuously laminated for more than 1.5 m.

The basal contact of LDO beds generally is sharp, and is characterized by an abrupt transition from F-NO to DO; for example, at the base of the 4.4 Ma interval (Figs 3 & 6). These abrupt changes from nannofossil ooze to LDO record the rapid onset of sustained mat-flux episodes (Kemp & Baldauf 1993).

Description and classification of laminae

The laminae have been classified according to observations made using SEM BSEI of polished

Table 1. *Classification of the laminae*

Lamina type	Composition	Thickness			Spacing	Regularity	Boundaries
		Range	Mean (μm)	Standard deviation (μm)			
T. longissima-rich	Assemblages of *T. longissima* Group diatoms	50 μm–4 mm	470	270	Sub-mm	The spacing of the laminae may vary from regular to irregular within a laminated diatom ooze interval	Abrupt to transitional between *T. longissima*-rich and mixed-assemblage laminae (Figs 8, 9). Abrupt between *T. longissima*-rich and nannofossil-rich laminae (Fig. 8)
Mixed-assemblage	Centric and pennate diatoms (including *T. longissima* Group species), nannofossils, foraminifera and radiolarians	25–900 μm	170	75	Sub-mm to mm	Irregular to regular	
Nannofossil-rich	Nannofossils with subordinate quantities of diatoms, foraminifera, and radiolarians	50 μm–1.35 mm	420	410	Sub-mm	The spacing of the laminae is predominantly regular within a laminated interval	

Fig. 5. (A) Section 138-850B-30X-5, 30–90 cm; effects of wire core splitting, which caused considerable deformation of the LDO making identification difficult, but creating a characteristic roughened surface. In Figures **(B)** and **(C)** saw-cut surfaces are shown. They allow for close examination of the millimetre to centimetre scale alternation between the paler (off-white) pure diatom ooze and the darker (green) mixed diatom and calcareous nannofossil ooze. **(B)** Section 138-851E-28X-2, 78–112 cm shows the centimetre-scale alternation between darker DO and lighter NDO, but with lamination visible throughout. **(C)** Section 138-851E-29X-5, 35–60 cm shows the continuous laminated ooze characteristic of the 11.1 Ma interval at sites 850 and 851. Scale: width of cores is 7 cm.

Fig. 6. Rapid onset of the 4.4 Ma event, Site 850 (Sample 138-850B-10H-7, 74–81 cm). The darkest high-porosity sediment is the LDO containing an assemblage of *T. longissima* Group diatoms. This is underlain by the paler background sediment type F-NO, comprising primarily of calcareous nannofossils, but with minor quantities of diatoms, foraminifera, and radiolarians. Planktonic and benthic foraminifera are represented by the black objects having a white outline (0.1–0.3 mm).

Table 2. *Details of laminated intervals observed during this study*

Interval sampled (cm)	Thickness of laminated interval(s)	Lamina type												Lamina associations
		T. longissima-rich				Mixed-assemblage				Nannofossil-rich				
		Number	Mean thickness	Standard deviation	Thickness range	Number	Mean thickness	Standard deviation	Thickness range	Number	Mean thickness	Standard deviation	Thickness range	
844B-12H-7, 24-32	1.85	6	0.165	0.070	.100-.300					5	0.110	0.050	.050-.200	B
	2.00	6	0.225	0.140	.100-.500					2	0.200	0.100	.100-.300	A, B
	2.76	10	0.130	0.090	.050-.350	4	0.050	0.000	0.050	4	0.245	0.165	.080-.500	A, B
	3.80	8	0.330	0.140	.150-.750	6	0.170	0.070	.075-.250					A
844B-12H-7, 75-79	4.70	6	0.425	0.140	.200-1.250	7	0.165	0.050	.100-.200					A
	1.60	5	0.180	0.070	.100-.300	6	0.325	0.140	.150-.400					A
	3.50	6	0.460	0.125	.250-.600	5	0.120	0.050	.050-.200					A
	4.35	8	0.430	0.155	.200-.700	5	0.120	0.070	.125-.300					A
844B-14H-6, 143-149	7.75	9	0.640	0.740	.150-2.40	8	0.080	0.030	.025-.100					A
844B-20H-5, 46-51	4.45	5	0.870	0.875	.150-2.50	8	0.250	0.200	.050-.650					A
	1.55	4	0.240	0.055	.100-.400	4	0.115	0.040	.050-.150					A
844C-8H-2, 125-130	2.75	6	0.210	0.105		4	0.080	0.020	.050-.100					A
844C-8H-3, 13-17	7.75	5	0.640	0.385	.100-1.25	5	0.280	0.135	.150-.550					A
	3.75	5	1.330	0.435	1.00-2.15	4	0.075	0.020	.050-.100					A
844C-8H-3, 142-150	3.75	5	0.490	0.220	.150-.750	6	0.125	0.040	.050-.150					A
844C-8H-5, 20-25	5.50	9	0.245	0.140	.100-.550	4	0.300	0.060	.200-.350					A
847B-15H-7, 39-49	3.46	11	0.365	0.265	.100-1.00	9	0.145	0.105	.050-.350					A
847C-15X-3, 110-115	6.30	6	0.900	0.780	.050-2.00	10	0.195	0.140	.100-.550					A
847C-15X-3, 134-144	2.40	7	0.260	0.155	.150-.550	5	0.180	0.350	.100-.450					A
849B-19X-6, 53-66	3.54	6	0.415	0.170	.250-.750	6	0.315	0.130	.175-.550					
	5.00	9	0.430	0.315	.050-.950	5	0.210	0.085	.100-.350					A
	5.62	9	0.500	0.215	.100-.800	9	0.125	0.115	.050-.400					A
	9.36	13	0.575	0.365	.100-1.45	8	0.140	0.010	.050-.350					A
	7.74	10	0.665	0.350	.200-1.40	12	0.155	0.010	.050-.350					A
849D-23X-6, 31-39	7.25	7	0.300	0.300	.100-1.00	9	0.115	0.080	.050-.250	7	1.130	0.870	.200-1.35	B
	7.78	8	0.155	0.100	.050-.400	7	0.265	0.105	.100-.400					A
850B-20X-7, 25-27	18.00	15	0.775	0.530	.150-2.00	15	0.125	0.060	.050-.200					A
851E-28X-2, 104-112	25.55	33	0.710	0.845	.050-4.00	29	0.115	0.120	.025-.700					A

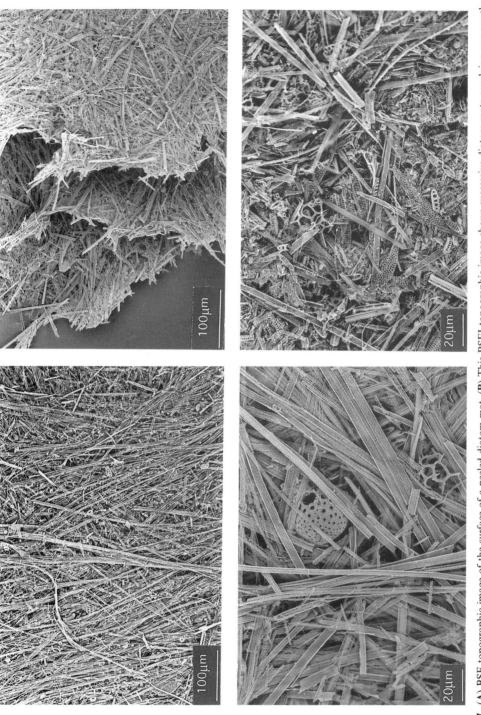

Fig. 7. (**A**) BSE topographic image of the surface of a peeled diatom mat. (**B**) This BSEI topographic image shows successive diatom mats overlying one another. (**C**) High-magnification topographic image of a mat surface (×500) with a radiolarian enclosed within the mat and a silicoflagellate on its surface. (**D**) Topographic image of a mixed-assemblage lamina from the 11.1 Ma interval (Site 844) showing fragmented *T. longissima* Group diatoms, other diatom species (including *Denticulopsis* and *Rossiella*), and calcareous nannofossils.

thin sections and broken sediment surfaces. All fossil assemblage percentage values have been calculated by area from the backscattered images.

The laminae can be subdivided into three types, in order of decreasing porosity: (1) assemblages dominated by *T. longissima* Group diatoms (95%+) (dark tone in BSEI); (2) mixed-assemblages of centric and pennate diatoms, (including specimens of the *T. longissima* Group), radiolarians, nannofossils, foraminifera and silicoflagellates (intermediate in BSEI); and (3) laminae containing more than 80% nanno-fossils with rare foraminifera and minor quantities of radiolarians and diatoms (bright in BSEI) (Tables 1 and 2).

T. longissima-rich laminae

Individual diatom mats, closely resembling modern examples of *Rhizosolenia* mats (Villareal & Carpenter 1989), have an average thickness of 20 μm (thickness range is 15 to 25 μm), and form the building blocks of the *T. longissima*-rich laminae (Figs 7A–B). Any one *T. longissima*-rich lamina will typically comprise of 25 of these microlaminae. Several varieties/subspecies of *T. longissima*, referred to as the *T. longissima* Group, make up between 95%–100% of the fossil assemblage. The *T. longissima* frustules are between 1.5 and 5 μm wide, up to 3.8 mm long, and form an interlocking meshwork (Fig. 7A through C). Between, and in some cases within, the individual mats are found rare calcareous nannofossils, planktonic and benthic foraminifera (e.g. *Cibicides* spp., *Gyroidina* spp., see King *et al.* 1995), non-*T. longissima* Group diatoms, silicoflagellates and radiolarians (see Fig. 7A through C). Bed-parallel partings within laminated sections have been interpreted as the contacts between successively deposited mats, as these are likely to be the zones of greatest weakness.

Mixed-assemblage laminae

Mixed-assemblage laminae are composed of fossil assemblages intermediate in composition between the *T. longissima*-rich and nannofossil-rich laminae types (Figs 7D & 9B). In detail, mixed-assemblage laminae are composed of variable quantities of diatoms (fragmented specimens of the *T. longissima* Group make up 10–60% of the assemblage, and other diatom species 10–50%), calcareous nannofossils (0–80%), and foraminifera, radiolarians and silicoflagellates (0–10%). *Denticulopsis* and *Rossiella* are typical non-*T. longissima* Group diatom species from the 12.8–12.5, 11.8–11.5, and 11.1 Ma mat intervals.

Nannofossil-rich laminae

Compositionally, nannofossil-rich laminae are F-NO and diatom-nannofossil oozes (DNO) (Fig. 8C). Thus, they are similar to the regionally prevalent background sedimentation. In addition to the dominance of nannofossils in these laminae (approximately 80%+), they are also characterized by an absence of diatom mats, although *T. longissima* Group diatoms may be present as a minor sediment component (0–10%). Subordinate quantities of non-*T. longissima* Group diatoms (<30%), foraminifera (<10%), and radiolarians (<10%) also occur. Foraminifera are significantly more abundant in these laminae than in either the mixed-assemblage or diatom-rich laminae.

Association of lamina types

Two types of lamina associations were observed, a mixed-assemblage/*T. longissima* lamination (Figs 8A & 9A), and a F-NO/*T. longissima* lamination (Fig. 8) (Table 3). The mixed-assemblage/ *T. longissima* lamina association is dominant

Table 3. *Classification of the LDO facies*

Lamina association	Number of intervals	Thickness			Occurrence
		Range (mm)	Average (mm)	Standard deviation	
Mixed-assemblage/ *T. longissima*	25	<2->25	5.95	5.21	Dominant within laminated intervals
F-NO/*T. longissima*	4	<1.9–7.7	3.45	2.21	Dominant at the transition between F-NO/LDO

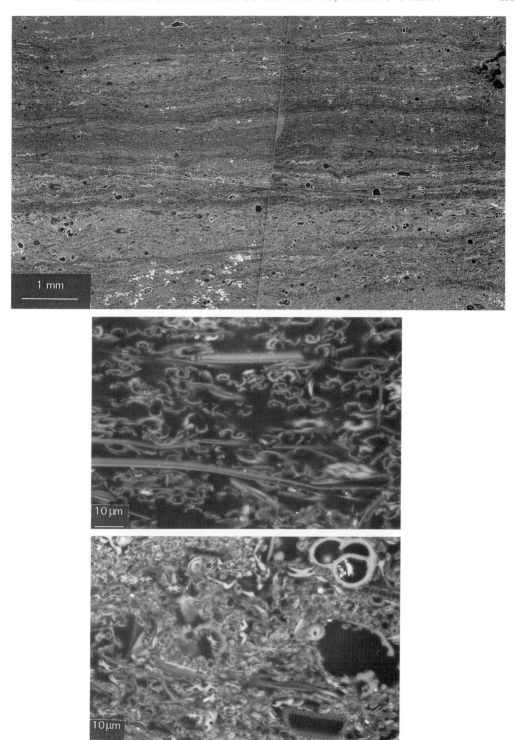

Fig. 8. (A) LDO from the 11.8–11.5 Ma interval (Site 844), showing the three types of laminae (see Fig. 4 for their identification). (B) High-magnification image of a typical *T. longissima*-rich lamina. (C) High-magnification image of a nannofossil-rich lamina containing scattered diatoms and foraminifera.

Fig. 9. (A) This laminated interval (Sample 138-849B-19X-6, 53–69 cm) exhibits thick well-defined *T. longissima*-rich laminae, that contain well-preserved foraminifera, interbedded with thin mixed-assemblage laminae. **(B)** This enlarged section of the lowermost mixed-assemblage lamina, 90 to 130 μm thick, shows a typical assemblage. It consists of calcareous nannofossils, centric and pennate diatoms, and a foraminifera. The pale gray areas above and below the foraminifera are composed of coccoliths and amorphous siliceous/calcareous matter.

within packets of LDO, and the transition between F-NO and LDO is most frequently marked by a F-NO/*T. longissima* lamina association. Site by site occurrences of the lamination types are summarized in Table 4.

Detailed sedimentological investigations of the 11.8–11.5, 11.1 and 4.4 Ma intervals indicates some slight variations in the lamina associations from those described above. The 11.8–11.5 Ma interval is characterized by more than 2 m of almost continuous LDO. Within this interval there are intermittent, but persistent occurrences of F-NO/*T. longissima* lamina associations. At the base of the 11.1 Ma interval, there are several occurrences of subcentimetre-sized LDO packets. These laminated sediments rarely contain a F-NO/*T. longissima* lamina association across the F-NO/LDO boundary, rather, there is normally an abrupt transition from F-NO to a mixed-assemblage/*T. longissima* lamina association. On a macroscopic scale, the 4.4 Ma interval is characterized by an abrupt transition from F-NO to LDO (see previous section). Microscopic evidence from sites 847 and 850 indicates that this change is marked by a transition from

F-NO to a mixed-assemblage/*T. longissima* lamina association, i.e., no nannofossil-rich laminae occur at this boundary.

As discussed in Kemp *et al.* (1995), the millimetre to centimetre scale alternations in the LDO intervals, record periods of more or less intense mat flux, these cycles may be analogous to periods of stronger/weaker El Niño/la Niña cycles. On the submillimetre scale, a similar signal is recorded by the mixed-assemblage/*T. longissima* lamina associations, and each lamina probably represents an individual event. The F-NO/*T. longissima* lamina associations represent periods of high mat flux alternating with periods of either low or negligible mat flux, and thus may indicate more extreme fluctuations in palaeoceanographical conditions.

A case study: micropalaeontology of three laminated diatom ooze intervals

Detailed SEM-led micropalaeontological investigations were undertaken on three laminated intervals from the middle Miocene *Actinocyclus*

Table 4. *The distribution and occurrence of the sedimentary facies at the Leg 138 sites*

LDO interval	Site	Lamina association(s)	Occurrence of lamina associations (location of samples)
14.6	844	Mixed-assemblage/*T. longissima*	In top 5 cm of a 15 cm packet of LDO
12.8–12.5	844	Mixed-assemblage/*T. longissima*; F-NO/*T. longissima*	F-NO/*T. longissima* lamina associations occur immediately above the boundary between F-NO and a 40 cm thick LDO packet. Mixed-assemblage/*T. longissima* lamina associations overly the nannofossil-rich/*T. longissima*-rich lamination
11.8–11.5	844	Mixed-assemblage/*T. longissima*; F-NO/*T. longissima*	Samples were taken from within a thick LDO sediment packet (>0.5 m thick). Mixed-assemblage/*T. longissima* lamina associations dominate, and there are intermittent, but persistent, occurrences of F-NO/*T. longissima* associations
11.1	844	Mixed-assemblage/*T. longissima*; F-NO/*T. longissima*	At the base of the LDO interval mixed-assemblage/*T. longissima* lamina associations occur as discrete LDO packets (sub cm scale) within F-NO sediments. Nannofossil-rich/*T. longissima*-rich lamination only rarely marks the transition between F-NO and LDO. Mixed-assemblage/*T. longissima* lamina associations only occur within
10–9.5	851	Mixed-assemblage/*T. longissima*	Less than 5 cm above the base of a 60 cm thick LDO packet, overlying 25 cm of intermittently laminated diatomaceous ooze.
6.3–6.1	850	Mixed-assemblage/*T. longissima*	Located within a 10 cm LDO packet.
	849	Mixed-assemblage/*T. longissima*; F-NO/*T. longissima*	At the F-NO/LDO transition of a 10 cm thick LDO packet, a F-NO/*T. longissima* lamina association occurs, and is succeeded by a mixed-assemblage/*T. longissima* lamina association.
5.8–5.1	847	Mixed-assemblage/*T. longissima*	An alternating mixed-assemblage and *T. longissima*-rich lamination occurs within a 20 cm thick LDO packet, the boundary between the laminated interval and the background sediment is not seen due to core disturbance.
5.1–4.98	849	Mixed-assemblage/*T. longissima*	A LDO packet approximately 10 m thick contains at least 9 cm of alternating mixed-assemblage and *T. longissima*-rich laminae. Transition to F-NO not seen.
4.4	847, 850	Mixed-assemblage/*T. longissima*	The 4.4 Ma interval consists of at least 50 m of LDO at its base. The boundary between the F-NO and the LDO is abrupt, a mixed-assemblage/*T. longissima* lamina association occurs in direct contact with the F-NO. Laminae are poorly-defined within the interval, but compositional data indicate that the *T. longissima*-rich and mixed-assemblage lamina types were probably dominant here.

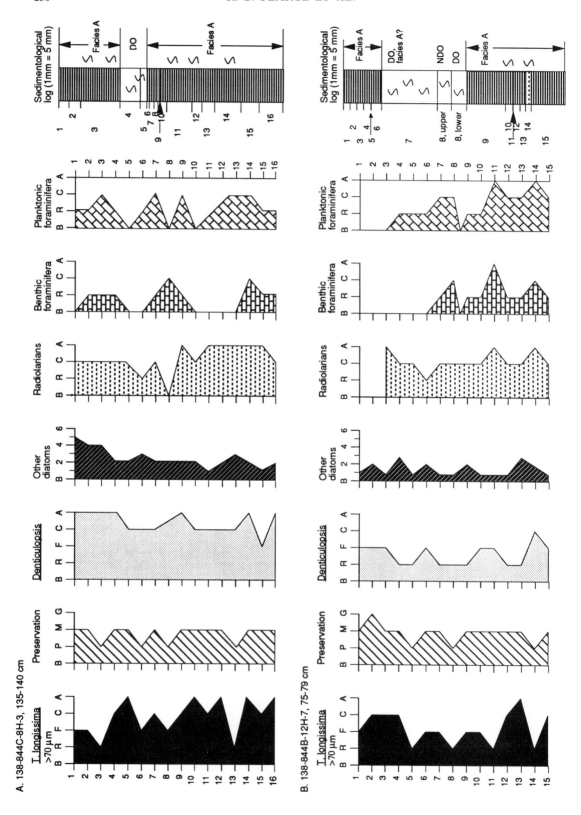

A. 138-844C-8H-3, 135-140 cm

B. 138-844B-12H-7, 75-79 cm

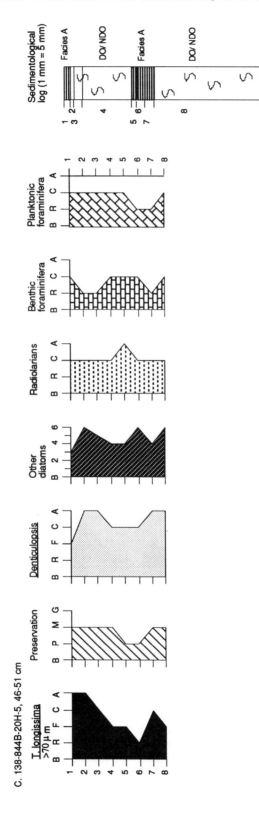

Fig. 10. Graphs of microfossil abundances. (**A**) Section 138-844C-8H-3, 135–140 cm. (**B**) Section 138-844B-12H-7, 75–79 cm. (**C**) Section 138-844B-20H-5, 46–51 cm.

moronensis, C. coscinodiscus, and *Cestodiscus peplum* zones, (138-844C-8H-3, 135–140 cm; 138-844B-12H-7, 75–79 cm; and 138-844B, 20H-5, 46–51 cm respectively) (Baldauf & Iwai 1995) (Figs 10 & 11).

The diatom assemblages are dominated by specimens of the *T. longissima* Group. Most of these specimens are fragmented, and have a length of less than about 40 μm. The occurrence of specimens of *T. longissima* longer than 70 μm that contain at least one apice varies from rare to abundant.

Diatom species other than *T. longissima* occur in each interval sampled. In Section 138-844C-8H-3, 135–140 cm, species consistently observed include *Denticulopsis simonseii* sl. and *Rossiella paeleacea.* After specimens of *T. longissima,* specimens of *Denticulopsis* are the next most abundant. *Denticulopsis* occurs in greatest numbers in Sample 3 where most of the specimens consist of girdle bands, rather than complete valves. In addition, rare specimens from the *Thalassiosira yabei* Group, *Hemidiscus cueniformis,* and *Asterolampra* sp. also were observed.

The abundance of 'Other Diatoms', including *Denticulopsis,* is lower in Section 138-844B-12H-7, 75–79 cm than in the other two intervals (Fig. 10B). Only scattered specimens of the *T. yabei* Group, *Thalassionema nitzschoides, T. nitzschioides var parva,* and *Triceratium* sp. were observed.

In Section 138-844B-20H-5, 46–51 cm, *Denticulopsis* specimens here are actually *Denticulopsis kanayae* rather than *D. simonseii. Rossiella* is extremely rare in this section.

The absolute abundance of diatoms throughout the sequences is in part controlled by preservation. Diatom preservation varies between poor and moderate, with one exception, Sample 2 from Section 138-844B-12H-7, 75–79 cm. This sample contains a well-preserved flora composed of the specimens mentioned previously, as well as specimens of *Azpeitia praenodulifer, A. moronensis, Nitzschia praereinboldii,* and *C. coscinodiscus.* Poorly-preserved diatom assemblages (e.g. Sample 3 in Section 138–844B-8H-3, 135–140 cm) are dominated by numerous girdle bands of *Denticulopsis,* and fragments (<70 μm long) of the *T. longissima* Group.

Smear slide analysis shows that radiolarians and silicoflagellates occur rarely, but consistently, throughout the samples. The silicoflagellate assemblage in the three sections consists of specimens of both *Dictyocha* and *Distephanus.* Specimens of *Dictyocha* generally outnumber those of *Distephanus* in Section 138-844B-8H-3, 135–140 cm, but in Section 138-844B-12H-7, 75–79 cm, *Distephanus* specimens are slightly more

numerous than *Dictyocha* specimens. The numbers of both species of silicoflagellate are equal in Section 138-844B-20H-5, 46–51 cm, with the exception of Sample 1 here *Dictyocha* is rarely observed.

Radiolarians are generally common, as determined by the BSEI method. The numbers of foraminifera (planktonic and benthic) are more variable, and planktonic forms generally outnumber benthic forms. Peaks in planktonic foraminifera abundance normally coincide with peaks in benthic foraminifera abundance.

Synthesis of the sedimentology and micropalaeontology

In Sections 138-844B-12H-7, 75–79 cm and 138-844B-20H-5, 46–51 cm, there generally is (1) an increased abundance of *T. longissima* fragments greater than 70 μm long (moderate to good preservation) within the *T. longissima*-rich laminae, accompanied, in some instances, by a reduction in the numbers of benthic foraminifera, and (2) an increased abundance of *T. longissima* fragments less than 70 μm long (poor to moderate preservation) in mixed-assemblage laminae, accompanied by a slight increase in benthic foraminifera. The abundance of 'Other Diatoms' does not show a positive correlation with mixed-assemblage laminae. Bioturbated horizons are dominated by fragmented (<70 μm in length), and poorly preserved *T. longissima* frustules. There is also some evidence to suggest that the abundance of 'Other Diatoms' increases in bioturbated laminae. No relationship between lamina type and fossil assemblage is seen in Section 138-844B-8H-3, 135–140 cm. The generally poor correlation between lamina type and fossil assemblage may be due to the relatively coarse sampling strategy.

Contemporary studies of benthic foraminifera within well-preserved LDO show a reduction in numbers relative to non-laminated sediment. This is attributed to the physical suppression of foraminifera benthic activity by the diatom meshwork (Kemp & Baldauf 1993; King *et al.* 1995).

Summary

It is clear that LDO occurs intermittently in time and space across the eastern equatorial Pacific Ocean, and in some instances these events can be correlated more closely.

The basic building blocks of all the LDO sediments are 20 μm-thick mats of the diatom

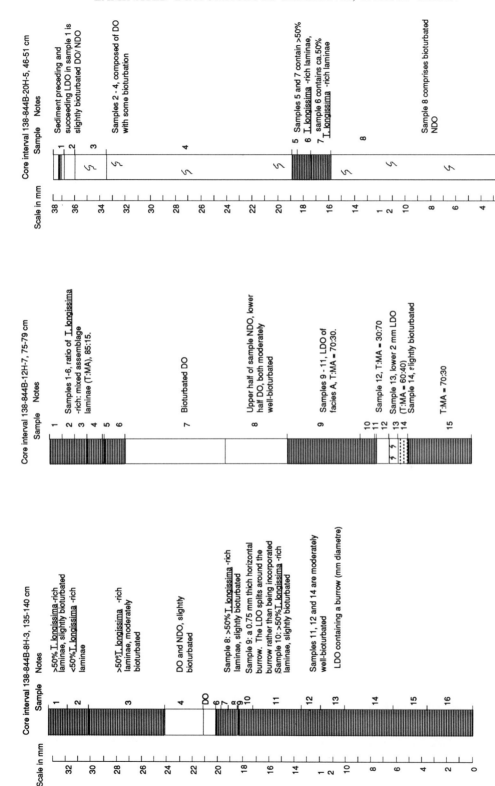

Fig. 11. Sedimentological logs of the intervals examined in the case study. All the LDO intervals belong to Facies A. T:MA = ratio of *T. longissima*-rich laminae to mixed-assemblage laminae. Symbols are the same as those used in Fig. 4.

Thalassiothrix longissima. Several varieties/sub-species of *T. longissima* occur within individual mats. Similar observations have been made of present-day *Rhizosolenia* mats in the central North Pacific gyre, where six species were found in ten different mat combinations (Villareal & Carpenter 1989). Three types of lamination have been identified in the LDOs. These are (1) an assemblage dominated by *T. longissima* Group diatoms, consisting of diatom mats; (2) a mixed-assemblage of centric and pennate diatoms, (including specimens of the *T. longissima* Group), radiolarians, silicoflagellates, nannofossils, and foraminifera, and (3) laminae containing more than 80% nannofossils with subordinate quantities of the fossil components mentioned above. These laminae occur in two lamina associations: a mixed-assemblage/*T. longissima* association and a F-NO/*T. longissima* association. The transition from F-NO to LDO is most frequently marked by a F-NO/*T. longissima* lamina association, and within packets of laminated diatom ooze, the mixed-assemblage/*T. longissima* lamina association dominates.

The average thickness of the *T. longissima*-rich laminae is 0.5 mm (25 layers of individual diatom mats, each about 20 μm thick) although, exceptionally, they range up to 4 mm. These laminae have been interpreted to result from individual episodes of multiple mat generation and flux, suggesting that, exceptionally, stacks of up to 200 mats may have been deposited from one mat-forming event (the likely origins of these mat-forming events is discussed in Kemp *et al.* 1995).

The microfossil assemblage within three laminated intervals at Site 844 was examined. Diatoms of the *T. longissima* Group occur within and between the mat intervals, whereas the various species of other diatoms, such as *Denticulopsis* (a common equatorial form) and *Rossiella*, are concentrated between the mats. Radiolarians and silicoflagellates, such as *Distephanus* and *Dictyocha*, occur in low numbers throughout the intervals sampled. Variations in the relative abundance of *Distephanus* and *Dictyocha* suggest systematic variation in temperature of the surface waters during episodes of mat flux. Both benthic and planktonic foraminifera are present in variable quantities. With increasing quantities of intact diatom mats, the abundance of benthic foraminifera decreases.

This work was conducted under NERC ODP Special Topic Grant GST/02/552 (R. Pearce). A. Kemp acknowledges support from the NERC ODP Special Topic. Helpful discussions with Leg 138 colleagues are gratefully acknowledged.

References

BALDAUF, J. G. & IWAI, M. 1995. Neogene diatom biostratigraphy for the eastern equatorial Pacific, Ocean Drilling Program Leg 138. *In*: MAYER, L. A., PISIAS, N. G., JANECEK, T. R., PALMER-JULSON, A. & VAN ANDEL, T. H. (eds) *Proceedings Ocean Drilling Program, Scientific Results*, **138**. College Station, TX (Ocean Drilling Program), 105–128.

BERGGREN, W. A., KENT, D. V., FLYNN, J. J. & VAN COUVERING, J. A. 1985. Cenozoic geochronology. *Geological Society America, Bulletin*, **96**, 1407–1418.

BLOOMER, S., MAYER, L. & MOORE, T. 1995. Seismic stratigraphy of the eastern equatorial Pacific: palaeoceanographic implications. *In*: MAYER, L. A., PISIAS, N. G., JANECEK, T. R., PALMER-JULSON, A. & VAN ANDEL, T. H. (eds) *Proceedings Ocean Drilling Program, Scientific Results*, **138**. College Station, TX (Ocean Drilling Program), 537–554.

BODEN, Per. 1991. Reproducibility in the random settling method for quantitative diatom analysis. *Micropalaeontology*, **37**, 313–319.

BRODIE, I. & KEMP, A. E. S. 1995. Variations in biogenic and detrital fluxes and formation of laminae in late Quaternary sediments from the Peruvian coastal upwelling zone. *Marine Geology*, **116**, 385–398.

BAUMGARTNER, T., FERREIRA-BARTRINA, V., SCHRADER, H. & SOUTAR, A. 1985. A 20-year varve record of siliceous phytoplankton variability in the central Gulf of California. *Marine Geology*, **64**, 113–129.

GOLDSTEIN, J. I., NEWBURY, D. E., ECHLIN, P., JOY, D. C., FIORI, C. & LIFSHIN, E. 1981. *Scanning Electron Microscopy and X-ray Microanalysis*. Plenum, New York.

GORSLINE, D. 1984. Studies of fine grained sediment transport processes and products in the California Continental Borderland. *Special Publication Geological Society, London*, **15**, 395–415.

GRIMM, A. K. 1992. High resolution imaging of laminated biosiliceous sediments and their paleoceanographic significance (Quarternary, Site 798, Oki Ridge, Japan Sea). *In*: PISCOTTO, K. A. *et al. Proceedings Ocean Drilling Program, Scientific Results*, **127/128**. College Station, TX (Ocean Drilling Program), 547–557.

KEMP, A. E. S. 1990. Sedimentary fabrics and variation in lamination style in Peru continental margin upwelling sediments. *In*: SUESS, E., VON HUENE, R. *et al.*, *Proceedings Ocean Drilling Program, Scientific Results*, **112**. College Station, TX (Ocean Drilling Program), 43–58.

——1995. Neogene and Quaternary pelagic sediments and depositional history of the eastern equatorial Pacific Ocean. *In*: MAYER, L. A., PISIAS, N. G., JANECEK, T. R., PALMER-JULSON, A. & VAN ANDEL, T. H. (eds) *Proceedings Ocean Drilling Program, Scientific Results*, **138**. College Station, TX (Ocean Drilling Program), 627–639.

—— & BALDAUF, J. G. 1993. Vast Neogene laminated diatom mat deposits from the eastern equatorial Pacific Ocean. *Nature*, **362**, 141–143.

——, —— & PEARCE R. B. 1995. Origins and paleoceanographic significance of laminated diatom ooze from the eastern equatorial Pacific (Leg 138). *In*: MAYER, L. A., PISIAS, N. G., JANECEK, T. R., PALMER-JULSON, A. & VAN ANDEL, T. H. (eds) *Proceedings Ocean Drilling Program, Scientific Results*, **138**. College Station, TX (Ocean Drilling Program), 641–645.

KING, S. C., KEMP, A. E. S. & MURRAY, J. W. 1995. Benthic foraminifer assemblages in Neogene laminated diatom ooze deposits in the eastern equatorial Pacific Ocean (Site 844). *In*: MAYER, L. A., PISIAS, N. G., JANECEK, T. R., PALMER-JULSON, A. & VAN ANDEL, T. H. (eds) *Proceedings Ocean Drilling Program, Scientific Results*, **138**. College Station, TX (Ocean Drilling Program), 665–673.

KRINSLEY, D. H., PYE, K. & KEARSLEY, A. T. 1983. Application of backscattered electron microscopy in shale petrology. *Geological Magazine*, **120**, 109–114.

MAYER, L., PISIAS, N., JANECEK, T. *et al.* 1992. *Proceedings Ocean Drilling Program, Initial Reports.* **138**, College Station, TX (Ocean Drilling Program).

SHACKLETON, N. J., CROWHAST, N., HAGELBERG, T. & PISIAS, N. G. 1995. A new Late Neogene timescale: application to Leg 138 sites. *In*: MAYER, L. A., PISIAS, N. G., JANECEK, T. R., PALMER-JULSON, A. & VAN ANDEL, T. H. (eds) *Proceedings Ocean Drilling Program, Scientific Results*, **138**. College Station, TX (Ocean Drilling Program), 73–101.

VILLAREAL, T. A. & CARPENTER, E. J. 1989. Nitrogen fixation, suspension characteristics, and chemical composition of *Rhizosolenia* mats in the Central North Pacific gyre. *Biological Oceanography*, **6**, 327–345.

Origins and palaeoceangraphic significance of laminated daitom ooze from the Eastern Equatorial Pacific Ocean

ALAN E. S. KEMP[1], JACK G. BALDAUF[2] & RICHARD B. PEARCE[1]

[1] *Department of Oceanography, University of Southampton,*
Southampton Oceanography Centre,
Southampton SO14 3ZH, UK
[2] *Department of Oceanography and Ocean Drilling Program,*
Texas A&M University, College Station, TX 77840, USA

Abstract: Laminated diatom ooze (LDO) has been recovered from several ODP Leg 138 sites and now is also recognized from several DSDP Leg 85 sites. These remarkable sediments are the result of massive and episodic flux of mats of the diatom *Thalassiothrix longissima*. By analogy with the *Rhizosolenia* diatom mat-forming events monitored by JGOFS (Joint Global Ocean Flux Study) in the equatorial Pacific Ocean during cooling conditions in late 1992, these episodes of massive flux of *T. Iongissima* mats may represent the 'fall out' from major frontal systems generated during La Niña (anti-El Niño) events. Laminations were preserved in the mat deposits because of the rapid mat deposition and high strength of the diatom mat meshwork, that subjugated benthic activity. This new mechanism of preservation of lamination in marine sediments has wide implications for other laminated sequences. The sustained periods of mat deposition documented in Neogene sediments of the eastern equatorial Pacific Ocean are part of the major cycles in the relative abundance of carbonate and silica in the region and, possibly, in the case of some intervals, also in the Atlantic Ocean.

The eastern equatorial Pacific Ocean is a region deemed responsible for up to half of global 'new' production and is a continuing focus for studies of modern oceanic and climatic processes, including most recently, the US JGOFS program (Wyrtki 1981; Chavez & Barber 1987; Chavez *et al.* 1990; Barber 1992; Yoder *et al.* 1994). Furthermore, changes in its sedimentary record have often recorded important palaeoceanographic events and heralded major reorganizations in ocean and climate history (van Andel *et al.* 1975; Mayer *et al.* 1985a, 1992). Leg 138 of the Ocean Drilling Program furnished a high-resolution record of the last 10 to 15 Ma depositional history of the eastern equatorial Pacific between 90° and 110° W (Fig. 1). The intervals of greatest diatom abundance are recorded by near-monospecific assemblages of the pennate diatom *Thalassiothrix longissima*, that occur in laminated diatom ooze (LDO), which have been interpreted as diatom mat deposits (Kemp & Baldauf 1993). The detailed sedimentology micropalaeontology and microfacies of the laminated diatom ooze are discussed elsewhere (Pearce *et al.* 1996). The

purpose of this study is to review evidence for the origins and palaeoceanographic significance of these remarkable sediments. The timescale used below is that of Berggren *et al.* (1985) as rendered in the Leg 138 'Explanatory Notes' chapter of Mayer *et al.* (1992).

The discovery of these laminated diatom mat deposits comes against an increasing awareness of both (1) the importance of relatively rare events in contributing a major proportion of flux to the seabed (e.g. Smith *et al.* 1992) and (2) the significance of buoyant diatom mats and individual large diatoms for nutrient cycling within the ocean surface layer (Villareal *et al.* 1993) and as agents of rapid flux to the sea floor (Sancetta *et al.* 1991).

Summary of geographic and temporal distribution of laminated diatom ooze

Laminated diatom ooze was recovered from Leg 138 Sites 844, 847, 849, 850 and 851. Investigation of DSDP Leg 85 cores established the presence of laminated diatom ooze in Sites 572, 573 and 574 (Fig. 2). Thus, the latitudinal extent of the LDO ranges from 90° to 133° West. LDO was deposited intermittently at the above sites from 15 to 4.4 Ma, but is concentrated at about 15, 13–12, 11, 10–9.5, 6.3–6.1 and 4.4 Ma. A less

Previously published in: PISIAS, N. G., MAYER, L. A., JANECEK, T. R., PALMER-JULSON, A. & VAN ANDEL, T. H. (eds) 1995. *Proceedings Ocean Drilling Program, Scientific Results*, **138**, College Station, TX (Ocean Drilling Program), 641–645.

From Kemp, A. E. S. (ed.), 1996, *Palaeoclimatology and Palaeoceanography from Laminated Sediments*, Geological Society Special Publication No. 116, pp. 243–252.

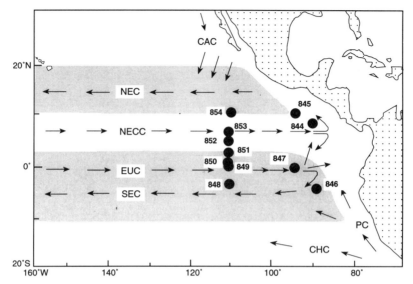

Fig. 1. Location of Leg 138 sites showing the configuration of major equatorial current systems: CAC, California Current; NEC, North Equatorial Current; NECC, North Equatorial Counter Current; EUC, Equatorial Under Current; SEC, South Equatorial Current; CHC, Chile Current; PC, Peru Current.

widespread interval confined to Sites 847 and 849 occurred between 5.8 and 5.1 Ma. No laminated sediments occur after 4.4 Ma. A plot of LDO intervals on a plate motion backtrack chart (Fig. 3) shows that almost all were deposited when sites lay between 0° and 2° S, although mat fragments are present in deposits farther north (Sites 851, 852). An exception is Site 844, which

contains several early to middle Miocene age LDO intervals. The laminated diatom ooze and mat fragments (see Kemp 1995) at Sites 844 and 845 may record the presence of a centre of mat formation, separated from the equatorial region.

Within laminated episodes, individual beds may be correlated over great distances. For

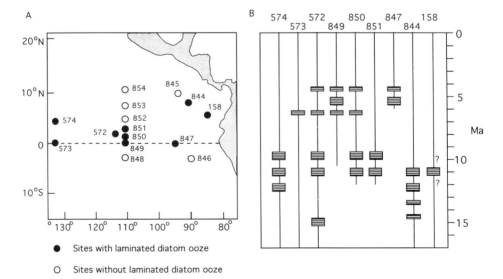

Fig. 2. Geographical (A) and temporal (B) extent of episodes of laminated diatom mat deposition in the eastern equatorial Pacific.

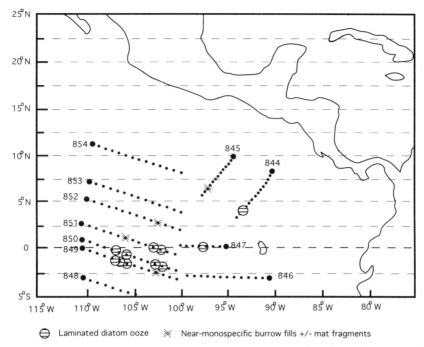

Fig. 3. Plate-motion backtracked diagram showing location, through time, of laminated diatom mat deposits and traces of diatom mat deposition (burrow fills and mat fragments).

example, the distinctive 60 cm bed of LDO that occurs at the base of the 4.4 Ma interval is present at Sites 847, 849, 850 and 572, and so may be correlated over a longitudinal extent of at least 2000 km. As well as the close similarity of the sequences, correlation is based on biostratigraphy and continuous gamma ray attenuation porosity evaluator (GRAPE) data (Mayer *et al.* 1992; Mayer 1991). The coincidence of the laminated intervals with abrupt changes in sediment density and, hence, seismic impedance contrast should provide a means of correlating depositional events over great distances from seismic records (Mayer *et al.* 1985; Bloomer *et al.* 1995) in a critical area of the ocean basins.

Ecological significance of *Thalassiothrix*

Micropalaeontological studies of the LDO indicate *Thalassiothrix longissima* is the dominant species, although several different varieties or subspecies occur (see Pearce *et al.* 1996). The presence of *Thalassiothrix* in Neogene sediments from the eastern equatorial Pacific has been widely used as an indicator of

significant upwelling and high primary production (Sancetta 1982, 1983). *Thalassiothrix longissima* is characterized by long straight to twisted cells, between 1.5 and 5 μm wide, but up to 4 mm long. Previous researchers have described it as a cosmopolitan species with a distribution ranging from the Southern Ocean to the North Pacific, North Atlantic, and the Norwegian Sea (Hendy 1937; Smayda 1958; Hasle & Semina 1987). However, this apparent wide-ranging distribution may result from differences in species definition and the assignment of other forms of *Thalassiothrix* sp. (e.g., *T. antarctica*, *T. acuta*, or *T. heteromorpha*) into *T. longissima* (Hasle & Semina 1987). Species definitions notwithstanding, micropalaeontological analysis of the laminated diatom oozes recovered during Leg 138 indicates the presence of different varieties of *T. longissima* within individual bundles of laminae (see Pearce *et al.* 1996). Whether these represent different varieties, subspecies, or species must await more detailed taxonomic study of the *Thalassiothrix* group and the laminated diatom ooze.

The few studies of *T. longissima* in the modern ocean have had a mainly taxonomic emphasis and contain little ecological information. *T. longissima* ranges in abundance up to 10^3

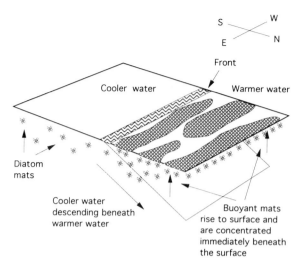

Fig. 4. Synoptic diagram showing the concentration of diatom mats at an oceanic frontal zone. The self-regulation of diatom mat buoyancy is disturbed as the colder water sinks beneath the warmer water. The mats rise to form a concentrated surface layer on the warm side of the front. These mats will subsequently die through nutrient exhaustion and overheating and sink. Thus, diatoms which need not represent a large portion of the primary production on on the cold side of the front may represent a large part of the export production beneath the frontal zone. The model is based on processes at the "great front" encountered by the 1992 Fall JGOFS cruises reported in Barber (1992) and Yoder *et al.* (1994).

to 10^6 cells per litre and occurs as single cells, colonies, or is arranged in bundle-shaped colonies and large tangled masses or mats (Hallegraeff 1986; Hasle & Semina 1987). The colonies or mats are formed, in part, by the attachment of several cells at the foot-pole resulting in a 'bunch of flowers' arrangement. No data are available on the sinking or flotation characteristics of mats of *T. longissima*, but several relevant recent studies of the mat-forming diatom *Rhizosolenia* have been used as an analogue.

Rhizosolenia: an analogous mat forming diatom

No information is available on sinking characteristics of *Thalassiothrix* mats; however new data from sediment trap studies have highlighted the significance of mass sinking of mats of rhizosolenid diatoms as agents of rapid flux to the seafloor (Sancetta *et al.* 1991). Concentrations of up to $4.4/m^3$ have previously been reported for *Rhizosolenia* mats (Martinez *et al.* 1983); however, massive concentrations of *Rhizosolenia* mats have been reported from the eastern equatorial Pacific during the Fall of 1992 (Barber 1992; Barber *et al.* 1994; Yoder *et al.* 1994).

The 1992 Fall US JGOFS experiment encountered a return to cold tongue (antit-El Niño or La Niña) conditions following the E1 Niño monitored during the Spring cruises. Residual warm water north of $2°$N led to the development of a sharp E–W front at $2°$N monitored by aircraft, satellite and the Space Shuttle Atlantis, which extended for over 1000 km (Barber 1992; Yoder *et al.* 1994). This front separated colder water to the south at 23.8°C which subducted to the north under the warmer water at 26.6°C. Great patches and streamers of mats of the diatom *Rhizosolenia* occurred on the north (warm) side of the front. Apparently, the mats (though not a major component of the plankton in the cold, highly productive waters south of the front) were so buoyant as to 'pop up' to form an almost contiguous layer a few centimeters from the surface on the warm side of the front. These layers of mats, which locally extended several kilometres away from the front, caused localized, near-surface, water column stratification; starved of nutrients and overheated, the mats appear to have subsequently died and sunk (Barber 1992, pers. comm.) (Fig. 4). Thus, the physics of the frontal system facilitates the concentration of the diatom mats, which might not otherwise be a significant component of either the phytoplankton or of 'new' or 'export' production.

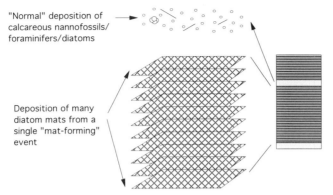

"Normal" deposition of calcareous nannofossils/ foraminifers/diatoms

Deposition of many diatom mats from a single "mat-forming" event

Fig. 5. Depositional model for the laminated mat deposits of the eastern equatorial Pacific. Stacks of generally around 25, but exceptionally, up to 200 mats, have been deposited from a single mat-forming event producing a *T. Longissima* lamina. These have been interlaminated with thinner laminae containing mixed microfossil assemblages, representing more normal equatorial deposition.

A depositional model for the *T. longissima* mat deposits

Individual, sub-millimetre laminae in all the laminated diatom ooze intervals in the eastern equatorial Pacific contain stacks of typically 20–30 individual 20 µm thick mats but occasionally contain up to 200 mats. These *T. longissima* laminae are interlayered with very thin laminae that contain mixed microfossil assemblages (Pearce *et al.* 1996) which represent 'normal' equatorial sediment. The *T. longissima* laminae probably represent the fall out from individual mat-forming/mat flux episodes, when large numbers of mats were concentrated, and sank *en masse* (Fig. 5). By analogy with the *Rhizosolenia* mat-forming events monitored by JGOFS, these episodes of mat flux may record La Niña (anti-El Niño) events when strong E–W frontal zones developed. The millimetre to centimetre scale alternations in the laminated diatom ooze, which record periods of more or less intense mat flux, represent time scales of several years to several tens of years and may record periods of stronger/weaker, El Niño/La Niña cycles. The origin of the decimetre-scale interbedding between LDO and nannofossil ooze (representing timescales from hundreds to a few thousand years) may relate to variations in intensity of circulation, with more vigorous surface circulation leading to the development of stronger frontal systems and, consequently, greater mat flux. Alternatively, these might relate to fluctuations in nutrient supply. Any explanation for these must provide a mechanism for the abrupt onset of mat deposition.

The presence of near-monospecific burrow fills and mat fragments outside the narrow 2° latitudinal limit of laminated diatom ooze suggests considerable dispersal of smaller quantities of diatom mats (Fig. 3).

Preservation of lamination in the *T. longissima* mat deposits: implications for the interpretation of laminated sediment sequences

The preservation of lamination in marine sediments (which forms a record of sequential flux events) has been universally regarded as resulting from the reduction or elimination of benthic animal activity by reduced bottom water oxygenation. Marine environments that are perceived to promote lamina preservation include the bottoms of anoxic silled basins such as the Black Sea (Hay 1988), or California borderland basins (Gorsline 1996; Schimmelmann & Lange 1996); zones of strong coastal upwelling, where an oxygen minimum layer intersects the shelf or slope, such as off Peru (Kemp 1990) or within the Gulf of California (Calvert 1966). In the light of these models, the presence of widespread laminated sediments on the deep seafloor of the eastern equatorial Pacific comes as a great surprise, as this region is not one associated with reduced oxygen levels (Kamykowski & Zentara 1990). Studies of benthic foraminifers (sensitive indicators of oxygenation) through these laminated intervals show no evidence for any increase in species characteristic of a low oxygen environment, but rather a decrease in tapered cylindrical and infaunal forms, attesting to the impenetrability of the diatom mat meshwork (King *et al.* 1995).

Independent evidence for strength of the diatom mat meshwork

Such is the tensile strength of individual laminations that some are seen to have been partially pulled out of the core during piston-coring or by the wire during core-splitting. Wire core-splitting caused considerable deformation of laminated intervals, making recognition of laminations difficult but producing a characteristic roughened surface. Only in the deeper intervals was a saw used for core-splitting, very nearly destroying the core but revealing obvious laminations on cut surfaces. During coring operations, the first penetration of even a thin bed of laminated sediment coincided with an abrupt increase in extraction overpull for the piston corer from typical values of 20 000 to 40 000 lbs to values of up to 150 000 lbs.

Thus, the diatom meshwork in the *T. longissima* laminae was of sufficient tensile strength and impenetrability to suppress benthic activity. During individual mat-flux events, with sedimentation, probably in days, of single layers up to 400 mats thick, the benthos was overwhelmed, resulting in complete preservation of the mats, which can be essentially regarded as prefabricated laminae. Laminae therefore were preserved by physical means rather than by reduced availability of dissolved oxygen. An important implication of this method of preservation is that deposition of small quantities of mats that did not overwhelm the benthos would not be preserved as laminae. Such intervals, containing abundant specimens of *T. longissima*, occur intermittently throughout the sequence, but are present in typical bioturbated pelagic sediments.

Implications for other deep sea near-monospecific and/or laminated giant diatom oozes

The physical suppression of bioturbation by diatom mats as a mechanism for preservation of lamination in the *T. longissima* oozes of the eastern equatorial Pacific Ocean has major implications for the origins of other enigmatic, near-monospecific and commonly laminated deepsea diatom oozes of Neogene and Quaternary age (Gardner & Burckle 1975; Gombos 1984; Muller *et al.* 1991). Because of the laminated

nature of these deposits, and given the existing preconceptions, origins have previously been ascribed to reduced oxygen conditions. For example, Muller *et al.* (1991) suggested that the presence of laminated oozes of the giant diatom *Bruniopsis mirabilis* in Messinian sediments from the South Atlantic relate to periods of (implausible) deepsea reduced bottomwater oxygenation. Gombos (1984) ascribed a similar origin to laminated oozes of the giant diatom *Ethmodiscus rex* at DSDP Site 520. Interestingly, *E. rex* is a diatom which, like the *Rhizosolenia* mats, is also capable of positive buoyancy. Furthermore, dead *Ethmodiscus* cells sink at speeds of up to 510 m/day. Whether the *E. rex* sank as aggregates or as single cells, the mass sinking of this giant diatom may have led to the preservation of lamination by physical suppression of benthos, and obviates the need to appeal to an unlikely and *ad hoc* period of reduced bottomwater oxygenation. Interestingly, many of the Atlantic Quaternary *E. rex* ooze deposits are in sites beneath or near localized, recurring frontal zones, and the timing of some of the major, late Quaternary equatorial Atlantic *E. rex* oozes, coincide with glacial build-up (Stabell 1986) when more intense frontal system development would be driven by increased wind strength.

Intriguingly, the timing of the *E. rex* ooze deposition at Site 520 and the *B. mirabilis* deposition at Site 701 appear to be coeval with the most intense period of mat deposition in the eastern equatorial Pacific: the 6.3 to 6.1 Ma event. This might suggest that general physical oceanic conditions favored mat production/flux at this time, perhaps implying high pole-to-equator gradients and enhanced formation of stronger oceanic frontal systems.

Implications for carbonate/silica variation in sediments

The *T. longissima* mat deposits, which are the direct result of surface processes, are clear evidence that surface processes, not dissolution at depth, controlled many of the major changes in the relative abundance of carbonate and silica during the Neogene, at least in the equatorial Pacific. After extensive SEM studies of the laminated mat deposits, we found no evidence that carbonate dissolution has occurred preferentially in the monospecific *T. longissima* diatom meshwork (Pearce *et al.* 1996). Individual, well preserved coccoliths (calcareous nannoplankton) occur occasionally, but *persistently*, through most of the mat deposits examined thus far. Planktonic and benthic foraminifers may show some effects of dissolution, but no more in the *T. longissima* laminations than in the carbonate-rich laminations. A synthesis of regional sedimentology shows that the episodes of diatom mat deposition do not coincide with major episodes of carbonate dissolution (Kemp 1995).

The episodic mass sinking of diatoms distorts the normal carbonate-silica cyclicity observed. Because the mats have been deposited much more rapidly than the surrounding nannofossil ooze, the usual assumptions of uniform sedimentation rates between age 'picks' upon which most palaeoceanographic techniques rely, are invalid. Indeed, the signal from the mat flux episodes may have to be filtered or 'squeezed'

out to permit conventional Milankovitch band analysis or production of 'tuned' GRAPE-based time scales (see Shackleton *et al.* 1995).

The GRAPE record (giving carbonate-opal variation: Mayer 1991) generally shows excellent correlation between equatorial sites (Lyle *et al.* 1992). However, when comparing intervals where one site has significant LDO and another has none, this correlation breaks down. For example, at two closely spaced sites (Sites 850 & 851; Fig. 6) the 4.4 Ma mat deposits are present at Site 850, but not at Site 851. Although the rest of the GRAPE signal correlates excellently, no record of the approximately 8 m thick diatom mat deposits can be seen at Site 851.

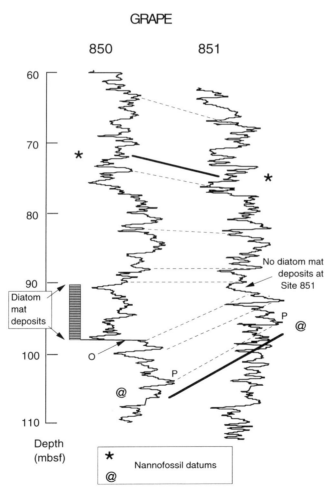

Fig. 6. Correlation of the GRAPE records of Sites 850 and 851 through the 4.4 Ma mat deposition episode. Although the rest of the GRAPE signal correlates excellently, no record is seen of the ~8 m thick diatom mat deposits of the 4.4 Ma interval at Site 851. Note that mat deposition is preceded by a carbonate maximum (P). O, onset of intense diatom mat deposition at Site 850.

This major difference in the sedimentary record of the two sites highlights the geographically restricted distribution of the laminated diatom ooze.

Discussion

Inspection of the GRAPE signals, (e.g. in Fig. 6), also reveals another intriguing aspect of the episodes of mat deposition. The peaks (indicated by 'P') represent the purest carbonate sediments having a minimum opal content. Each major interval of laminated diatom mat deposits is underlain by such a carbonate maximum. These carbonate maxima do not immediately precede the abrupt onset of mat deposition ('O' in Fig. 6), but appear to be consistently followed by the same distinctive pattern of decrease then increase, then abrupt decrease in carbonate content. Together, these major cycles represent the most abrupt and largest magnitude changes in carbonate–silica cyclicity in the equatorial Pacific. These major cycles currently are being investigated using a range of interdisciplinary sedimentological, micropalaeontological and isotopic techniques.

The intervals of mat deposition between 14.6 and 9.5 Ma occur during a period of many major episodes of ocean cooling and reorganization including a shift in the main locus of silica deposition from the Atlantic to the Pacific Ocean (Baldauf & Barron 1990; Barron & Baldauf 1990). The 6.3–6.1 Ma interval represents the most rapid burst of mat deposition and opal export in the eastern equatorial Pacific. The timing of this (during the early Messinian) and the apparently synchronous deposition of LDO at high sedimentation rates in the South Atlantic (see above) raise questions about major variations in the silica cycle and the relation of these to major oceanographic events. Attempts to quantify opal fluxes during this period currently are under way.

The cessation of laminated sediment deposition after 4.37 Ma coincides with a shift in silica production from the equatorial Pacific to the Antarctic circum-polar ocean (Leinen 1979; Brewster 1980) with the continued growth of the ice sheets and approximates the closure of the Pan-American Seaway (Keigwin 1982). At present, one can only speculate as to the cause of this cessation. Possible mechanisms range from a change in nutrient supply relating to differences in communication between the Atlantic and Pacific Oceans, to changes in circulation, or a combination of the effects of these on diatom ecology.

Conclusions

From the combined evidence of the the Leg 138 results and the Fall JGOFS cruise, the best model for the generation of deepsea diatom mat deposits (manifest as near-monospecific laminated diatom ooze) is one of generation at sharp oceanic frontal zones. Such frontal zones provide mechanisms both for concentrating buoyant diatom aggregates or mats, and then killing and sinking them. The evidence of the *Thalassiothrix* mat deposits suggests that such frontal systems may have generated flux at rates one or two orders of magnitude faster than 'normal' high equatorial sedimentation rates. The mat deposits appear to be a key part of major carbonate–silica variations in the equatorial Pacific, whose significance and origins demand further study.

The contributions of Pearce and Kemp were supported by a NERC ODP Special Topic research grant and the Maurice Hill fund of the Royal Society (Kemp). Baldauf acknowledges support from USSAC. We are grateful to Dick Barber, Connie Sancetta, and many of our Leg 138 colleagues for stimulating discussions.

References

BALDAUF, J. A. & BARRON, J. A. 1990. The distribution of Eocene through Quaternary biosiliceous sediments: A distribution resulting in part from Polar cooling. *In*: BLIEL, U. & THIEDE, J. (eds) *Geological History of the Polar Oceans: Arctic versus Antarctic*. Kluwer, Dordrecht, 575–607.

BARRAN, J. A. & BALDAUF, J. G. 1990. Development of Biosiliceous sedimentation in the North Pacific during the Miocene and Early Pliocene. *In*: TSUCHI, R. (ed.) *Pacific Neogene Events: Their Timing, Nature and Interrelationship*. University of Tokyo Press, 43–63.

BARBER, R. T. 1992. Fall survey cruise finds cooling conditions in equatorial Pacific. *U.S. JGOFS News*, **4**, no. 1:1–6.

——, MURRAY, J. W. JR & MCCARTHY, J. J. 1994. Biogeochemical interactions in the equatorial Pacific. *Ambio*, **23**, 62–66.

BERGGEN, W. A., KENT, D. V., FLYN, J. J. & VAN COUVERING, J. A. 1985. Cenozoic geochronology. *Geological Society of America Bulletin*, **96**, 1407–1418.

BLOOMER, S. F., MAYER, L. A. & MOORE, T. J. JR. 1995. Seismic stratigraphy of the eastern equatorial Pacific Ocean: paleoceanographic implications. *In*: PISIAS, N. G., MAYER, L. A., JANECEK, T. R., PALMER-JULSON, A. & VAN ARDEL, T. H. (eds) *Proceedings of the Ocean Drilling Program (Scientific Results)*, **138**. College Station, TX. (Ocean Drilling Program) 537–554.

BREWSTER, N. A. 1980. Cenozoic biogenic silica sedimentation in the Antarctic Ocean. *Geological Society of America Bulletin*, **91**, 337–347.

CALVERT, S. E. 1966. Origin of diatom-rich varved sediments from the Gulf of California. *Journal of Geology*, **74**, 546–565.

CHAVEZ. F. P. & BARBER, R. T. 1987. An estimate of new production in the equatorial Pacific. *Deep-Sea Research*, **34**, 1229–1243.

——, BUCK, K. R. & BARBER, R. T. 1990. Phytoplankton taxa in relation to primary production in the equatorial Pacific. *Deep-Sea Research*, **37**, 1733–1752.

GARDNER, J. V. & BURCKLE, L. H. 1975. Upper Pleistocene *Ethmodiscus rex* oozes from the eastern equatorial Atlantic. *Micropaleontology*, **21**, 234–242.

GOMBOS, A. M. 1984. Late Neogene diatoms and diatom oozes in the central South Atlantic. *In*: HSÜ, K. J., LABRECQUE, J. L. *et al. Initial Reports of the Deep Sea Drilling Project*, **73**, 487–494.

GORSLINE, D. S., NAVA-SANCHEZ, E. & MURILLO DE NAVA, J. A study of occurrences of Holocene laminated sedimants in California borderland basins: Products of a variety of depositional proceses. *This volume*.

HALLEGRAEFF, G. M. 1986. Taxonomy and morphology of the marine planktonic diatoms *Thalassionema* and *Thalassiothrix*. *Diatom Research*, **1**, 57–80.

HASLE, G. R. & SEMINA, H. J. 1987. The marine planktonic diatoms *Thalasslothrix longissima* and *Thalassiothrix antarctica* with comments on *Thalassionema* spp. and *Synedra Reinboldii*. *Diatom Research*, **2**, 175–192.

HAY, B. J. 1988. Sediment accumulation in the central eastern Black Sea over the last 5100 years. *Paleoceanography*, **3**, 491–508.

HENDY, N. I. 1937. The plankton diatoms of the Southern Seas. *Discovery Reports*, **16**, 151–364.

KAMYKOWSKI, D. & ZENTARA, S.-J. 1990. Hypoxia in the world ocean as recorded in the historical data set. *Deep-Sea Research*, **37**, 1861–1874.

KEIGWIN, L. D. 1982. Isotopic Paleoceanography of the Caribbean and East Pacific: Role of Panama Uplift in Late Neogene Time. *Science*, **217**, 350–353.

KEMP, A. E. S. 1990. Sedimentary fabrics and variation in lamination style in Peru continental margin upwelling sediments. *In*: SUESS, E., VON HUENE, R. *et al.* (eds) *Proceedings of the Ocean Drilling Program, (Scientific Results)*, **112**, College Station, TX. (Ocean Drilling Program), 43–58.

——1995. Neogene and Quaternary pelagic sediments and depositional history of the eastern equatorial Pacific Ocean (ODP Leg 138). *In*: PISIAS, N. G., MAYER, L. A., JANECEK, T. R., PALMER-JULSON, A. & VAN ANDEL, T. H. (eds) *Proceedings of the Ocean Drilling Program, (Scientific Results)*, **138**. College Station, TX. (Ocean Drilling Program), 627–639.

—— & BALDAUF, J. 1993. Vast Neogene laminated diatom mat deposits from the eastern equatorial Pacific Ocean. *Nature*, **362**, 141–144.

KING, S. C., KEMP, A. E. S. & MURRAY, J. W. 1995. Changes in benthic foraminifer assemblages through Neogene laminated diatom ooze deposits in the eastern equatorial Pacific Ocean (ODP Site 844). *In*: PISIAS, N. G., MAYER, L. A., JANECEK, T. R., PALMER-JULSON, A. & VAN ANDEL, T. H. (eds) *Proceedings of the Ocean Drilling Program, (Scientific Results)*, **138**. College Station, TX. (Ocean Drilling Program), 665–673.

LEINEN, M., 1979. Biogenic silica accumulation in the central equatorial Pacific and its implications for Cenozoic Paleoceanography. *Geological Society of America Bulletin*, **90**, 801–803.

LYLE, M., MAYER, L., PISIAS, N., HAGELBURG, T., DADEY, K. & BLOOMER, S. 1992. Downhole logging as a paleoceanographic tool on Ocean Drilling Program Leg 138: interface between high-resolution stratigraphy and regional synthesis. *Paleoceanography*, **7**, 691–700.

MARTINEZ, L., SILVER, M. W., KING, J. & ALLDREDGE, A. L. 1983. Nitrogen fixation by floating diatom mats: a source of new nitrogen to oligotrophic ocean waters. *Science*, **221**, 152–154.

MAYER, L. A. 1991. Extraction of high-resolution carbonate data for palaeoclimate reconstruction. *Nature*, **352**, 148–150.

——, THEYER, F., THOMAS., E. *et al.* 1985a. Initial Reports of the Deep Sea Drilling Project, **85**, Washington (U.S. Govt. Printing Office) .

——, SHIPLEY T. H., THEYER, F., WILKENS. R. W. & WINTERER, E. L. 1985b. Seismic modelling and paleoceanography at DSDP Site 574. *In*: MAYER, L. A., THEYER, F., THOMAS, E. *et al. Initial Reports of the Deep Sea Drilling Project*, **85**, Washington (U.S. Govt. Printing Office), 947–970.

——, SHIPLEY, T. H. & WINTERER, F. I. 1986. Equatorial Pacific Seismic Reflectors as indicators of Global Oceanographic events. *Science*, **233**, 761–764.

——, PISAS, N. G., JANECEK, T. R. *et al.* 1992. *Proceedings of the Ocean Drilling Program, (Initial Reports)*, **138**, College Station, TX (Ocean Drilling Program).

MULLER, D. W., HODELL, D. A. & CIGIELSKI, P. F. 1991. Late Miocene to earliest Pliocene (9.8–4.5 Ma) paleoceanography of the subantarctic southeast Atlantic: stable isotopic, sedimentologic, and microfossil evidence. *In*: CIESIELSKI, P. F., KRISTOFFERSEN, Y, *et al. Proceedings of the Ocean Drilling Program, (Scientific Results)*, **114**, College Station, TX (Ocean Drilling Program), 459–474.

PEARCE, R. B., KEMP, A. E. S., BALDAUF, J. G. & KING, S. C. 1996. High-resolution sedimentology and micropalaentology of laminated diatomaceous sediments from the eastern equatorial Pacific Ocean (Leg 138). *This volume*.

SANCETTA, C., 1982. Diatom biostratigraphy and paleoceanography, Deep Sea Drilling Project Leg 68. *In*: PRELL, W. L., GARDNER, J. V. *et al. Initial Reports of the Deep Sea Drilling Project*, **68**, Washington (U.S. Govt. Printing Office), 301–309.

——1983. Biostratigraphic and paleoceanographic events in the eastern equatorial Pacific: Results of Deep Sea Drilling Project Leg 69. *In*: CANN,

J. R., LANGSETH, M. G., HONOREZ, J., VON HERZEN, R. P., WHITE, S. M. *et al. Initial Reports of the Deep Drilling Project*, **69**, Washington (U.S. Govt. Printing Office), 311–320.

——, VILLAREAL, T. & FALKOWSKI, P. 1991. *Rhizosolenia* mats. *Limnology and Oceanography*, **36**, 1452–4157.

SCHIMMELMANN, A. & LANGE, C. 1996. Tales of 1001 varves: a review of Santa Barbara Basin sedimant studies. *This volume*.

SHACKLETON, N. J., CROWHURST, S., HAGELBERG, T., PISIAS, N. G. & SCHNEIDER, D. A. 1995. A new late Neogene time scale: application to Leg 138 Sites. *In*: PISIAS, N. G., MAYER, L. A., JANECEK, T. R., PALMER-JULSON, A. & VAN ANDEL, T. H. (eds) *Proceedings of the Ocean Drilling Program (Scientific Results)*, **138**, College Station, TX (Ocean Drilling Program).

SMAYDA, T. J. 1958. Biogeographical studies of marine phytoplankton. *Oikos*, **9**, 158–191.

SMITH, K. L., BALDWIN, R. J. & WILLIAMS, P. M. 1992. Reconciling particulate organic carbon flux and sediment community oxygen consumption in the deep North Pacific. *Nature*, **359**, 313–316.

STABELL, B. 1986. Variations of diatom flux in the eastern Equatorial Atlantic during the last 400,000 years ("Meteor" Cores 13519 and 13521). *Marine Geology*, **72**, 305–423.

VAN ANDEL, T. H., HEATH, G. R. & MOORE, T. C., JR 1975. Cenozoic tectonics, sedimentation and paleoceanography of the central equatorial Pacific. *Memoir of the Geological Society of America*, **143**.

VILLAREAL, T. A., ALTABET, M. A. & CULVER-RYMSZA, K. 1993. Nitrogen transport by vertically migrating diatom mats in the North Pacific Ocean. *Nature*, **363**, 709–712.

WYRTKI, K. J. 1981. An estimate of Equatorial Upwelling in the Pacific. *Journal of Physical Oceanography*, **11**, 1205–1214.

YODER, J. A., ACKLESON, S., BARBER, R. T., FLAMANT, P. & BALCH, W. A. 1994. A line in the Sea. *Nature*, **371**, 689–692.

Index